AF701482

THE DOMESTIC DYNAMICS OF CHINA'S ENERGY DIPLOMACY

Series on Contemporary China (ISSN: 1793-0847)

Series Editors: Joseph Fewsmith *(Boston University)*
Zheng Yongnian *(East Asian Institute, National University of Singapore)*

---

*Published**

Vol. 26 Looking North, Looking South: China, Taiwan, and the South Pacific
edited by Anne-Marie Brady

Vol. 27 China's Industrial Development in the 21st Century
edited by Mu Yang & Hong Yu

Vol. 28 Cross-Taiwan Straits Relations Since 1979: Policy Adjustment and Institutional Change Across the Straits
edited by Kevin G. Cai

Vol. 29 The Transformation of Political Communication in China: From Propaganda to Hegemony
by Xiaoling Zhang

Vol. 30 The Great Urbanization of China
edited by Ding Lu

Vol. 31 Social Structure of Contemporary China
edited by Xueyi Lu

Vol. 32 East Asia: Developments and Challenges
edited by Yongnian Zheng & Liang Fook Lye

Vol. 33 China and East Asia: After the Wall Street Crisis
edited by Peng Er Lam, Yaqing Qin & Mu Yang

Vol. 34 The World Turned Upside Down:
The Complex Partnership between China and Latin America
by Alfredo Toro Hardy

Vol. 35 Township Governance and Institutionalization in China
by Shukai Zhao

Vol. 36 Health Policy Reform in China: A Comparative Perspective
by Jiwei Qian & Åke Blomqvist

Vol. 37 Understanding Chinese Society: Changes and Transformations
by Eileen Yuk-ha Tsang

Vol. 38 The Domestic Dynamics of China's Energy Diplomacy
by Chi Zhang

*To view the complete list of the published volumes in the series, please visit:
http://www.worldscientific.com/series/scc

Series on Contemporary China – Vol. 38

THE DOMESTIC DYNAMICS OF CHINA'S ENERGY DIPLOMACY

Chi Zhang

National Defense University, China

World Scientific

NEW JERSEY • LONDON • SINGAPORE • BEIJING • SHANGHAI • HONG KONG • TAIPEI • CHENNAI • TOKYO

Published by

World Scientific Publishing Co. Pte. Ltd.
5 Toh Tuck Link, Singapore 596224
USA office: 27 Warren Street, Suite 401-402, Hackensack, NJ 07601
UK office: 57 Shelton Street, Covent Garden, London WC2H 9HE

Library of Congress Cataloging-in-Publication Data
Zhang, Chi (College teacher)
The domestic dynamics of China's energy diplomacy / Chi Zhang (National Defense University, China).
pages cm. -- (Series on contemporary China ; 38)
ISBN 978-9814696739
1. Energy policy--China. 2. Petroleum industry and trade--China. I. Title.
HD9502.C62Z43 2015
333.790951--dc23
2015021589

British Library Cataloguing-in-Publication Data
A catalogue record for this book is available from the British Library.

Copyright © 2016 by World Scientific Publishing Co. Pte. Ltd.

All rights reserved. This book, or parts thereof, may not be reproduced in any form or by any means, electronic or mechanical, including photocopying, recording or any information storage and retrieval system now known or to be invented, without written permission from the publisher.

For photocopying of material in this volume, please pay a copying fee through the Copyright Clearance Center, Inc., 222 Rosewood Drive, Danvers, MA 01923, USA. In this case permission to photocopy is not required from the publisher.

In-house Editors: Dong Lixi/Prathima

Typeset by Stallion Press
Email: enquiries@stallionpress.com

For my parents Zhang Peixiang & Chi Geling

Contents

List of Abbreviations

ASEAN	Association of Southeast Asian Nations
AU	African Union
BE	Bureau of Energy
BP	British Petroleum
bpd	Barrels per day
CACF	China–Arab Cooperation Forum
CCP	Chinese Communist Party
CDB	China Development Bank
CEIB	China Export–Import Bank
CELAC	Community of Latin American and Caribbean States
CICIR	China Institute of Contemporary International Relations
CIIS	China Institute of International Studies
CNPC	China National Petroleum Corporation
CNODC	China National Oil and Gas Exploration and Development Corporation

CNOOC	China National Offshore Oil Corporation
CPI	Consumer price index
DPP	Democratic Progressive Party
EC	State Economic Commission
EIA	U.S. Energy Information Administration
EOR	Enhanced oil recovery
FDI	Foreign direct investment
FELSG	Financial and Economic Leading Small Group
FOCAC	Forum on China–African Cooperation
FTA	free trade area
G4	Group of Four
GCC	Gulf Cooperation Council
GDP	Gross domestic product
IEA	International Energy Agency
IOCs	International oil companies
IPE	International political economy
KMT	Kuomintang
LAS	League of Arab States
LNG	Liquefied natural gas
M&A	Mergers and acquisitions
ME	Ministry of Energy
MFA	Ministry of Foreign Affairs
MFI	Ministry of Fuel Industry
MLR	Ministry of Land and Resources
MOC	Ministry of Commerce
MOU	memorandum of understanding
MPI	Ministry of Petroleum Industry
NEA	National Energy Administration
NDRC	National Development and Reform Commission
NGOs	Non-Governmental Organisations
NOCs	National oil companies
ODI	Outward Direct Investment
OECD	Organisation for Economic Cooperation and Development
OPEC	Organisation of the Petroleum Exporting Countries
PLA	People's Liberation Army

PRC	People's Republic of China
PSA	Production sharing agreement
PSC	Production sharing contract
RMB	Renminbi
SASAC	State-Owned Assets Supervision and Administration Commission
SCO	Shanghai Cooperation Organisation
SDPC	State Development Planning Commission
SEC	State Energy Commission
SELSG	State Energy Leading Small Group
SETC	State Economic and Trade Commission
Sinopec	China Petrochemical Corporation
SIOEDC	Sinopec International Oil Exploration and Development Company Ltd.
SLOC	Sea Lanes of Communication
SOEs	State-Owned enterprises
SPC	State Planning Commission
SPR	Strategic Petroleum Reserves
TNCs	Transnational Corporations
UN	United Nations
UNCTAD	United Nations Conference on Trade and Development
USD	US dollars
WHO	World Health Organisation

Preface

China's economic rise has led to the dramatic expansion of its appetite for energy resources, in particular oil and natural gas. Consequently, the country's domestic energy production is not able to satisfy its consumption. During the past two decades, China has been increasingly dependent on foreign energy supply. Particularly, its dependence on foreign oil and natural gas has surpassed 60 per cent and 30 per cent respectively.

In accordance, China's energy diplomacy has been developing fast recently, searching oil and natural gas resources worldwide. The Chinese government has been actively engaging energy exporting states and Chinese national oil companies (NOCs) have spared no effort in conducting outward direct investment (ODI) and transnational operation in many energy rich areas in the world. Thus, Beijing is playing a more and more prominent role on the world stage; and Chinese NOCs have become significant players in the international energy arena. This new phenomenon is one of the most important developments in the international political economy. It coincides with the rise of China and continuous turbulences in the international energy market, therefore receiving growing attention globally.

This study examines the key domestic dynamics of China's energy diplomacy, especially the interaction between national and corporate interests behind the movement. I argue that the Chinese government and Chinese NOCs are two equally important players in the game; and neither of them should be underestimated. There are both national and corporate interests behind the phenomenon. On the one hand, energy diplomacy improves China's energy security and international relations, contributing to the maintenance of the Chinese Communist Party (CCP) government's political legitimacy which is identified as the state's essential national interests. On the other hand, this movement enables Chinese NOCs to access new investment markets abroad, generate greater profits and fulfil their long-term development strategies, which are their critical corporate interests. In this way, the convergence of national and corporate interests is realised, providing the most important and sustainable momentum for China's energy diplomacy.

Also, the government and the NOCs are natural partners in the movement, as they need each other's activities to reach their respective interests. Beijing's engagement with energy exporting states are beneficial for the NOCs' overseas business operation; while the NOCs' ODI and transnational operation contribute to China's energy security and provide Beijing with a new platform to strengthen its relations with many countries. In the recent years, the two players have been cooperating with each other to advance China's energy diplomacy. The partnership between the government and the NOCs are the guarantee of such rapid progress.

Furthermore, behind China's energy diplomacy, the government and the NOCs interplay with each other domestically. The government–NOC relationship has been evolving with China's economic and enterprise reform. The decentralisation process has led to a shift of power from the government to the NOCs, enabling them to become increasingly autonomous and powerful. Generally, the government manages the NOCs and oversees their business activities through various oil supervising agencies in the State Council, the nomenclatural system and the investment approval process; while the NOCs have substantial material, institutional and ideological resources to exert considerable impact upon the government's policy making. The recent years have witnessed the growth of the NOCs' economic and political capability *vis-à-vis* the waning of the state's leverage on them.

Although China's energy diplomacy advances quickly, two issues might affect the future of this movement. One is the occasional divergence of national and corporate interests; the other is the difficulties for the government to effectively manage the NOCs. These two issues are interrelated with each other, posing a big challenge in front of the Chinese leadership. The outcome of the interaction between the government and the NOCs will determine how national and corporate interests can be coordinated and realised. The development of such domestic dynamics will eventually have a critical influence on China's energy diplomacy.

This book is based on my PhD thesis that I completed in March 2009 at the University of Nottingham. In the past six years, significant progress has been achieved by China's energy diplomacy and new developments have been added to the domestic dynamics of this movement. Therefore, this book provides an updated analysis of the phenomenon.

In particular, here I would like to express my gratitude to some individuals and institutions. First and foremost, I am very grateful to my parents who have supported me all the way. To them I dedicate this book. Meanwhile, I am grateful to my wife. Without her understanding and support, it would be difficult for me to finish this book. Then, my special thanks go to Professor Zheng Yongnian, my mentor and first Ph.D. supervisor, who has consistently provided me with valuable guidance, strong intellectual support and steady encouragement. Also, I must thank Professor Adam Morton, my second Ph.D. supervisor, who provided insightful advice for my research. Besides, I would like to acknowledge the support from my colleagues at

National Defence University of China. Finally, I wish to thank the University of Nottingham where I received my PhD degree; and East Asian Institute, National University of Singapore for offering me a three-month Visiting Research Fellowship, during which I finished writing up this book.

Chapter

1

Introduction: Why Domestic Dynamics Matter

One of the most important events in the international political economy (IPE) at the beginning of the 21st century is China's energy diplomacy. This movement coincides with the rise of China and turbulence in the world energy market. Thanks to its economic reform, China has been experiencing fast socio–economic development for about three decades and become the second largest economy in the world. Accordingly, the country's energy consumption has been rapidly growing and its domestic production no longer satisfies its appetite. In consequence, China is increasingly dependent on overseas energy supply to keep its economic prosperity. Recently, the Chinese government has been actively engaging energy exporting countries around the world and Chinese national oil companies (NOCs) are expanding their outward direct investment (ODI) in overseas energy assets. Thus, Beijing is playing a more prominent role on the world stage; and Chinese NOCs are emerging significant players in the international energy arena. This new phenomenon has received increasing attention worldwide.

1.1 Why do domestic dynamics matter?

Some observers are eager to find out the implication of China's energy diplomacy for the international community. Does this development pose a challenge to the world political and economic order? Does China follow a mercantilist approach in competing with the other players in the energy market? What is the impact of energy diplomacy on China's international relations? How should the West respond to China's energy diplomacy? To answer questions of these kinds, there is a small but growing literature on China's energy diplomacy. Many analysts have examined the issue on the inter-state level to discuss China's energy security, China's energy diplomacy, Chinese NOCs' overseas investment, as well as the economic, political and strategic implication of this movement for the rest of the world.[1]

---

[1] See, for example, Erica S. Downs, *China's Quest for Energy Security* (Santa Monica, CA: Rand Corporation, 2000); Amy Myers Jaffe and Steven W. Lewis, 'Beijing's oil diplomacy', *Survival* 44: 1, 2002, pp. 115–134; Philip Andrews-Speed, Xuanli Liao and Roland Dannreuther, *The Strategic Implications of China's Energy Needs*, the International Institute for Strategic Studies, Adelphi Paper 346 (Oxford: Oxford University Press, 2002); Roland Dannreuther, 'Asian security and China's energy needs', *International Relations of the Asia–Pacific* 3, 2003, pp. 197–219; Henry J. Kenny, 'China and the competition for oil and gas in Asia', *Asia–Pacific Review*, 11: 2, 2004, pp. 36–47; Pak K Lee, 'China's quest for oil security: Oil (wars) in the pipeline?', *Pacific Review* 18: 2, 2005, pp. 265–301; Robert E. Ebel, *China's Energy Future: The Middle Kingdom Seeks Its Place in the Sun* (Washington DC: The CSIS Press, Center for Strategic and International Studies, 2005); David Zweig and Bi Jianhai, 'China's global hunt for energy', *Foreign Affairs* 84: 5, 2005, pp. 25–38; Ian Taylor, 'China's oil diplomacy in Africa', *International Affairs* 82: 5, 2006, pp. 937–959; Xuecheng Liu, 'China's energy security and its grand strategy', *Policy Analysis Brief*, The Stanley Foundation, September 2006; Yuanming Alvin Yao, 'China's oil strategy and its implications for U.S.–China relations', *Issues & Studies* 42: 3, 2006, pp. 165–201; James Tang, 'With the Grain or Against the Grain?:

Although these studies are informative, their focus on state to state interaction has not explained an important question — what is the key momentum of China's energy diplomacy? As to the answer of this question, conventional wisdom tends to regard China's energy security concerns as the motivation and view Chinese NOCs as agencies of the Chinese state. Its assumption is often like this: Beijing's energy security considerations drive energy diplomacy and ODI in energy assets abroad, which affects the international community. However, such mercantilist logic is too simple a characterisation, or even a misperception. In particular, it overlooks the critical domestic dynamics of the phenomenon — the interaction between China's national interests and Chinese NOCs' corporate interests.

Indeed, it is outdated to adopt a state-centric view to explain the current Chinese foreign policy, without paying attention to various domestic forces. Marc Lanteigne noted that the conventional perception of Chinese foreign policy being decided by a centralised leadership in Beijing is no longer as valid as it used to be. The number of actors who participate in the formulation of Beijing's foreign policy has grown within the Chinese government as well as increasingly outside it.[2] Robert G. Sutter also observed that the number of people in and outside of the Chinese government with an interest and influence in China's foreign policy making has grown substantially since the Maoist period.[3] In addition, Zhao Suisheng wrote that although China's foreign policy making is still highly centralised, the process of

Energy Security and Chinese Foreign Policy in the Hu Jintao Era', the Brookings Institution, October 2006, http://www.brookings.edu/~/media/Files/rc/papers/2006/10china_tang/tang2006.pdf, accessed 20 June 2008; Aaron L. Friedberg, '"Going out": China's pursuit of natural resources and implications for the PRC's grand strategy', NBR Analysis, 17: 3, the National Bureau of Asian Research, September 2006, http://www.nbr.org/publications/analysis/pdf/vol17no3.pdf, accessed 20 June 2008; and Suisheng Zhao, 'China's Global Search for Energy Security: Cooperation and competition in Asia–Pacific', *Journal of Contemporary China* 17: 55, 2008, pp. 207–227.

[2] Marc Lanteigne, *Chinese Foreign Policy: An Introduction* (London and New York: Routledge, 2009), p. 19.

[3] Robert G. Sutter, *Chinese Foreign Relations: Power and Policy since the Cold War* (Plymouth, UK: Rowman & Littlefield Publishers, Inc., 2008), p. 53.

foreign policy information processing, deliberation and decision making and the management of foreign relations are no longer controlled by a handful of individual leaders. Non-governmental actors like foreign policy think tanks, social groups and public opinion have become increasingly important in the process of China's foreign policy making.[4] Notably, Chinese NOCs are among the social groups that exert considerable influence on Beijing's foreign policy making. However, the conventional wisdom about China's energy diplomacy underestimates the role of these firms while overestimating that of the state. Thus, it does not provide a comprehensive explanation of the motivation of China's energy diplomacy.

As a result, many existing analyses and policy suggestions on this phenomenon are based on misleading perceptions, which may cause unconstructive consequences. As Kenneth Lieberthal and Mikkal Herberg pointed out, Washington's response to China's energy rise has been relatively counterproductive. 'Compounded by China's own lack of transparency, U.S. reactions have suffered from a poor understanding of China on many levels', such as the complex interests motivating the country's worldwide quest for energy, the goals and relationships characterising China's energy institutions and Chinese NOCs and the linkage between energy and other issues in China.[5] Erica Downs also noted that the mercantilist perception of China's energy issues may exacerbate the bilateral friction that both Beijing and Washington want to avoid; and 'treating China like a mercantilist state may prompt it to behave like one'.[6]

Therefore, it is necessary to have a more comprehensive understanding of the momentum of China's energy diplomacy, particularly

---

[4] Suisheng Zhao, 'The Making of Chinese Foreign Policy: Actors and Institutions', in Kweku Ampiah and Sanusha Naidu eds., *Crouching Tiger, Hidden Dragon?: Africa and China* (South Africa: University of KwaZulu-Natal Press, 2008), pp. 39–52.

[5] Kenneth Lieberthal and Mikkal Herberg, 'China's search for energy security: Implications for U.S. policy', NBR Analysis, 17: 1, the National Bureau of Asian Research, April 2006, http://www.nbr.org/publications/analysis/pdf/vol17no1.pdf, accessed 20 June 2008.

[6] Erica Downs, 'China's Quest for Overseas Oil', *Far Eastern Economic Review*, September 2007, p. 56.

the long-neglected domestic dynamics — the interaction of national and corporate interests. This research attempts to make a contribution in this regard.

1.2 The contribution of this study

In brief, this study explains the Chinese government's national interests and Chinese NOCs' corporate interests behind China's energy diplomacy and examines the measures and resources with which the government and the NOCs interact with each other, which have a profound influence on the development of the movement. In this way, this research aims to make an informed contribution in the following six areas.

1.2.1 *The political economy of contemporary China*

The Chinese government's energy diplomacy is a type of political behaviour, while Chinese NOCs' overseas expansion is an economic or business activity. How the state and the firms' respective interests can be realised and how the two players interact with each other are important questions of the political economy of contemporary China. These two issues are the focus of this study.

1.2.2 *Chinese foreign policy*

Energy diplomacy is a new but increasingly important component of Chinese foreign policy. This research will not only show the development of this phenomenon, but also take a further step towards explaining the key national and corporate interests behind it.

1.2.3 *China's global business*

Chinese NOCs are among the Chinese transnational corporations (TNCs) that have emerged in the recent years to increasingly participate in global competition. This study will trace the NOCs' history and demonstrate the motivation and performance of their ODI and transnational operation, shedding light on China's expanding global business. Notably, there are similar corporate interests and considerations between the NOCs and some other big Chinese state-owned enterprises (SOEs) regarding their involvement in global business.

1.2.4 *State–SOEs relationship in China*

During the economic reform period, both the Chinese state and Chinese SOEs have undergone transformation. The analysis of interaction between the government and the NOCs in this research will add to the understanding of relationship between the Chinese state and Chinese SOEs. Although this study focuses on the government–NOC relationship, there are similar dynamics for the state's interaction with other big SOEs in China.

1.2.5 *China's energy issues*

This research will explore the roles of the government and the NOCs in China's energy policy making, as well as the interplay between them on some energy issues. Apart from the NOCs, there are other important energy SOEs in China, such as those dominant in the coal, electricity and nuclear industries. They play more or less similar roles to the NOCs in China's energy affairs.

1.2.6 *The link between IPE theories and the study of China*

This research will test the empirical evidence of the domestic dynamics of China's energy diplomacy against the three mainstream IPE theories, so as to show their merits and shortcomings in explaining the phenomenon, before discussing the new diplomacy idea's inspiration for conceptualisation of the interaction between national and corporate interests behind China's energy diplomacy.

1.3 Some key terms

There are some key terms, such as energy diplomacy, the domestic dynamics, the government and the NOCs, which need clarifying here.

1.3.1 *Energy diplomacy*

Energy includes many categories like coal, oil, natural gas, hydropower, solar power, wind power, nuclear power, etc. This research will focus on oil and natural gas, as they are China's most deficient and needed energy resources, as well as the main resources that China searches through its energy diplomacy.

As for diplomacy, it can be understood in the narrow or the broad sense. In the narrow sense, diplomacy is exclusively a state or governmental behaviour. This is the traditional definition. In the broad sense, however, diplomacy goes beyond the state's domain and includes transnational activities of non-governmental actors such as corporations. This is a relatively new and expanded definition.

Accordingly, in the traditional and narrow sense, China's energy diplomacy only means the Chinese government's diplomatic activities to engage energy exporting states. In the newer and broader sense, China's energy diplomacy includes two aspects. One is the traditional aspect of the government's energy related diplomatic activities. The other is Chinese NOCs' ODI and transnational operation. Such a broad sense of diplomacy is more comprehensive than the traditional definition, as it sees both the role of the state and that of the firms in China's global search for energy resources. It will also be reflected by the conceptualisation of this study in the next chapter. This research will draw on Susan Strange's idea of the new diplomacy as the general theoretical framework to examine the interaction between the Chinese government and Chinese NOCs. This idea also views diplomacy as a broad arena including state–state, state–firm and firm–firm interactions.

Therefore, this research will distinguish between the broad and the narrow sense of energy diplomacy. This book will refer to the broad sense as China's energy diplomacy or energy diplomacy, which includes both the government and the NOCs' behaviours; and the narrow sense as the Chinese government's energy diplomacy or Beijing's energy diplomacy, which is restricted to traditional state behaviour.

1.3.2 *The domestic dynamics*

This study's exploration of the domestic dynamics of China's energy diplomacy will focus on the interaction between the Chinese government's national interests and Chinese NOCs' corporate interests.

1.3.3 *The government*

There are the central government and various levels of local governments in China. This research refers to the Chinese government or the government as China's central government led by the Chinese Communist Party (CCP).

1.3.4 *The NOCs*

There are different oil companies in China, including SOEs and other types of firms. Among the SOEs, some are owned by the central government, the others by local governments. This study refers to the NOCs as the oil companies owned by the central government. It will focus on the three leading NOCs — China National Petroleum Corporation (CNPC), China Petrochemical Corporation (Sinopec) and China National Offshore Oil Corporation (CNOOC).

1.4 The argument

China's economic rise has led to the dramatic expansion of its appetite for energy resources, especially oil and natural gas. Consequently, China's domestic energy production is not able to satisfy its demand; and the country is increasingly dependent upon foreign energy supply. In accordance, China's energy diplomacy has been rapidly developing recently, searching for oil and natural gas resources worldwide. The Chinese government has been actively engaging energy exporting states; and Chinese NOCs have been fast conducting ODI and transnational operation in many energy rich areas in the world.

The author argues that the convergence of national and corporate interests is the key momentum of China's energy diplomacy.

Both the role of the Chinese government and that of Chinese NOCs are important and neither should be underestimated. On the one hand, energy diplomacy contributes to China's national interest, as it improves China's energy security and international relations, which are beneficial for the maintenance of the CCP's political legitimacy. On the other hand, energy diplomacy contributes to Chinese NOCs' corporate interests, as it helps these firms access overseas investment markets, generate greater profits and fulfil their long-term development strategies. The government and the NOCs are two partners in the movement. They need each other's critical activities to realise their respective interests. Beijing's energy diplomacy and good relations with many energy rich states are beneficial for the NOCs' operation there; while the NOCs' transnational operation improves China's energy security and gives Beijing a new platform to enhance its ties with energy rich countries. The convergence of national and corporate interests and the partnership between the government and the NOCs have ensured the rapid progress of China's energy diplomacy.

Moreover, the government–NOC relationship has been evolving with China's economic and enterprise reform. The decentralisation process has led to power devolution from the government to the NOCs, enabling them to become increasingly autonomous and powerful. Currently, the CCP government administrates the NOCs and oversees their business activities at home and abroad through various oil supervising agencies in the State Council, the nomenclatural system and the investment approval process; while the NOCs have substantial material, institutional and ideological resources to influence the government's policy making and secure the state's diplomatic and financial support for their global business. The recent years have witnessed the rise of these firms' economic and political capability *vis-à-vis* the decline of the state's leverage over them. Although China's energy diplomacy is advancing quickly, two problems will affect the future of this movement. One is the occasional divergence of national and corporate interests; the other is the difficulties for the government to manage the NOCs. 'How to address these issues', is a big challenge in front of the Chinese leadership.

1.5 The organisation

This book is divided into eight chapters. Following this introduction, Chapter 2 provides a review of recent literature on domestic dynamics especially the relationship between national and corporate interests behind China's energy diplomacy, before dealing with the conceptualisation of this research. There are two major problems with the existing explanations of the movement. The first is the conventional wisdom that underestimates the role of the NOCs and their corporate interests. The second is an inadequate attention to the government's role and national interests. Actually, the government and the NOCs are two equally important and indispensable actors in the phenomenon and neither of them should be downplayed. As to the conceptualisation, this study draws on the new diplomacy idea for the overall theoretical framework and incorporates useful elements from the three mainstream IPE theories — realism, liberalism and Marxism to explain some aspects in the phenomenon.

Chapter 3 describes the conditions of China's energy demand and supply. There are six reasons leading to the country's surging appetite for energy resources, in particular oil and natural gas. Meanwhile, the capacity for its domestic energy production is limited. These factors have caused China's increasing dependence on energy import.

Chapter 4 demonstrates the development of the Chinese government's energy diplomacy across various regions in the world and Chinese NOCs' ODI and transnational operation in those regions, highlighting the futures and new developments of this movement.

Chapter 5 explores the Chinese government's national interests and Chinese NOCs' corporate interests behind China's energy diplomacy. It explains that energy diplomacy serves both national and corporate interests. The major national interests are improving energy security and promoting Beijing's international relations and multilateral diplomacy, which contribute to China's economic development, social stability and national unity, thus helping sustain the CCP's political legitimacy. The major corporate interests include seeking overseas investment markets, generating more profits and fulfilling corporate development strategies. The convergence of national and corporate interests is the fundamental and sustainable motivation of China's energy diplomacy.

Chapter 6 discusses the interaction of the government and the NOCs, especially their approaches and abilities to engage and influence each other. During China's economic and enterprise reform, the state has decentralised much power and authority to the NOCs. Nowadays, the government manages the oil sector and the NOCs mainly through various oil supervising agencies in the State Council, the nomenclatural system and the investment approval process; while the NOCs possess significant material, institutional and ideological resources to exert considerable influence upon the government's policy making.

Chapter 7 examines Chinese discourses of energy security, which not only justify but may also influence China's energy diplomacy. This chapter focuses on two questions — what are the key challenges to China's energy security and what major measures should be adopted to safeguard China's energy security? It identifies three major Chinese academic discourses of energy security — the oil supply-focused discourse, the strategic ability-focused discourse and the energy efficiency-focused discourse. The first is the mainstream voice in China. There is also a Chinese official discourse of energy security that is similar to some points in the first and the third academic discourses. Among them, the oil supply-focused discourse and the official discourse provide critical justification for China's energy diplomacy.

In conclusion, Chapter 8 touches on two emerging issues that will increasingly influence China's energy diplomacy in the future. One is the occasional divergence of national and corporate interests. The other is the government's difficulties in managing the NOCs. How to coordinate different domestic interests and deal with these powerful firms are severe challenges for the Chinese leadership.

Chapter

2

Conceptualising the Interaction of National and Corporate Interests

This chapter provides a review of some recent literature on domestic dynamics especially the relationship between national interests and corporate interests behind China's energy diplomacy, before dealing with the conceptualisation of this research. The chapter is divided into two sections. The first section identifies two problems with the existing literature on the issue, each of which underestimates the role of either corporate or national interests. The second section tests the author's general observation of the domestic dynamics of China's energy diplomacy against the three mainstream international political economy (IPE) theoretical perspectives — realism, liberalism and Marxism, in order to indicate their merits and shortcomings in explaining the phenomenon, before providing the alternative theoretical perspective for the conceptualisation of this research — the new diplomacy idea as the overall framework that incorporates useful elements from realism, liberalism and Marxism.

2.1 Existing explanations of the domestic dynamics of China's energy diplomacy

Some scholars have made initial efforts to explain the domestic dynamics of China's energy diplomacy, and touched upon the relationship between national and corporate interests behind the phenomenon. However, there are two major problems with their explanations. The first is the conventional and widespread perception that underestimates the importance of corporate interests behind China's energy diplomacy and the national oil companies' (NOCs) self-motivation in conducting outward direct investment (ODI) and transnational operation; while the second problem is an underestimation of the importance of national interests behind the movement and the role of the government. The following paragraphs discuss the two problems in detail.

In the first place, conventional wisdom tends to view Chinese NOCs as agencies of the Chinese government; and their overseas expansion as a highly coordinated governmental strategy to secure energy supply from abroad. According to this logic, the NOCs are in the grip of the government or merely puppets of the state, simply because they are state-owned or state-controlled in theory. Their mission is to carry out the Chinese leadership's decisions. Since China increasingly relies on foreign energy supply, the government is nervous about the potential threats to China's energy security. Thus, it orders or directs the NOCs to secure energy supply around the world through ODI in and mergers and acquisitions (M&A) of oil and natural gas assets abroad; these firms' overseas expansion is under the government's instruction and planning. Also, Chinese NOCs' corporate interests are usually considered to be subordinated to the Chinese state's national interests; and their transnational business is often regarded as more likely for the purpose of realising China's national interests, compared with pursuing their corporate interests. Due to the fact that prior to China's economic reform, the country's economy was centrally planned, it is understandable that many people might continue to adopt a state-centric view on the political economy of China.

For example, Zhao Suisheng maintained that China has adopted a state-centred approach towards energy security and launched the state-led search for energy resources around the world, deepening political and commercial relationships with all energy producing nations and aggressively investing in oil fields and pipelines abroad.[1] Similarly, Philip Andrews-Speed, Xuanli Liao and Roland Dannreuther contended that China cannot afford to be heavily dependent on the international market for oil supply, hence the government directs the NOCs to invest in overseas projects to reduce this dependence; and domestic political and foreign policy concerns trump economic and efficiency concerns.[2] Kang Wu and Shair Ling Han argued that because Chinese NOCs are state-owned, economic concerns and commercial interests may not be priorities in their overseas investment.[3] Also, Robert E. Ebel mentioned that Beijing is concerned about its rising dependence on the sustainability of crude oil import, and has sent its NOCs on a worldwide search for equity oil.[4] Ian Taylor held that Chinese NOCs continue to be fundamentally state-owned and their administration functions are largely assumed by the government. Thus, their operation is likely to be in accordance with China's national strategy regarding resources and foreign policy.[5]

---

[1] Suisheng Zhao, 'China's global search for energy security: Cooperation and competition in Asia–Pacific', *Journal of Contemporary China* 17: 55, 2008, pp. 207–227.

[2] Philip Andrews-Speed, Xuanli Liao and Roland Dannreuther, *The Strategic Implications of China's Energy Needs*, the International Institute for Strategic Studies, Adelphi Paper 346 (Oxford: Oxford University Press, 2002), p. 25 and p. 101.

[3] Kang Wu and Shair Ling Han, 'State-company goals give China's investment push unique features', *Oil & Gas Journal*, April 18 2005, p. 20.

[4] Robert E. Ebel, *China's Energy Future: The Middle Kingdom Seeks Its Place in the Sun* (Washington DC: The CSIS Press, Center for Strategic and International Studies, 2005), p. 13.

[5] Ian Taylor, 'China's oil diplomacy in Africa', *International Affairs* 82: 5, 2006, pp. 937–959.

Stephanie Kleine-Ahlbrandt and Andrew Small maintained that China's economic expansion leads to increasing energy demand; and 'at the instigation of some authoritarian governments eager to count Beijing as a sponsor, China sent its state-controlled companies to make massive investments, sweetening the deals with significant loans and military assistance'.[6] Joshy M. Paul wrote that Beijing's worldwide search for energy resources for its economic success has created a notion of 'scramble for energy' that portrays China's increasing investment and equity stake in the energy sector considered as a political strategy by the Chinese elites.[7]

Such conventional perceptions largely stem from the theoretical state ownership of Chinese NOCs. Of course, the NOCs are state-owned enterprises (SOEs) in theory. Nonetheless, those regarding the NOCs as a tool of the government adopt an overly simplistic view of the complex government–NOC relationship. Some analysts have frequently stressed *state-owned* to the neglect of *enterprises*. Actually, in today's China where a market economy has been initially established, SOEs and NOCs are first and foremost *enterprises* and then *state-owned*. Just like many conventional firms, the NOCs consider their own corporate interests instead of the state's national interests as the priority of their business operations.

In particular, they are not agencies of the state. Thanks to China's economic and enterprise reform, Chinese NOCs have been transformed into transnational corporations (TNCs). More importantly, the Chinese state has decentralised a significant part of its administrative power to some leading SOEs including the NOCs and greatly reduced its intervention in these firms' operation, in accordance with the country's transformation towards a market economy. Therefore, the NOCs have enjoyed a high degree of autonomy and become quite independent players on the domestic and international stages, while maintaining their monopoly in China's energy market. The first

---

[6] Stephanie Kleine-Ahlbrandt and Andrew Small, 'China's New Dictatorship Diplomacy: Is Beijing Parting With Pariahs?', *Foreign Affairs* 87: 1, 2008, p. 41.

[7] Joshy M. Paul, 'The role of energy security in China's foreign policy: A maritime perspective', *Maritime Affairs* 6: 2, 2010, pp. 49–71.

section of Chapter 6 will describe the decentralisation process in detail. As a matter of fact, all of the ODI initiatives lie with the NOCs themselves, instead of the government.

With the rise of the Chinese economy in the recent years, the NOCs' financial capacities have been expanding fast. Their growing accumulation of surplus capital is a fundamental factor pushing them to look for investment opportunities abroad to seek further profit. They also have the ambition and long-term corporate strategy to internationalise themselves in order to become world-class TNCs. These corporate interests are the most important motivation for Chinese NOCs to conduct ODI in and M&A of overseas energy assets. To a great extent, these firms' corporate interests and behaviours objectively converge with the government's national interests (safeguarding China's energy security and improving China's international relations) and are in line with the Chinese official energy diplomacy activities. Such convergence of national and corporate interests is the key momentum of China's energy diplomacy. Chapter 5 will provide detailed analysis of this point.

Also, the NOCs have inherited significant political authority from China's previous bureaucratic system under the planned economy and further developed such advantages during the reform era. Currently, they possess powerful political status within the Chinese Communist Party (CCP) and high bureaucratic levels within the Chinese state, enjoy patronage from China's top leaders, benefit from their intimate connections with various energy administrative organisations and have various approaches to directly participate in or exert significant influence on the government's energy decision making and policy formulation. Their political advantages and growing economic capability have been reinforcing each other for years, making them strong players in China's economic and political life. Thus, to a certain extent, these firms are able to secure the state's support to realise their corporate interests such as ODI and transnational operation. Chapter 6 will demonstrate the CCP and the Chinese government's administration and supervision over Chinese NOCs as well as these firms' leverage over the state's policy making. The chapter will explain that the government's management and control of these firms are not as strong as often assumed; while the NOCs are able to exert considerable impact on the government's policy making.

Notably, there are two major differences between Chinese NOCs and conventional international oil companies (IOCs) especially those in Western countries. The first is that Chinese NOCs are theoretically state-owned or state-controlled, while they actually have much autonomy in their business operation. The second is that Chinese NOCs enjoy high political status and bureaucratic ranking (ministerial level) in China and have strong leverage over the Chinese government's policy formulation.

Therefore, too much emphasis on the NOCs' theoretical ownership by the government, as in the conventional wisdom, may lead to an underestimation of the role of their corporate interests and self-motivation. Without a strong incentive to gain corporate advantages, it would be very difficult for the NOCs to be so active in their overseas expansion and achieve significant success in the past two decades.

Moreover, Chapter 9 will offer a few examples of the NOCs' resistance to the government's preferences and the state's insufficient capability to rein these firms in, which are good examples to refute the conventional wisdom that regards these firms as puppets of the state. The NOCs' activities and interests sometimes are not in accordance or even conflict with the government's preferences. Such occasional divergence of national and corporate interests will affect China's energy diplomacy in the future.

The second problem with the existing literature on the domestic dynamics of China's energy diplomacy is an exaggeration of the NOCs' capability and an underestimation of national interests in the movement. Some scholars view China's energy diplomacy and the government–NOC relationship in a nearly opposite way. Although they have noted the co-existence of national and corporate interests in the phenomenon, they tend to downplay the role of the government and national interests and see the NOCs as so powerful that their corporate interests are the dominant motivator of China's energy diplomacy.

For instance, McGregor argued that China's foreign policy has been 'hijacked' by big companies like the NOCs.[8] Also, Downs wrote that China's energy projects and agenda are often driven by the NOCs' corporate interests rather than by China's national interests.[9] Similarly, Jiang and Sinton maintained that Chinese NOCs' overseas investment is the result of a complex interplay between individuals and groups associated with the NOCs; and commercial incentives play the largest part in the movement.[10]

However, Chapter 5 will show that the Chinese government does have its vital interests behind energy diplomacy — improving energy security and promoting Beijing's diplomacy, which contributes to China's economic development, social stability and national unity, helping sustain the CCP's political legitimacy. Both national interests and corporate interests are indispensable and neither should be underestimated. The convergence of the two national and corporate interests is the key and sustainable motivation of China's energy diplomacy, although sometimes there is divergence between them.

Furthermore, as will be explained in Chapter 6, while the government's authority over the NOCs is not as strong as expected by some people, it is able to manage the NOCs through various energy supervising agencies in the State Council as well as the nomenclatural system and oversee these firms' ODI activities largely through the investment approval process.

Therefore, a better understanding of the domestic dynamics of China's energy diplomacy is necessary. This study attempts to provide

---

[8] Richard McGregor, 'Chinese diplomacy 'hijacked' by big companies', Financial Times on-line, 16 March 2008, http://www.ft.com/cms/s/0/28b21418-f386-11dc-b6bc-0000779fd2ac.html?nclick_check=1, accessed 18 March 2009.

[9] Erica S. Downs, 'China', *Energy Security Series*, the Brookings Foreign Policy Studies, the Brookings Institution, December 2006, http://www.brookings.edu/~/media/Files/rc/reports/2006/12china/12china.pdf, accessed 8 April 2008.

[10] Julie Jiang and Jonathan Sinton, *Overseas Investments by Chinese National Oil Companies: Assessing the Drivers and Impacts*, International Energy Agency, February 2011, http://www.iea.org/publications/freepublications/publication/overseas_china.pdf, accessed 24 August 2014.

a more comprehensive explanation of the motivation of China's energy diplomacy, in particular the interaction between national and corporate interests.

2.2 Conceptualisation of the domestic dynamics of China's energy diplomacy

The fundamental domestic dynamics of China's energy diplomacy is the interaction between China's national interests and Chinese NOCs' corporate interests. The government and the NOCs are two equally important and indispensable players in the game. They have respective essential interests behind the movement. The convergence of national and corporate interests is the momentum behind China's energy diplomacy.

As to the conceptualisation of the phenomenon, none of the three mainstream IPE theories — realism, liberalism and Marxism — is adopted in this research as the overall framework, because they have respective shortcomings in explaining the domestic dynamics of China's energy diplomacy: The realist perspective emphasises or even exaggerates the role of the state and national interests while downplaying the role of firms and corporate interests; the liberalist perspective focuses on the role of corporations but underestimates that of the state and national interests; and the Marxist perspective neglects the state and national interests and its interpretation of class is not in accordance with Chinese politics today. Alternatively, this research draws on the idea of the new diplomacy or the triangular diplomacy (government–government, firm–firm and government–firm relationships) to examine the overall interaction between national and corporate interests that motivates China's energy diplomacy. Under the general dynamics, there are aspects like the government's concern about energy security and consideration of national interests, the importance and self-motivation of the NOCs, a more pluralised energy policy making process in China and the NOCs' resources and capacity to influence the government's policy making. These aspects of the domestic dynamics can be explained by realism, liberalism and Marxism. Therefore, as Figure 2.1 illustrates, when adopting the new

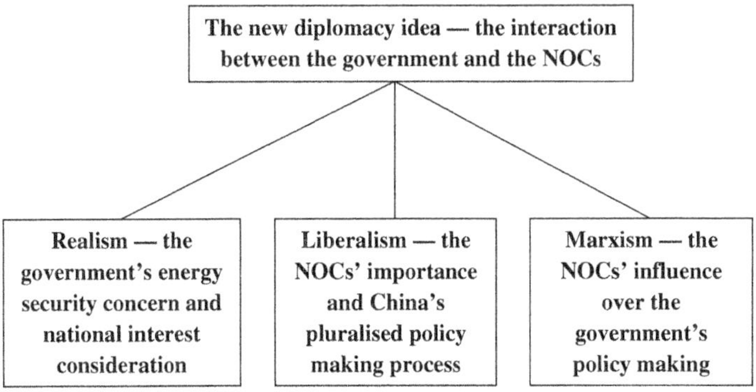

Figure 2.1: Conceptualisation of the interaction between national and corporate interests behind China's energy diplomacy

Source: Author's compilation.

diplomacy idea as the general theoretical framework to explore the interaction between the government and the NOCs, this research also incorporates useful elements from realism, liberalism and Marxism to explain some aspects in the phenomenon.

The rest of this section consists of four sub-sections. The first, second and third sub-sections respectively test the author's observation of the domestic dynamics of China's energy diplomacy against the three mainstream IPE theories — realism, liberalism and Marxism, respectively identify their advantages and disadvantages in explaining the phenomenon and explain why none of them is used as the overall theoretical perspective of this study, as well as how they can contribute to the conceptualisation of this research. The last sub-section explains why the new diplomacy idea is adopted as the general theoretical framework and how it explains the interaction between the government and the NOCs behind China's energy diplomacy.

2.2.1 *Realism*

Realism focuses on the role of the state and the importance of power in shaping outcomes in the IPE. Its core idea is that economic activities are and should be subordinate to the goal of state building and national interests. This perspective adopts a zero-sum view of international relations and contends that the nature of the international system is anarchic and the duty of each state is to safeguard its own national interests. It considers the struggle for economic resources among states as pervasive and inherent in the nature of the international system itself. National security is, and always will be, the principal concern of the state; and states should try to maintain self-sufficiency and protect their strategic industries against foreign competition. Moreover, the state is the dominant actor at both domestic and international levels. Although some recent forms of realism recognise the importance of non-state or market-based actors such as TNCs, they subordinate the importance of TNCs to that of the state. According to realists, the economic power of TNCs is limited; they are subject to the dictates of the state. The reason that corporations have become prominent economic actors is that states have abandoned regulation or loosened controls over the movement of capital. For example, Robert Gilpin wrote that states use their power to implement policies to channel economic forces in ways favourable to their own national interests. Using the case of the United States' power and TNCs, he argued that 'the multinational corporation has prospered because it has been dependent on the power of, and consistent with the political interests of, the United States.' This theoretical perspective prefers state control of key economic activities. In addition, realism or statism regards the state as an autonomous actor pursuing goals associated with power and the general interests of the society.[11]

The most significant contribution of realism to this research is its reflection of the fierce international competition for energy resources,

---

[11] Robert O'Brien and Marc Williams, *Global Political Economy: Evolution and Dynamics* (New York: Palgrave Macmillan, 2004), pp. 14–17; Robert Gilpin, *The Political Economy of International Relations* (Princeton: Princeton University Press, 1987), pp. 31–32; Robert Gilpin, *Global Political Economy: Understanding the International Economic Order* (Princeton: Princeton University Press, 2001), pp. 17–21; Robert Gilpin, 'U.S. Power and the Multinational Corporation', in Nikolaos Zahariadis ed., *Contending Perspective in International Political Economy* (Beijing: Peking University Press and Thomson Learning, 2004), pp. 25–26; and Stephen D. Krasner, *Defending the National Interest: Raw Materials Investments and U.S. Foreign Policy* (Princeton: Princeton University Press, 1978), pp. 5–34.

power politics with regard to energy issues as well as some states' concern about energy security. As realists believe, the international system is anarchic and every country should protect its own national interests; national security is naturally the major concern of the state. In China's energy diplomacy, the role of the state is undoubtedly critical. The realist perspective can explain the Chinese government's consideration of energy security and national interests.

To begin with, as Chapter 5 will explain, Beijing is worried about the country's energy security situation, which is directly related to economic development, social stability and national unity of the country. These three aspects are the pillars of the political legitimacy of the CCP. This is the biggest national interest consideration behind China's energy diplomacy which helps satisfy China's demand for overseas energy resources and safeguard the country's energy security. Whether China can guarantee its energy security determines whether the CCP government is able to sustain its political legitimacy and rule over China.

However, China's energy security is threatened by five (potential) external factors. The first is the unstable geo-political situation in some major energy exporting regions in the world, such as the Middle East and West Africa, which affects these regions' ability to export oil and the sufficient supply in the international oil market. The second threat is the fluctuation of the international oil prices, which may cause worldwide energy or economic crises that may affect the economic development of some developing countries like China. The third factor is the threats to China's energy import transportation. Most of the country's oil and liquefied natural gas (LNG) imports are transported by sea, which faces the risks of natural disaster, shipping accidents, piracies, terrorist attacks, etc.; and China's simplistic energy import shipping routes are negative for its energy import security. The fourth is the international energy containment against China, including strategic, business and ideological containment. The fifth factor is the pressure or threat from the United States, because the United States is the only country capable of implementing an oil embargo against China; and Beijing is in a passive position in China–U.S. relations and the two countries have some divergent strategic interests.

The Chinese government's concern about energy security and its wariness of some external uncertainties are in accordance with the realist view that there is constant competition among states for controlling the world's natural resources in an anarchic international society; and each country must safeguard its own national interests and national security. Chapter 5 will show Beijing's energy security concerns in detail and explain that China's energy diplomacy contributes to China's energy security in four ways: First, it helps increase and diversify the sources of China's energy import. Second, it increases the volume of overseas energy supply to China. Third, it diversifies the transportation methods through which China imports energy. Fourth, it helps stabilise the prices for China's domestic oil products and the Chinese economy.

Furthermore, Chapter 5 will argue that energy diplomacy provides a new platform for Beijing to engage many energy rich countries, which helps enhance China's relationship with them. It is beneficial for the People's Republic of China (PRC) to win more political support in multilateral institutions such as the United Nations (UN) and diplomatic backup on some critical issues like the Taiwan issue, human rights, the Beijing Olympic Games, etc. The examples of the competition between Beijing and Taipei for diplomatic recognition, African states' support for China regarding the dispute over the proposed expansion of UN Security Council in 2005 and the election of new Director-General of the World Health Organisation (WHO) in 2006 will be provided in this chapter. Also, energy diplomacy is in accordance with the implementation of the Chinese government's recently declared strategy of the Silk Road Economic Belt and the 21st Century Maritime Silk Road ('*sichou zhilu jingjidai*' and '*21 shiji haishang sichou zhilu*', or '*yidai yilu*' for short). In other words, apart from defending energy security, China's energy diplomacy helps realise the PRC's other national interests such as diplomatic and international strategic interests. Thus, realism is appropriate in explaining the Chinese government's energy security concern and national interest consideration.

However, the realist perspective merely focuses on the inter-state level and easily neglects the critical domestic dynamics of China's energy diplomacy and the essential role of the NOCs. It views the state as a unitary actor and the interests and activities of the NOCs are subordinated to the interests and rule of the Chinese state. Such a state-centric view is not in line with the reality of today's Chinese politics, because China has been undergoing a transformation from planned economy towards market economy for years and the political and economic life in the country has been significantly pluralised. Many domestic actors, especially those increasingly powerful corporations (state-owned, private-owned, shareholding and foreign invested) are exerting considerable impact over the policy making process at various levels of the government. To a certain extent, some of the government's policies are largely made to serve the interests of those companies. The rise of corporations is a new characteristic of the political economy of contemporary China. For instance, Chapter 4 and Chapter 6 will discuss the emergence and development of Chinese NOCs as well as their expanding economic and political clout.

Another shortcoming of realism is its lack of an attention to the role of ideas or ideology. As will be demonstrated in Chapter 6 and Chapter 7, some Chinese official slogans or strategies such as the 'going global' (*Zou chuqu*) strategy and the mainstream voice in Chinese discourses of energy security that calls for implementing energy diplomacy to safeguard China's energy security, serve as a critical theoretical justification for China's energy diplomacy.

In the end, although realism provides a useful approach to explaining Beijing's energy security concerns and national interest considerations, this is only one aspect of the phenomenon. The NOCs and their corporate interests play an indispensable role in China's energy diplomacy, whereas the realist perspective tends to neglect this important factor. Therefore, realism is not adopted for the general conceptualisation of this research.

2.2.2 *Liberalism*

Liberalism focuses on individuals and interest groups. Unlike realism which sees the state as an independent and unitary actor, liberalism views the state as a referee among competing social groups. It imagines politics as a vector diagram in which the state is under a series of pressures and moves in the direction preferred by the strongest societal forces. Moreover, instead of emphasising the inevitability of conflict, liberalism regards international relations as essentially cooperative and believes in the positive-sum of economic relations. It maintains that all can benefit from participating in a system of free markets and trade; states and peoples can cooperate for mutual benefit. Also, liberal institutionalists view the state as enmeshed in a web of interdependence and international organisations. Interdependence is assumed to be both a description of events and a prescription for the conflict resolution. Within this approach, TNCs play an important and positive role.[12]

Unlike the realist perspective which easily overlooks the domestic dynamics of China's energy diplomacy, the liberal perspective adopts a pluralist view on the issue. It is helpful for overcoming the problems with the conventional state-centric interpretation of the motivation of China's energy diplomacy and examining the critical role of the NOCs in the phenomenon. It can also be used to describe the pluralised processes of the Chinese government's energy policy making.

In the first place, the liberal perspective's stress on the role of TNCs is useful for unveiling the importance of the NOCs in the movement. Chapter 5 will argue that the NOCs have vital interests in carrying out ODI in and M&A of overseas oil and natural gas assets. Through such activities, they can generate more profit and fulfil their long-term corporate development strategies. Their corporate interests and self-motivation, rather than the state's national interests or direction, are the basic motivator of their transnational operation and overseas expansion. Chapter 6 will also contend that these firms have been becoming increasingly powerful both economically and politically in the recent years, due to the growth of the Chinese economy and China's economic and enterprises reform. They are able to exert

[12] Robert O. Keohane and Joseph S. Nye, *Power and Interdependence* (Boston: Scott, Foresman and Company, 1989), pp. 3–22; O'Brien and Williams, *Global Political Economy*, pp. 18–19; and Thomas D. Lairson and David Skidmore, *International Political Economy: The Struggle for Power and Wealth* (Belmont, CA: Wadsworth/Thomson Learning, 2003), p. 12.

important influence on Beijing's energy policy formulation and gain official endorsement for their actions to seek corporate interests such as transnational operations.

Compared with the situation under the previous planned economy, today's energy policy making process in China has been greatly pluralised. In the past, the predecessors of the NOCs were part of China's central government, such as the Ministry of Petroleum Industry (MPI). The country's energy administration authority was concentrated with merely a couple of governmental departments like the MPI and the State Planning Commission (SPC).

However, China's economic reform has led to the pluralisation of interests within the country. Many new interest groups have appeared and are actively participating in the political process. Among them, some large SOEs especially those monopoly enterprises are powerful interest groups in their respective sectors, exerting a significant impact upon the government's policy formulation. For example, the NOCs have become critical players in energy affairs and energy policy making. They not only retain their monopoly in China's energy market, making them actual market regulators, but also possess significant resources and channels to participate in the government's energy policy making. Thus, they have become critical stakeholders in the game.

Today, there are many stakeholders in China's energy policy formulation, including the NOCs, the central government, local governments, the Chinese People's Liberation Army (PLA), etc. Within the central government or the State Council which is the focus of this research's examination of China's energy administration agencies, there are various departments involved in energy management and supervision as well as energy diplomacy. The structure of the State Council's administration over the oil sector, such as those supervising bodies holding important authorities over the oil industry and the NOCs, have continuously changed with the seven rounds of bureaucratic restructurings throughout the reform era. Chapter 6 will demonstrate that since the economic reform, there have been more energy supervising bodies in the State Council.

At present, there are three principal oil supervising agencies — the National Development and Reform Commission (NDRC) with the

National Energy Administration (NEA) under it, the State-owned Assets Supervision and Administration Commission (SASAC) and the Ministry of Land and Resources (MLR). Indeed, oil affairs are related to the interests of many social groups, and more and more governmental departments or interest groups have become involved in the administration of this very profitable sector. Since there are so many governmental bodies sharing responsibility of oil affairs and they may have different preferences regarding energy issues, divergence or conflicts of interests among them is understandable, affecting the government's administration over the oil sector. These issues will be tackled in Chapters 6 and 8.

Besides, China's foreign policy making has also been significantly pluralised; and most of the decisions and policies are made through ministerial and bureaucratic consultation. Moreover, non-governmental actors such as business and lobby groups have been transformed from irrelevant actors to strong players in Chinese foreign policy.[13]

For example, the current CCP Central Committee's Financial and Economic Leading Small Group (*Zhongyang caijing lingdao xiaozu*, FELSG), which is China's top decision making body over energy affairs, includes not only President Xi Jinping as the head and Premier Li Keqiang as the deputy head, but also another 24 senior officials as group members, such as Liu Yunshan and Zhang Gaoli who are members of the Standing Committee of the CCP's 18th Central Committee's Political Bureau, Liu Yandong, Wang Yang, Ma Kai, Wang Huning, Li Zhanshu, Yang Jiechi and Yang Jing who are members of the CCP's 18th Central Committee's Political Bureau, some ministers in the State Council and the PLA General Chief of Staff Fang Fenghui; and the FELSG Office is headed by Liu He, Vice-Chairman of the NDRC.

With specific regard to energy diplomacy, some other governmental bodies apart from those principal supervising agencies, such as the Ministry of Commerce (MOC) and the Ministry of Foreign Affairs (MFA), also hold important leverage over the situation. Each actor

---

[13] Marc Lanteigne, *Chinese Foreign Policy: An Introduction* (London and New York: Routledge, 2009), p. 24.

has its own interests and considerations which might converge or conflict with others'. For instance, it is said that there has been friction between the MFA and the NDRC and between the MFA and the NOCs.[14] Such conflicting interests need coordination by higher level governmental bodies. Since there is a more pluralised policy making process, it is likely that a diversity of opinions will reach China's top leadership and lead to more informed decision making. The implementation of actual policies involves the interaction and negotiation among key stakeholders.[15]

In order to coordinate the differences among those stakeholders, the Chinese leadership had to establish high-level coordination organs. For example, the State Energy Leading Small Group (SELSG) was established in 2005 to coordinate different domestic interests in the energy sector. It was replaced in the 2008 bureaucratic restructuring by the State Energy Commission (SEC).

Notably, during China's economic and enterprises reforms, there has been a power shift from the government to the NOCs, as will be discussed in Chapter 6. In accordance with China's effort to transfer itself from planned to market economy, such reforms have not only created the three major NOCs, but also empowered them and given them increasing autonomy from governmental intervention. Before the reforms, state enterprises were part of the bureaucracy and operated under state plans. In comparison, today the NOCs have become much more autonomous and powerful, in charge of their own business operations. Therefore, the government's leverage over these enterprises has been greatly reduced.

In brief, China's energy policy making process has been pluralised. There are not only various governmental bodies with different preferences involved in energy supervision, but also increasingly independent and powerful NOCs exercising effective authority over China's energy market and exerting considerable influence on the government. Therefore, compared with the realist perspective that

---

[14] Downs, 'China', p. 19.

[15] Erica S. Downs, 'The Chinese Energy Security Debate', *The China Quarterly*, 177, 2004, pp. 22–31.

views the state as a unitary actor, the liberalist perspective that notices different interest groups within a state and the significant role of corporations has an advantage in explaining some aspects in this study.

Nevertheless, there are also some limits with liberalism in explaining the domestic dynamics of China's energy diplomacy. Firstly, the liberal perspective focuses on interest groups or social groups instead of the state. It tends to downplay the importance of the state and national interests. But this research identifies the government and the NOCs as two equally important and indispensable players in the movement, and examines the interaction between national and corporate interests behind the phenomenon. Thus, an analysis of the state and national interests is necessary.

Secondly, liberalism is not compatible with the factor of constant international competition for the control of the world's energy resources and the common phenomenon of states' intervention in energy business. It tends to see the cooperative feature of the interstate relations and prefers a free market with minimum state intervention. However, energy is never a pure commercial issue in the market; it is always related to geo-political competition.

Thirdly, liberalism does not pay enough attention to the importance of ideas or ideology. Chapters 6 and 7 will indicate that some Chinese official slogans and Chinese discourses of energy security are used as the essential justification for China's energy diplomacy.

In a word, liberalism's advantage lies in its great attention to the role of corporations, as well as its pluralist view of China's energy management system and policy formulation. However, it downplays the role of the state and takes an overly optimistic or even unrealistic perception of the international energy competition. Therefore, the liberal perspective is not chosen for the general conceptualisation of this study.

2.2.3 *Marxism*

Marxism focuses on class and views class as the main actor in the global political economy. It rejects the individualism of the liberal perspective and embraces the collectivism of the realist perspective.

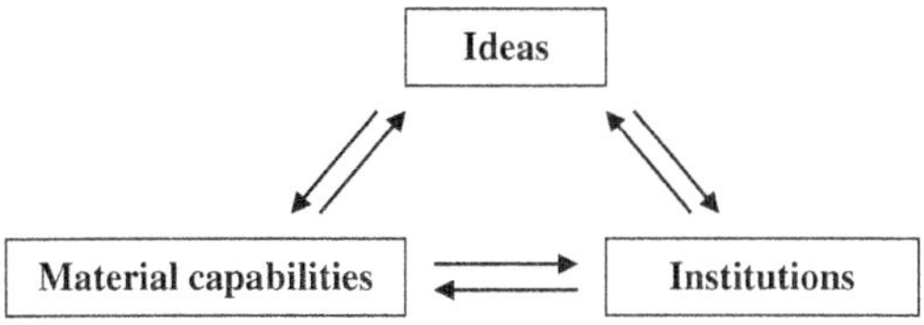

Figure 2.2: Three elements of the hegemony

Source: Robert W. Cox with Timothy J. Sinclair, *Approaches to World Order* (Cambridge: Cambridge University Press, 1996), p. 98.

However, it rejects statism and stresses the importance of class. It regards the state as the representative of class interests rather than the expression of the harmony of communal interests as realism maintains. Moreover, the neo-Gramscian perspective which evolves out of Marxism emphasises 'the role of transnational classes and ideology' in understanding the world economy.[16]

Within this perspective, Robert Cox's idea of the three elements of hegemony (see Figure 2.2) offers an inspiration in explaining why the NOCs are able to exert great influence over the government's energy policy making, although this research does not argue that the NOCs are the hegemony in the domestic dynamics of China's energy diplomacy.

Cox divided the forces supporting the hegemony into three categories — material capabilities, ideas and institutions — which interact in a structure. The question of which way the lines of force run depends on particular cases.

Firstly, material capabilities refers to productive and destructive potentials, such as technological and organisational capabilities, natural resources, industries, armaments, wealth, etc. Secondly, ideas means thought patterns. Thirdly, institutions tend to stabilise and perpetuate a particular order.[17]

[16] O'Brien and Williams, *Global Political Economy*, pp. 21–22.

[17] Robert W. Cox with Timothy J. Sinclair, *Approaches to World Order* (Cambridge: Cambridge University Press, 1996), pp. 97–99.

To some extent, as will be discussed in detail in Chapter 6, Chinese NOCs are armed with all of the three elements, which promote one another to continuously empower these firms. In the first place, their material capabilities are largely derived from their expanding profit and financial capacity. Three factors have contributed to their windfall profit. First and foremost, the NOCs maintain their monopoly status in China's energy market, while gaining increasing autonomy throughout the economic reform. Thus, with the development of the Chinese economy and the increase of China's energy consumption, the NOCs have fast expanded their business across the country and earned huge profits. They have become the leading enterprises and the most profitable SOEs in China. Second, like many other SOEs, the NOCs still enjoy soft budget. Due to China's tax reform, SOEs, including the NOCs, kept all of their after-tax profit between 1994 and 2008. The Chinese government's implementation of the new corporate income tax rate since 2007 has also directly alleviated their tax burdens and contributed to the growth of their profits. Although these firms do not remit their profits to the state, the government has been subsidising them for years. When they suffer losses, the state is always ready to subsidise them, enabling them to earn much profit each year, as Chapter 5 will show. Third, some of the NOCs' subsidiaries have been listed on overseas stock markets. Therefore, they are able to secure more financial resources.

Moreover, the NOCs' institutional advantage is based on their monopoly in China as well as their intimate ties with the CCP and the Chinese government. Their status as the *de facto* controllers and regulators of China's energy market gives them key resources to bargain with the government for the realisation of their corporate interests. Besides, they have special political status in China. The three leading NOCs — China National Petroleum Corporation (CNPC), China Petroleum and Chemical Corporation (Sinopec) and China National Offshore Oil Corporation (CNOOC) — were successively separated from the former MPI, inheriting much clout of this previously powerful ministry. They not only enjoy high bureaucratic levels and status in the government and the CCP systems, but are also supported by the personal patronage of China's top leadership. Some (former) top

Chinese leaders, such as Zeng Qinghong, Zhou Yongkang and Zhang Gaoli, all of whom were/are members of the Standing Committee of the CCP Central Committee's Political Bureau, started their careers in the oil sector or have long-term experiences in this sector. Many other officials from the oil industry have also become important figures on China's political stage. In addition, the NOCs have close personal and financial connections with China's energy administration bodies like the NDRC. Together with their widespread and powerful networks within the party-state, they frequently participate in and exert direct influence on the government's energy studies and policy making.

Last but not least, as will be addressed in Chapter 6 and Chapter 7, the NOCs' ideological resources largely come from two aspects. One is some Chinese official slogans and strategies that legitimise these firms' overseas expansion and the government's support for their global business. A typical example is the Chinese government's call for capable Chinese enterprises to implement the so-called 'going global' strategy. In fact, such official advocacy came about one decade later than the NOCs' own initiatives to conduct ODI and transnational operation. In other words, the NOCs themselves have already gone global even without this official strategy. In this sense, the slogan provides a justification for these firms' ongoing behaviour.

The other important ideological resource for the NOCs is the mainstream voice in China's academic discourses of energy security and the Chinese official discourse of energy security. The former holds that energy diplomacy contributes to China's energy security and national interests; and calls for the government to promote its energy diplomacy and provide assistance to the NOCs' overseas business. It not only offers a key theoretical justification for China's energy diplomacy, but may also influence this movement in the future. The Chinese official discourse of energy security echoes this mainstream academic voice, adding to the legitimacy of China's energy diplomacy.

Nevertheless, the Marxist perspective's focus on class is its weakness in explaining the domestic dynamics of China's energy diplomacy. This research tackles the interaction between national and corporate interests behind the movement; both the state and the NOCs are indispensable players in the game. But like liberalism, Marxism deals with groups within a society rather than the state. Also, the Marxist idea that the state represents the interests of the ruling class is not in line with today's Chinese politics.

Today, it is difficult to follow Marxism to identify which class is the ruling class and which classes are those being ruled in China. As the only ruling party in China, the CCP has been undergoing a transformation from a traditional communist party to a new type of organisation that incorporates interests of various social groups. Originally, during the contemporary Chinese revolution (1921–1949), the membership of the CCP was based upon the working class and the peasantry. However, since China's economic reforms in the late 1970s, more and more people from other classes and social strata such as the intelligentsia and entrepreneurs have joined the party. Especially, since the beginning of the 21st century, the CCP has begun to embrace entrepreneurs. Usually, according to Marxism, the working class and the *bourgeoisie* are opponents and foes. But now in China, they are comrades within the CCP. Considering the Chinese government's long time focus on economic development and its pro-business tendency during the past three decades or so, as well as the popular combination of the interests of businessmen and that of many officials in China, entrepreneurs or the *bourgeoisie* (including Chinese capitalists and foreign capitalists) have become powerful and exert great leverage on policy making of governmental institutions at various levels across China. This phenomenon is also reflected by the composition of the National People's Congress, China's top legislative, as more businessmen are there, at the cost of the number of representatives from the working class and the peasantry.

However, it is improper to say that the CCP has been changed into a capitalist party, because the capitalist members in the party are an absolute minority in number although they are influential; and it is impossible for the CCP leadership to only serve the interests of the business circle and overlook that of workers and peasants. In fact, how to realise fast and sustainable socio–economic development, in order to reduce poverty and satisfy the growing need of the grass roots for living a better life, is always at the centre of the CCP government's work. But one thing is certain, the CCP is transforming from a traditional revolutionary party to a modern ruling party, or changing from a closed and exclusive socialist party to an increasingly open and inclusive organisation that incorporates interests of as many social groups as possible, so as to maintain its political legitimacy. China's situation is unique. If adhering to traditional Marxism, since the CCP is the ruling party, the class it represents could be assumed as the ruling class; as a result, almost all classes in China — the working class, the peasantry, intelligentsia, entrepreneurs, soldiers, etc. — are participating in the CCP and can be viewed as ruling classes. In this sense, the CCP and the Chinese state have gone beyond traditional Marxism to represent interests of different social strata or the Chinese nation as a whole, which is closer to realism's description of the state.

In brief, Marxism, in particular Cox's idea about the three power sources of hegemony, offers a useful inspiration to see the NOCs' capability to effectively influence the government's policy making, especially to gain an insight into the ideological justification for China's energy diplomacy. However, Marxism focuses on class instead of the state and corporations, while this research deals with the government-NOC interplay behind China's energy diplomacy. Also, the Marxist interpretation of the state and the ruling class does not reflect today's Chinese politics. Thus, Marxism is not used for the general conceptualisation of this study.

2.2.4 *The alternative theoretical framework*

None of the three mainstream IPE theories described above is employed as the overall theoretical perspective to explore the interaction of national and corporate interests behind China's energy diplomacy. But they are useful in explaining some aspects of the phenomenon. Therefore, they need to be incorporated into a general theoretical framework which can be used to address the overall interaction between national and corporate interests and that between the government and the NOCs.

The argument about the so-called new diplomacy or triangular diplomacy proposed by John Stopford and Susan Strange in their co-authored book *Rival States, Rival Firms: Competition for World Market Shares* offers an inspiration for the general conceptualisation of this research. The key argument of this book, as Strange concluded in another book, is that outcomes in the IPE are determined by the triangular diplomacy between states, between states and firms, and between firms and other firms. Some diplomatic bargains are carried out within the countries and look like domestic politics; some are reached between government representatives and look like international politics; and somewhere firms alone are engaged can be categorised as transnational politics.[18] Particularly, the two authors contended that the traditional perception of diplomacy as interstate activities or government–government relationship had to be expanded to include government–firm and firm–firm interactions. Especially, national boundaries can no longer define the rules. The negotiation and action is conducted on a triangular basis. 'The traditional players in the embassies and foreign ministries are still in business, but they have been joined by members of other government ministries and by the executives of firms, both local and multinational.'[19]

Also, in her later book *The Retreat of the State: The Diffusion of Power in the World Economy*, Strange provided a further clarification of her perspective and the power shift from states to TNCs. She argued that over the post-war period, the impersonal forces of world markets have been integrated more by private enterprises in finance, industry and trade than by the cooperative decisions of governments. These forces are now more powerful than the states to which ultimate political authority over society and economy is supposed to belong. She maintained that in the last half of the 20th century, the power shift from states to markets would be the biggest change in the IPE. It is most marked with regard to production, trade, investment and finance. One of the major shifts has been the increased power and

[18] Susan Strange, *The Retreat of the State: The Diffusion of Power in the World Economy* (Cambridge: Cambridge University Press, 1996), p. xiv.

[19] John M. Stopford and Susan Strange, *Rival States, Rival Firms: Competition for World Market Shares* (Cambridge: Cambridge University Press, 1991), pp. 19–22.

influence of TNCs. Such a shift has made political players of TNCs. The domain of state authority in society and economy is shrinking. What once were domains of authority exclusive to state authority are now being shared with other sources of authority. Although TNCs do not take over from the governments of states, they have certainly encroached on the governments' domain of power. 'They are increasingly exercising a parallel authority alongside governments in matters of economic management affecting the location of industry and investment, the direction of technological innovation, the management of labour relations and the fiscal extraction of surplus value.' Those powerful transnational cartels formulate systems of governance and actively exercise direct authority over markets.[20]

Strange's perspective is special. She pointed out that politics is larger than what politicians do; and power is exercised by non-state authorities as well as by governments. The international politics cannot be fully understood or analysed without paying attention to international business. Conversely, the international business cannot be fully understood without paying attention to international and domestic politics. There is no basic distinction between domestic politics and international politics; and national economies cannot be analysed in isolation from the world economy. She wrote that 'the new realism of the Stopford–Strange analysis of corporate strategies and state development policies makes it imperative to look seriously at the power exercised by authorities other than states.'[21] In general, her theoretical perspective can be called 'unorthodox realism'.[22]

---

[20] Strange, *The Retreat of the State*, pp. 4–94.

[21] Strange, *The Retreat of the State*, p. xv.

[22] Strange's idea is based on realism but also incorporates important elements of liberalism to explain the development in the international political economy. Thus, how to define her distinguished perspective is another question. The term of unorthodox realism used by Robert O'Brien and Marc Williams in describing the theoretical perspective of Strange is adopted here. They concluded that Strange was an unorthodox realist. Her theoretical approach contains a strong element of power politics, but is also unorthodox because she 'urged observers to take account of the growing role of markets, corporations and technological innovations in changing the environment in which state operates.' O'Brien and Williams, *Global Political Economy*, p. 27.

Obviously, the new diplomacy defined by Stopford and Strange is a kind of diplomacy in a broad sense. It includes not only the traditional inter-state diplomacy, but also the interaction between firms and states, and that among firms. To a great extent, it is in accordance with this research's focus on the domestic dynamics of China's energy diplomacy, especially the interaction of national and corporate interests. The Chinese government and Chinese NOCs are two equally important players in the movement. China's energy diplomacy can also be viewed in the broad sense of the new diplomacy. Generally, it includes not only the government's active engagement with energy exporting countries in various parts of the world, but also the NOCs' fast development of their transnational operation particularly ODI in and M&A of oil and natural gas assets abroad. Both players have critical interests behind the movement. More importantly, the interaction between the government's national interests and the NOCs' corporate interests, which is the focus of this research, is largely examined within this broad definition of diplomacy. The convergence of national and corporate interests is the key momentum of this joint movement by the two players. Therefore, this study draws on the new diplomacy idea as the general theoretical framework.

Another issue that needs clarifying here is that among the three aspects of the new diplomacy idea (government–government, firm–firm and government–firm interactions), why only one aspect — the interaction between the government and the firm is stressed in this research. The reason is that the other two aspects, to which much attention has been paid traditionally, have been relatively well explored in prior academic research. The government–government interaction is prioritised in realism, which focuses on the role of state and government; and the firm–firm interaction has been tackled by liberalism, which emphasises the role of the market and TNCs. But the government–firm interaction is often overlooked and there is not much literature in this regard. Thus, academic research in this area has much room to improve.

Basically, the advantage of the new diplomacy idea in conceptualising general interaction between national and corporate interests behind China's energy diplomacy is that it attaches great importance to both the states and TNCs. While recognising the importance of the states, it noted TNCs' increasing role in the world economy and politics, which is a significant new development in the IPE after the Second World War. Similar to the idea of the triangular diplomacy, David Zweig and Bi Jianhai wrote that China's traditional foreign policy making bodies still make the key decisions, 'but a more pluralistic environment is emerging and allowing business leaders to help shape foreign policy.' For example, the China Institute for International Studies which is a governmental think tank under the MFA has held numerous conferences bringing together representatives from the academic circle, business, the government, etc. to devise strategies for China's top leadership.[23]

As to the interaction between national and corporate interests behind the movement, Chapter 5 will explain that the government's key national interests are safeguarding China's energy security and improving Beijing's international relations, which are beneficial for the country's economic development, social stability and national unification, hence contributing to the CCP's political legitimacy. The NOCs' major corporate interests are seeking new investment markets, generating further profits and fulfilling their development strategies. Thus, there is a convergence of national and corporate interests, which is the fundamental momentum of China's energy diplomacy.

Furthermore, the phenomenon is an outcome of the interaction between the government's and the NOCs'. Chapter 6 will show that China's economic and enterprise reform has created the three leading Chinese NOCs and transformed them into modern TNCs; and the Chinese state has decentralised significant authority to the NOCs. Therefore, these firms enjoy high degree of autonomy in business operation, close linkages with the party-state and monopoly in the domestic market, which boosts their rapid development. In particular, they have significant material, institutional and ideological resources that enable them to directly participate in and effectively influence the government's policy formulation. They have actually become

[23] David Zweig and Bi Jianhai, 'China's global hunt for energy', *Foreign Affairs* 84: 5, 2005, pp. 25–38.

important players in China's policy making process alongside various governmental institutions. They are increasingly powerful both economically and politically to the detriment of the state's leverage over them. The government oversees the NOCs and the oil sector mainly through oil supervising agencies in the State Council, the nomenclatural system and the investment approval process. But these traditional methods are not as effective as often assumed. Chapter 8 will mention that sometimes, there is divergence between national and corporate interests, because the NOCs always regard their corporate interests or profits as the priority, while the government has to take into consideration of national interests like energy security and the overall diplomatic and international strategies. Of course, however, there is more convergence than divergence of national and corporate interests. That is why China's energy diplomacy has made so much progress in the past two decades, as will be demonstrated in Chapter 4.

Although the new diplomacy idea is useful for the overall theoretical framework of this research, it does not explain every aspect is the interaction of national and corporate interests behind China's energy diplomacy, such as the government's national security concerns and national interest consideration, the long-neglected but critical role of the NOCs, China's pluralised policy making process and the NOCs' resources to exert influence over the government. Whereas, some ideas from the three mainstream IPE theories can be used to explain these sub-issues. Therefore, while adopting the new diplomacy idea as the general theoretical framework, this study also incorporates useful elements from realism, liberalism and Marxism for the conceptualisation.

2.3 Conclusion

As to the existing explanations of the relationship between national and corporate interests behind Chinas' energy diplomacy, there are two major problems. One is the conventional logic that underestimates the importance of the NOCs and their corporate interests and self-motivation in the movement. The other is an underestimation of the role of the government and national interests. In fact, both the government and the NOCs are indispensable and neither should be downplayed.

The conceptualisation of this research draws on the new diplomacy idea as the overall theoretical framework and incorporates useful elements from the three mainstream IPE theories — realism, liberalism and Marxism. This study does not choose the three mainstream theories for the overall conceptualisation; all of them have merits and shortcomings in explaining the issue. The realist perspective can explain the Chinese government's concern about energy security and its national interest consideration, but downplays the role of the NOCs and corporate interests; the liberalist perspective attaches importance to the long-neglected role of the NOCs and their corporate interests and is useful to see the pluralised policy making process in China, but it underestimates the role of the state and national interests; and the Marxist perspective has an advantage in explaining the NOCs' ability to exert influence on the government's policy making, but neglects the state and national interests and its interpretation of class does not reflect the current situation in China. Therefore, although all of them explain parts of the phenomenon, they are not suitable for exploring the overall interaction between national and corporate interests behind China's energy diplomacy. The new diplomacy idea that pays attention to the state-firm interaction can be adopted to examine the general situation. Under this general framework, the three mainstream IPE theories are included to explain those important aspects.

The next chapter discusses the basic background of China's energy diplomacy, i.e. China's energy demand and supply, explaining why the country is facing energy shortage, in particular its growing dependence on oil and natural gas supply from abroad.

Chapter

3

China's Energy Shortage

This chapter offers a brief discussion of China's energy shortage, in particular the demand and supply situation of oil and natural gas. China's fast socio–economic development and industrialisation have been built upon rapidly growing consumption of energy resources. The country's energy consumption has increased substantially and continuously since its economic take off in the early 1990s. There are six major factors leading to China's expanding appetite for energy resources, namely the economic growth pattern, China's role as a major manufacturer in the world economy, the low energy utilisation level, the rapid urbanisation process, the changing energy consumption structure, and the country's increasing use of transport facilities. As a result, although China itself is a major energy producer, its domestic energy supply is not able to meet its demand. In the recent years, the gap between the domestic energy production and consumption has been widening; and the country has been increasingly dependent on overseas oil and natural gas supply. Such fundamental energy demand–supply situation forms the basic background of China's energy diplomacy.

This chapter consists of three sections. The first section demonstrates the substantial growth of China's energy consumption, in particular oil and natural gas, before providing six reasons of this phenomenon. The second section deals with the country's domestic energy supply condition and capability. The third section illustrates the expanding gap between China's energy consumption and production, as well as the country's increasing reliance on energy import.

3.1 China's energy demand

Thanks to the fast development of the Chinese economy, China's energy consumption has expanded for more than five times since the 1980s and reached 4,260 million tonnes of coal equivalent in 2014, according to the statistics from China's National Bureau of Statistics (see Figure 3.1).[1] The country has been the largest energy consumer in the world since 2010 when its overall energy consumption surpassed that of the United States (Figure 3.2 shows the major energy consumers in the world in 2013). In 2013, according to the statistics from British Petroleum (BP), China's total energy consumption reached 2,852.4 million tonnes of oil equivalent, 4.7 per cent increase from the previous year, and accounting for 22.4 per cent of the world's total energy consumption. Figure 3.3 illustrates the shares of the five leading energy consumers — China, the United States, Russia, India and Japan's energy consumption in the world's total energy consumption in 2013. Also, China's energy consumption structure was diversified, with coal taking up 67 per cent of the energy portfolio, oil 18 per cent, hydroelectricity 7 per cent, natural gas

[1] Guojia tongji ju (National Bureau of Statistics, the People's Republic of China), '2014 guomin jingji he shehui fazhan tongji gongbao' (Statistical Communiqué of the People's Republic of China on the 2014 National Economic and Social Development), 26 February 2015, http://www.stats.gov.cn/tjsj/zxfb/201502/t20150226_685799.html, accessed 8 March 2015.

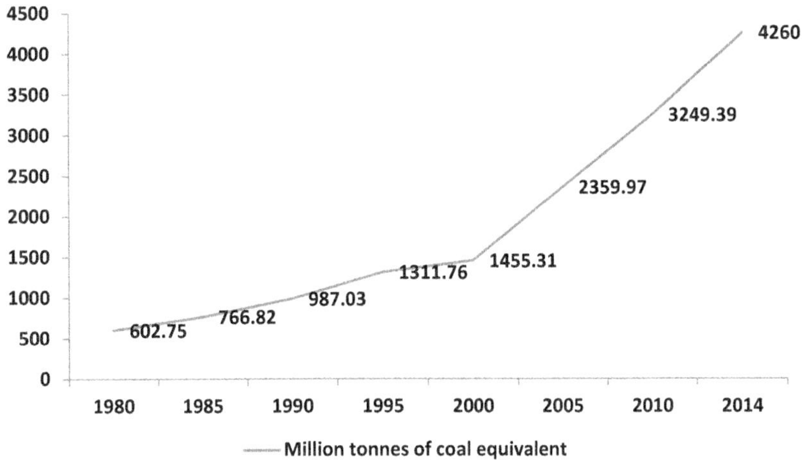

Figure 3.1: China's overall energy consumption, 1980–2014 (Million tonnes of coal equivalent)

Source: Guojia tongjiju nengyuan tongjisi (Department for Energy Statistics, National Bureau of Statistics, the People's Republic of China), *Zhongguo nengyuan tongji nianjian 2013* (*China Energy Statistical Yearbook 2013*), (Beijing: Zhongguo tongji chubanshe (Beijing: China Statistics Press), 2013), p. 55; and Guojia tongjiju (National Bureau of Statistics, the People's Republic of China), '2014 guomin jingji he shehui fazhan tongji gongbao' (Statistical Communiqué of the People's Republic of China on the 2014 National Economic and Social Development), 26 February 2015, http://www.stats.gov.cn/tjsj/zxfb/201502/t20150226_685799.html, accessed 8 March 2015.

5 per cent, renewable energy 2 per cent and nuclear energy 1 per cent, respectively.[2]

While China is the top energy consuming country in the world, the level of its per capita energy consumption is still relatively low. According to the Chinese official white paper *China's Energy Policy 2012*, the country's per capita energy consumption is only one third of the average amount of developed countries. 'But as the economy and society progress and living standards improve, China's energy

[2] British Petroleum, *BP Statistical Review of World Energy June 2014*, http://www.bp.com/content/dam/bp/pdf/Energy-economics/statistical-review-2014/BP-statistical-review-of-world-energy-2014-Bob-Dudley-Group-Chief-Executive-introduction.pdf, accessed 24 August 2014.

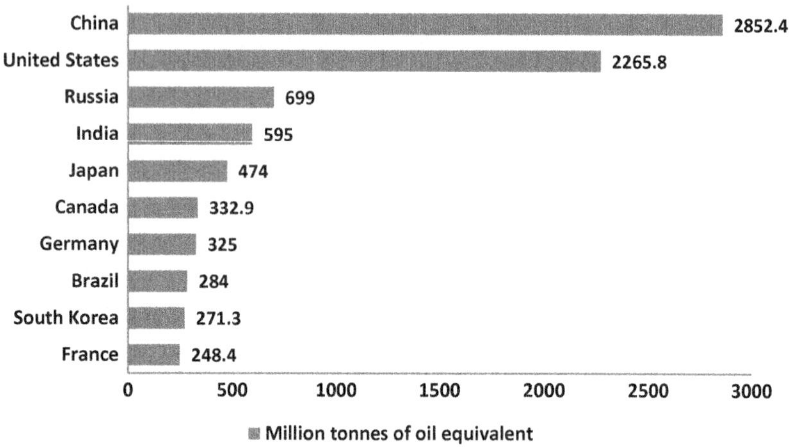

Figure 3.2: Top ten energy consuming countries in the world, 2013 (Million tonnes of oil equivalent)

Source: British Petroleum (BP), *BP Statistical Review of World Energy June 2014*, http://www.bp.com/content/dam/bp/pdf/Energy-economics/statistical-review-2014/BP-statistical-review-of-world-energy-2014-Bob-Dudley-Group-Chief-Executive-introduction.pdf, accessed 24 August 2014.

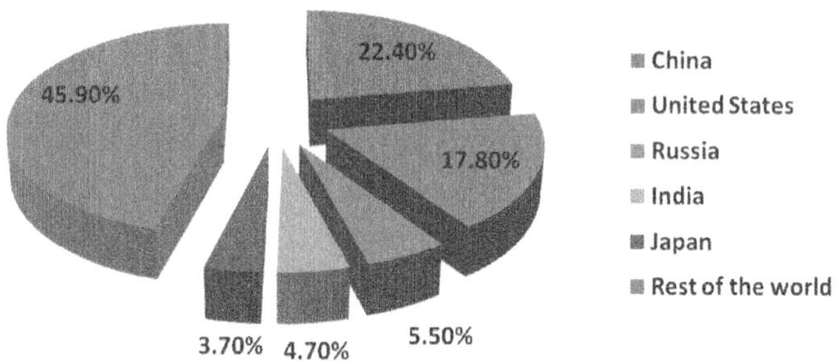

Figure 3.3: Shares of China, the United States, Russia, India and Japan's energy consumption in the world's total consumption, 2013 (Percentage)

Source: BP, *BP Statistical Review of World Energy June 2014.*

consumption will continue to rise sharply, and there will be a growing restraint on resources.'[3]

[3]The State Council Information Office, the People's Republic of China, *China's Energy Policy 2012*, October 2012, http://english.gov.cn/archive/white_paper/2014/09/09/content_281474986284499.htm, accessed 14 January 2015.

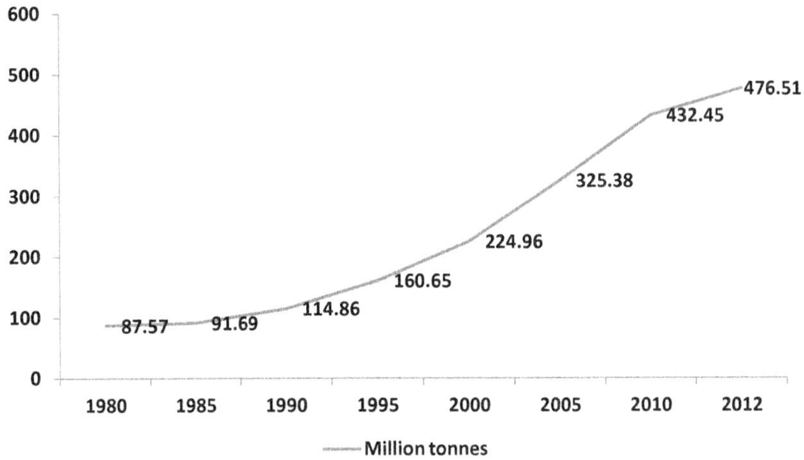

Figure 3.4: China's oil consumption, 1980–2012
(Million tonnes)

Source: Guojia tongjiju nengyuan tongjisi (Department for Energy Statistics, National Bureau of Statistics, the People's Republic of China), *Zhongguo nengyuan tongji nianjian 2013* (*China Energy Statistical Yearbook 2013*), pp. 118–119.

With specific regard to China's oil consumption, it has increased for nearly four times during the past 35 years, from 87.57 million tonnes in 1980 to 476.51 million tonnes in 2012 (Figure 3.4).[4] Moreover, China surpassed Japan in 2003 to become the world's second largest oil consumer, only after the United States (see Figures 3.5 and 3.6). In 2013, the country consumed 507.4 million tonnes or 10,756 thousand barrels per day (bpd) of oil, accounting for 12.1 per cent of the world's total oil consumption.[5]

[4] Guojia tongjiju nengyuan tongjisi (Department for Energy Statistics, National Bureau of Statistics, People's Republic of China), *Zhongguo Nengyuan Tongji Nianjian 2013* (*China Energy Statistical Yearbook 2013*), pp. 118–119.

[5] British Petroleum, *BP Statistical Review of World Energy June 2008*, http://www.bp.com/liveassets/bp_internet/globalbp/globalbp_uk_english/reports_and_publications/statistical_energy_review_2008/STAGING/local_assets/downloads/pdf/statistical_review_of_world_energy_full_review_2008.pdf, accessed 25 December 2008; and British Petroleum, *BP Statistical Review of World Energy June 2014*.

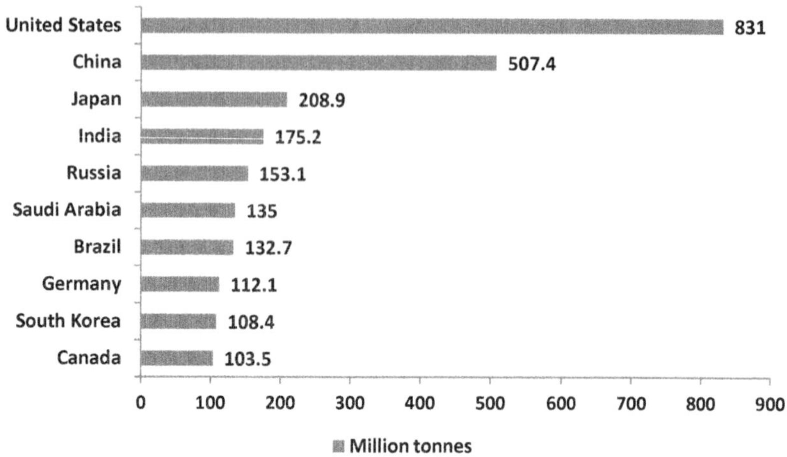

Figure 3.5: Top 10 oil consuming countries in the world, 2013 (Million tonnes)

Source: BP, *BP Statistical Review of World Energy June 2014*.

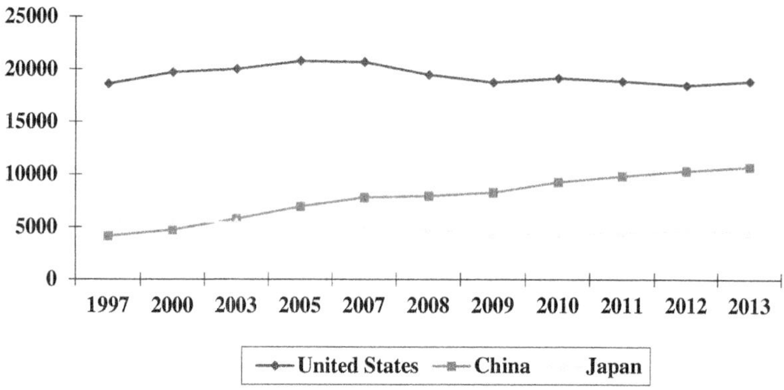

Figure 3.6: The oil consumption in the United States, China and Japan, 1997–2013 (1000 bpd)

Sources: BP, *BP Statistical Review of World Energy June 2008*, http://www.bp.com/liveassets/bp_internet/globalbp/globalbp_uk_english/reports_and_publications/statistical_energy_review_2008/STAGING/local_assets/downloads/pdf/statistical_review_of_world_energy_full_review_2008.pdf, accessed 25 December 2008, and BP, *BP Statistical Review of World Energy June 2014*.

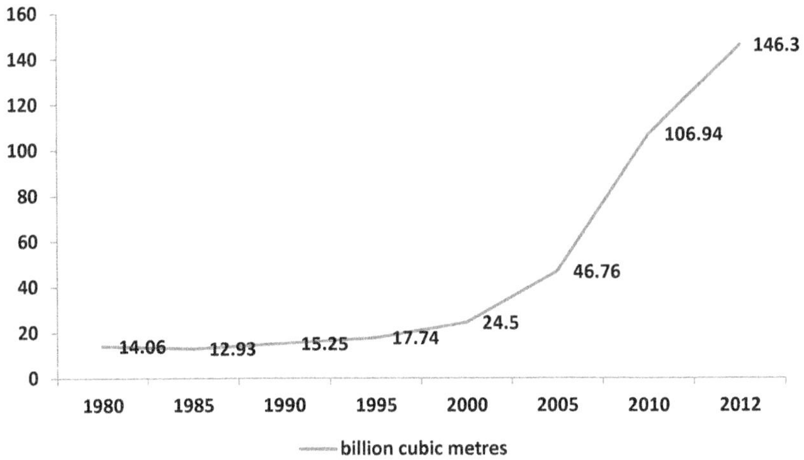

Figure 3.7: China's natural gas consumption, 1980–2012 (Billion cubic metres)

Source: Guojia tongjiju nengyuan tongjisi (Department for Energy Statistics, National Bureau of Statistics, the People's Republic of China), *Zhongguo nengyuan tongji nianjian 2013* (*China Energy Statistical Yearbook 2013*), pp. 128–129.

It is estimated that China's oil consumption in 2014 and 2015 will increase by 360 thousand bpd on average.[6]

As to China's natural gas consumption, it has grown for about seven times since the beginning of this century (Figure 3.7).[7] Indeed, China's natural gas market is one of the fastest growing in the world. The annual increase rate of the country's natural gas consumption from 2000 to 2013 was 16 per cent.[8] In 2013, the country consumed 161.6 billion cubic metres or 145.5 million tonnes of oil equivalent

[6] U.S. Energy Information Administration, 'Short-Term Energy Outlook', December 2014, http://www.eia.gov/forecasts/steo/pdf/steo_full.pdf, accessed 13 December 2014.

[7] Guojia tongjiju nengyuan tongjisi (Department for Energy Statistics, National Bureau of Statistics, the People's Republic of China), *Zhongguo Nengyuan Tongji Nianjian 2013* (*China Energy Statistical Yearbook 2013*), pp. 128–129.

[8] Xinhua, 'Woguo tianranqi duiwai yicundu shengzhi 32.2 per cent' (China's natural gas dependency on foreign supply has rose to 32.2 per cent), 19 January 2015, http://news.xinhuanet.com/energy/2015-01/19/c_127398476.htm, accessed 8 February 2015.

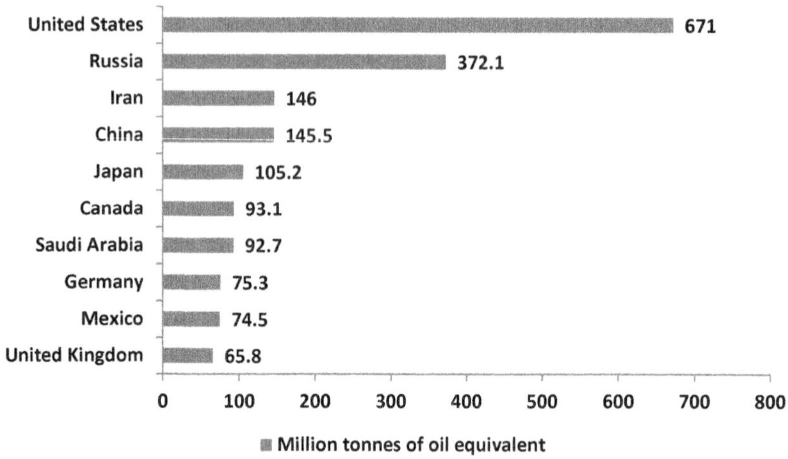

Figure 3.8: Top 10 natural gas consuming countries in the world, 2013 (Million tonnes of oil equivalent)

Source: BP, *BP Statistical Review of World Energy June 2014.*

of natural gas, accounting for 4.8 per cent of the world's overall consumption of natural gas, keeping its position as the world's fourth largest natural gas consumer (Figure 3.8) after the United States (73.72 billion cubic metres), Russia (41.35 billion cubic metres) and Iran (16.26 billion cubic metres).[9] The International Energy Agency (IEA) forecasted that China's natural gas consumption would increase by 12 per cent annually from 2013 to 2019. The fast expansion of China's appetite for natural gas is primarily driven by the higher living standards of the country's middle class, strong industrial consumption and China's desire to clean up air pollution by switching from coal to gas for power generation.[10]

Nevertheless, in accordance with the Chinese official white paper which pointed out the country's per capita energy consumption is

[9] British Petroleum, *BP Statistical Review of World Energy June 2014.*

[10] Julie Jiang and Chen Ding, *Updates on Overseas Investments by China's National Oil Companies: Achievements and Challenges since 2011*, International Energy Agency, Partner Country Series, 2014, http://www.iea.org/publications/freepublications/publication/PartnerCountrySeriesUpdateonOverseasInvestmentsbyChinasNationalOilCompanies.pdf, accessed 24 August 2014.

Figure 3.9: Per capita oil consumption in various parts of the world, 2013 (Tonnes)

Source: BP, *BP Statistical Review of World Energy June 2014.*

very low, China's per capita consumption of both oil and natural gas are among the lowest levels in the world, as shown by two figures in *BP Statistical Review of World Energy June 2014* (Figures 3.9 and 3.10). This means that there is still much room for the future growth of China's energy consumption, especially with the country's continuous socio-economic development and the improvement of the living standard of the Chinese people. Just as the White Paper stated, Although China's energy consumption has experienced rapid growth in the past few years, the country's per capita energy consumption is low — only about one third of the average of developed countries. Moreover, 'China will still be in a stage featuring accelerated industrialisation and urbanisation for a long time to come, facing the challenging tasks of developing its economy and improving its people's livelihood. Its energy needs will go on to increase in the future'.[11]

[11]The State Council Information Office, the People's Republic of China, *China's Energy Policy 2012.*

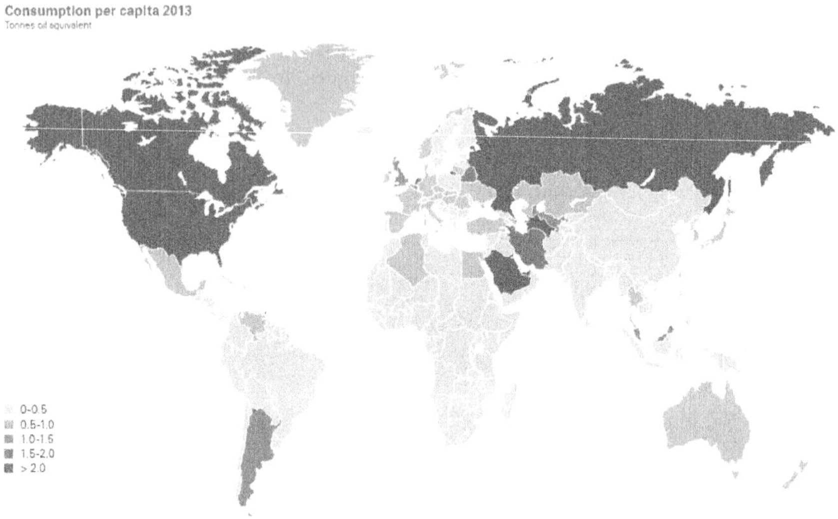

Figure 3.10: Per capita natural gas consumption in various parts of the world, 2013 (Tonnes of oil equivalent)

Source: BP, *BP Statistical Review of World Energy June 2014.*

Although it is easy to reach a conclusion that China's spectacular economic growth since the 1980s had led to the surge of its energy demand, a more important question needs answering — why and how does China's economic development boost its appetite for energy so quickly? The rest of this sub-section offers six major reasons for the country's expanding energy consumption, including the economic growth pattern, China's role as a major manufacturer in the world economy, the low energy utilisation level, the fast-expanding urbanisation, the transformation of China's energy consumption structure, and the country's growing use of transport facilities.

3.1.1 *Economic growth pattern*

Since its economic reforms began in the late 1970s, China has experiencing double-digit rates of rapid economic development for about three decades; its gross domestic product (GDP) has increased by more than 30 times, although the country's per capita GDP is still very low. In 2010, the GDP of China overtook that of Japan, making the country the second largest economy in the world only behind the United States. China's GDP was USD 9.3 trillion in 2013 and is expected to surpass USD 10 trillion in 2014. In accordance, large amounts of energy resources have been consumed to support the rise of the Chinese economy especially the country's fast expanding industrialisation.

More importantly, the way that energy resources are consumed in China is determined in a great extent by the country's economic growth pattern. Rather than increasing in an unchanged way throughout its economic reform, China's energy consumption has been changing with the shift of the economic growth pattern. As Barry Naughton pointed out, its economic growth patterns were different during two periods. In the first period from the late 1970s to the mid-1990s, China's industrial development was focused on light industry, with relatively less energy consumption; while the second period since 1995 has witnessed a fast growth of energy-intensive industries. In the former period, China's industrial structure was diversified and the light industry that had been neglected previously was developed rapidly, providing it with opportunity to create more value with given inputs of capital and energy. The most energy-intensive sectors grew more slowly than the overall industry. The energy intensity of the Chinese economy dropped substantially during this period. However, the share of light industry in China's total industrial output declined dramatically between 1995 and 2004. Meanwhile, several energy-intensive sectors grew more rapidly than the overall industry. Because of the ongoing industrialisation and a sustained investment effort, China's energy demand surged.[12]

In particular, in the first a few years of this century, there was a new round of fast economic growth in China, which was dominated by the development of heavy industry. The newly emerging demand for infrastructure construction caused by the country's fast economic growth and urbanisation has stimulated investment in heavy industry.

[12] Barry Naughton, *The Chinese Economy: Transition and Growth* (Cambridge, MA: The MIT Press, 2007), pp. 329–333.

Such economic growth patterns have boosted, and are expected to continue boosting energy consumption in China's industrial sectors.

According to the Chinese official White Paper, industry is the major energy consuming sector in China, as its energy consumption accounts for more than 70 per cent of China's total energy consumption. Within it, energy consumption by four major energy-intensive industrial sectors, namely steel, non-ferrous metals, chemicals and building materials, accounts for 40 per cent of the country's total energy consumption.[13] While China's economic growth was much faster than its energy consumption for most of the time, especially during the 1990s, this situation was reversed in 2003 and 2004 (Figure 3.11). The annual increase rate of China's energy consumption in 2003 and 2004 was above 15.3 per cent and 16.1 per cent respectively, much higher than the country's GDP growth rate which was 10 per cent and 10.1 per cent respectively.[14] Also, as Figure 3.11 demonstrates, the gap between the growth rate of China's GDP and that of its energy consumption in the 2000s was relatively narrow than that in the 1990s.

3.1.2 *'The workshop of the world'*

The second factor influencing China's energy consumption is the country's role as 'the workshop of the world'. As an export-oriented economy and an important part of the global production chain, China imports and consumes substantial energy resources in order to produce manufactured goods for export. Thus, many consumers of those manufactured goods are not people in China but people in other parts of the world especially North America and Europe. That is to say, the ultimate consumers of those energy resources are often people purchasing 'made in China' goods outside China. 'Much of

---

[13]The State Council Information Office, the People's Republic of China, *China's Energy Policy 2012.*

[14]Guojia tongjiju nengyuan tongjisi (Department for Energy Statistics, National Bureau of Statistics, the People's Republic of China), *Zhongguo Nengyuan Tongji Nianjian 2013* (*China Energy Statistical Yearbook 2013*), p. 3.

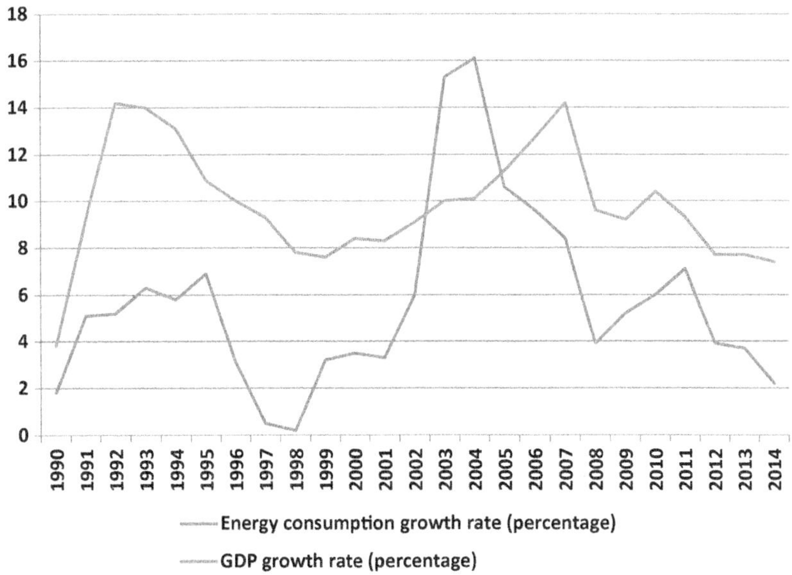

Figure 3.11: A comparison of China's GDP growth and energy consumption growth, 1990–2014 (Percentage)

Source: Guojia tongjiju nengyuan tongjisi (Department for Energy Statistics, National Bureau of Statistics, the People's Republic of China), *Zhongguo nengyuan tongji nianjian 2013* (*China Energy Statistical Yearbook 2013*), p. 3; Guojia tongjiju (National Bureau of Statistics, the People's Republic of China), *Zhongguo tongji nianjian 2014* (*China Statistical Yearbook 2014*), online edition, http://www.stats.gov.cn/tjsj/ndsj/2014/indexch.htm, accessed 8 March 2015; and Guojia tongjiju (National Bureau of Statistics, the People's Republic of China), '2014 guomin jingji he shehui fazhan tongji gongbao' (Statistical Communiqué of the People's Republic of China on the 2014 National Economic and Social Development), 26 February 2015, http://www.stats.gov.cn/tjsj/zxfb/201502/t20150226_685799.html, accessed 8 March 2015.

the energy China consumes is used to make products sold to the rest of the world, thus replacing energy demand in other countries'.[15]

[15] Daniel H. Rosen and Trevor Houser, 'China Energy: A Guide for the Perplexed', *China Balance Sheet*, A Joint Project by the Center for Strategic and International Studies and the Peterson Institute for International Economics, May 2007, http://www.petersoninstitute.org/publications/papers/rosen0507.pdf, accessed 3 May 2007, p. 5.

Therefore, when producing many cheap manufactured goods for consumers all over the world, China not only suffers from some negative effects such as pollution and environmental degradation, but is also widely blamed as a chief energy consumer and major polluter in the world. Although there have not been any detailed statistics showing the proportion of energy used to manufacture goods for export in China's total energy consumption, some other statistics might be useful for reference. For example, a United Kingdom government-funded body found that a quarter of China's greenhouse gas emissions in 2004 were caused by manufacturing goods exported to the West.[16]

China's role of doing such a hard but thankless job is due to its position in global production chain (see Figure 3.12). To some extent, such position shifts a great number of the energy demand from developed countries to China. In the past three decades or so, many Western industrialised countries have transferred, usually through foreign direct investment (FDI), a significant proportion of their manufacturing sectors with labour intensive technique to China — a country with abundant cheap labour and a very favourable environment for FDI inflow.[17]

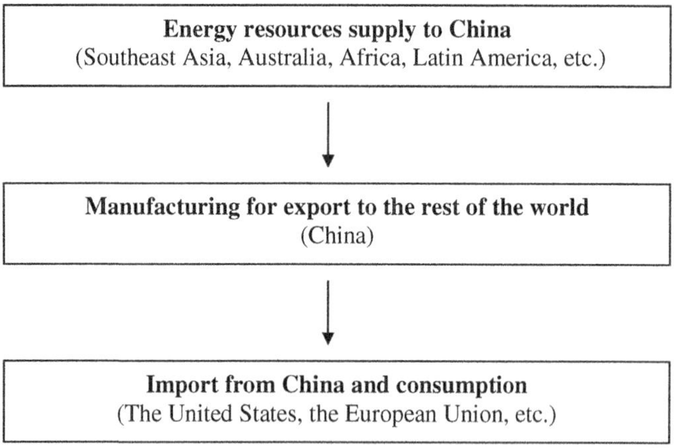

Figure 3.12: China as a major manufacturer in the global production chain
Source: Author's compilation.

[16] BBC News, 'Exports fuel China's CO2 output', 19 October 2007, http://news.bbc.co.uk/1/hi/uk/7052115.stm, accessed 20 December 2007.
[17] Naughton, *The Chinese Economy*, pp. 401–423.

As a result, while becoming an essential part of the global production chain, China has encountered the challenge of insufficient domestic energy supply. It has to import more and more energy resources from the Middle East, Southeast Asia, Africa, Australia, Latin America, etc., and then produce manufactured goods for export to developed countries like the United States and some European countries.[18] Although China does not ultimately consume such huge amounts of energy, it has to import these resources in order to keep the 'the workshop of the world' running and the Chinese economy booming.

3.1.3 *Low energy utilisation level*

The third factor is China's inefficient use of energy resources. China has been improving its energy efficiency for years, but it still has a long way to go. In the past two decades or so, the Chinese government has been making great efforts in promoting energy conservation by formulating a series of energy conservation policies and measures, and the country has realised notable achievements in energy conservation drive, with its energy efficiency continuously improving.[19]

Indeed, the energy efficiency of the Chinese economy has increased substantially since the late 1970s. From 1978 to 2005, the country's energy consumption grew at an annual rate of 5.2 per cent, while its real GDP grew at almost 10 per cent each year, indicating that China produces 10 times the real GDP it produced in 1978, with an energy consumption of only three and half times as much energy as it consumed in 1978. In this sense, the Chinese economy has become more energy efficient.[20]

---

[18] For a detailed discussion of China at the centre of global and regional production networks, see John Wong, 'China's Economic Growth in East Asian Context', in John Wong and Wei Liu ed., *China's Surging Economy: Adjusting for More Balanced Development* (Singapore: World Scientific Publishing Co. Pte. Ltd., 2007), pp. 35–38.

[19] Zhou Fuqiu, 'Energy conservation policy development in China', in Elspeth Thomson, Youngho Chang and Jae-Seung Lee eds., *Energy Conservation in East Asia*, (Singapore: World Scientific Publishing Co. Pte. Ltd., 2011), pp. 143–159.

[20] Naughton, *The Chinese Economy*, p. 329 and pp. 336–337.

Similarly, the official Chinese White Paper maintained that China's energy-saving progresses are obvious. According to this Paper, during 1981 and 2011, the annual increase rate of China's energy consumption was 5.82 per cent, while the country's GDP grew by 10 per cent annually. Between 2006 and 2011, China's energy consumption for every RMB 10,000 of GDP declined by 20.7 per cent, saving energy equivalent to 710 million tonnes of standard coal. [21]

Nevertheless, China is still an inefficient energy user in many aspects, compared with developed economies.[22] For instance, the official Chinese White Paper also pointed out that due to the shortcomings of China's industrial structure and economic growth pattern, China's energy consumption for per unit GDP is much higher than that of developed countries and even some industrialising countries. China's energy-intensive industries suffer from a technological underdevelopment. The share of energy consumption by the secondary industries, especially those energy-intensive sectors, is too large in China's overall energy consumption. China's low energy efficiency results in the country's high energy consumption for per unit GDP.[23]

Specifically, to produce every USD 1 million of GDP, the energy consumption of China is 2.5 times as that of the United States, 5 times as that of some European Union countries and 9 times as that of Japan. China's energy utilisation ratio in average is 32 per cent, nearly 10 per cent lower than the level in developed countries. The overall efficiency of China's energy system (*nengyuan xitong de zongxiaolu*) is merely 9 per cent, less than half of that in developed countries. This indicates that about 90 per cent of energy was wasted during extraction, processing, transformation, transportation, distribution and final utilisation. Among them, the efficiency of extraction was 32 per cent,

---

[21] The State Council Information Office, the People's Republic of China, *China's Energy Policy 2012.*

[22] Zhou Fuqiu, 'Energy conservation policy development in China'.

[23] The State Council Information Office of the People's Republic of China, *China's Energy Policy 2012*, October 2012, http://english.gov.cn/archive/white_paper/2014/09/09/content_281474986284499.htm, accessed 14 January 2015.

that of intermediate steps (processing, transformation, transportation and distribution) was 70 per cent, and that of final utilisation was 41 per cent. The so-called energy efficiency (*nengyuan xiaolu*) refers to the product of the extraction efficiency and the intermediate steps efficiency mentioned above. China's energy efficiency is about 29 per cent, 10 per cent lower than the advanced level in the world; and its final utilisation efficiency is more than 10 per cent lower than the advanced level. Also, the country's average oil consumption in per unit GDP growth is 30 per cent higher than the U.S. level and double the Japanese level. According to a research report produced by the Development Research Centre of the State Council of the PRC, China's oil density (oil consumption in per unit GDP) is 2.35 times (nominal rate of exchange) as much as the average level of Organisation for Economic Cooperation and Development (OECD) countries.[24]

3.1.4 *Urbanisation*

Another factor boosting China's energy demand is the country's ongoing urbanisation. Thanks to China's rapid industrialisation and gradual relaxation of control over internal immigration, an increasing number of people, about 20 million each year, have been moving from rural areas to urban areas to search for better jobs, livelihood and permanent or temporary settlements. While China's urbanisation rate was merely 18 per cent in 1978 when the country reformed and opened up, it has increased to 52.6 per cent in 2012, reaching the average level in the world (Figure 3.13 illustrates China's increasing rate of urbanisation since the beginning of this century). In 2014, there were 142 cities in China with a population of above one million, including six super-cities with a population of above 10 million. In the recent years, three mega city clusters, namely Jingjinji (Beijing–Tianjin–Hebei) region cluster, the Yangtze River delta cluster and the

---

[24] Cui Minxuan, ed., *2007 Zhongguo nengyuan fazhan baogao (The Energy Development Report of China 2007)* (Beijing: Shehui kexue wenxian chubanshe, March 2007), pp. 24–25, p. 100 and p. 134.

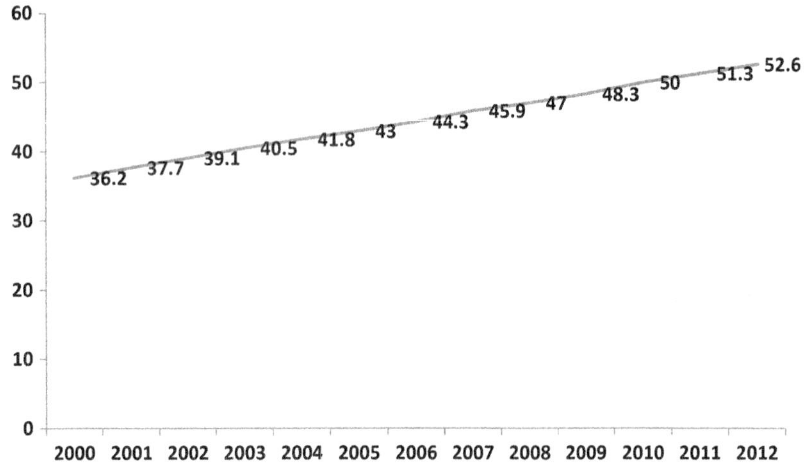

Figure 3.13: China's urbanisation rate, 2000–2012
(Percentage)

Source: Guojia tongjiju nengyuan tongjisi (Department for Energy Statistics, National Bureau of Statistics, the People's Republic of China), *Zhongguo nengyuan tongji nianjian 2013* (*China Energy Statistical Yearbook 2013*), pp. 4–5.

Pearl River delta cluster, have emerged in Eastern part of China which is relatively developed.[25]

China's urbanisation process is still going on. According to the Chinese Academy of Social Science, the country's urbanisation rate is expected to reach at least 55 per cent by 2020.[26] This will significantly change the proportion of rural population and that of urban population in China. According to reports published by the Organisation of

[25] Xinhua, 'Zhongguo baiwan yishang renkou chengshi da142ge, 6ge chengshi renkou chaoqianwan' ('The number of cities in China with a population of above one million reaches 142, and the number of cities with a population of above 10 million reaches 6'), 19 March 2014, http://news.xinhuanet.com/fortune/2014-03/19/c_119840576.htm, accessed 17 January 2015.

[26] Souhu, 'Zhongguo 2020 nian chengshihualv jiangda 55 per cent, chenggshibing jiang jizhong baofa' ('China's urbanisation rate will reach 55 per cent in 2020 and there will be an outbreak of city diseases'), 9 February 2012, http://news.sohu.com/20120209/n334228775.shtml, accessed 17 January 2015; and Cui Minxuan, ed., *The Energy Development Report of China 2007*, p. 23.

the Petroleum Exporting Countries (OPEC), nowhere in the world is the process of urbanisation as vivid as in China. While in 1950 nearly 90 per cent of the country's population lived in rural areas, the urbanisation process began to accelerate after 1980. By 2011, for the first time, there were more people living in urban areas in China than in rural areas, and by 2040 the share in urban areas is expected to have reached 73 per cent. In particular, between 2013 and 2040, China will be the region that will experience the most dramatic migration from rural to urban areas in the world. Although China's rural population was 254 million more than its urban population in 2006, the country's urban population was 82 million more than its rural population in 2013, and it is estimated that there will be 672 million more urban residents than rural residents in 2040 in China (see Figure 3.14).[27]

The continuous urbanisation process and expanding urban population have directly led to China's increasing appetite for energy. For example, the development of China's urbanisation is accompanied by expanded prosperity, productivity gains in the move away from the agricultural sector, and higher levels of car ownership as well. Urban development will also contribute to the swift expansion of the middle class, which will have an implication for the country's consumption patterns. China's evolving middle class, which is already the second largest in the world only after the United States, could be a key source of long-term energy demand.[28] Indeed, in China, the per capita residential energy consumption in urban areas is much higher than that in rural areas, although the gap is narrowing down as the latter is catching up recently. Figure 3.15 provides a comparison of China's per capital residential energy consumption in urban and rural areas between 1980 and 2012.

---

[27] OPEC, *World Oil Outlook 2008* (Vienna: OPEC Secretariat, 2008), http://www.opec.org/library/World%20Oil%20Outlook/pdf/WOO2008.pdf, accessed 6 January 2009, p. 21; and OPEC, *2014 World Oil Outlook* (Vienna: OPEC Secretariat, 2014), http://www.opec.org/opec_web/static_files_project/media/downloads/publications/WOO_2014.pdf, accessed 22 January 2015, pp. 39–41.
[28] OPEC, *2014 World Oil Outlook*, pp. 41–42.

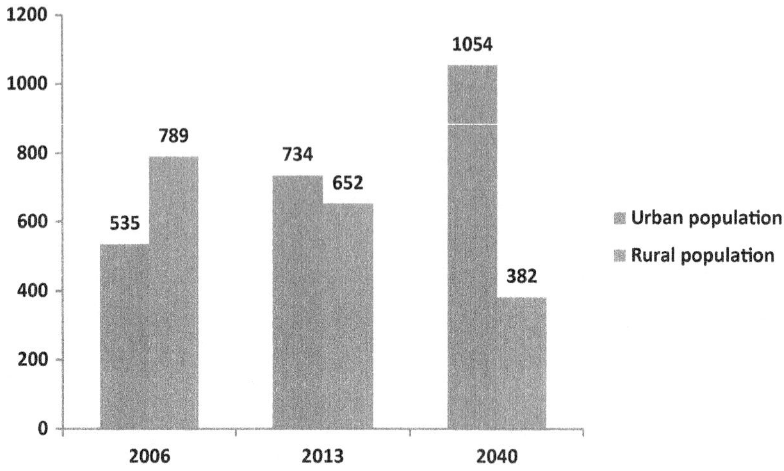

Figure 3.14: Change of China's population by urban/rural classification, 2006–2040 (Millions)

Source: Organisation of Petroleum Exporting Countries (OPEC), *World Oil Outlook 2008* (Vienna: OPEC Secretariat, 2008), http://www.opec.org/library/World%20Oil%20Outlook/pdf/WOO2008.pdf, accessed 6 January 2009, p. 21; and OPEC, *2014 World Oil Outlook* (Vienna: OPEC Secretariat, 2014), http://www.opec.org/opec_web/static_files_project/media/downloads/publications/WOO_2014.pdf, accessed 22 January 2015, p. 39.

In 2013, China's urbanisation rate reached 53.7 per cent. The new generation of Chinese leadership had singled out urbanisation as the main driver of the country's economic development in the coming years and introduced New Urbanisation Programme at the Chinese Communist Party (CCP) 18th National Congress in November 2012. In March 2014, the Chinese central government followed this up with the launch of a compendium to guide the New Urbanisation Programme, which is expected to serve as a strategic, comprehensive and instructional compendium for China's urbanisation in the future.[29]

[29] Zhou Zhihua, 'China Launches New Urbanization Plan (2014–2020)', EAI Background Brief, No. 910, East Asian Institute, National University of Singapore, 16 April 2014.

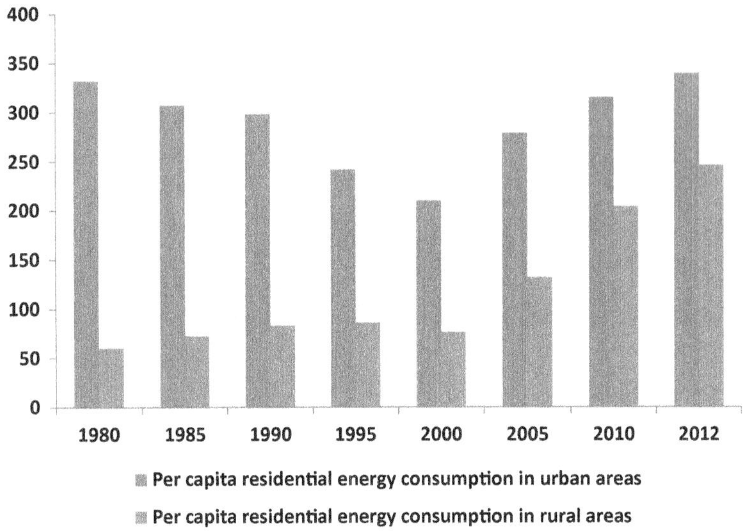

Figure 3.15: China's per capita residential energy consumption, 1980–2012 (Kilogram standard coal)

Source: Guojia tongjiju nengyuan tongjisi (Department for Energy Statistics, National Bureau of Statistics, the People's Republic of China), *Zhongguo nengyuan tongji nianjian 2013* (*China Energy Statistical Yearbook 2013*), p. 9.

3.1.5 *The change of energy consumption structure*

In the recent years, the proportions of oil and natural gas in China's total energy consumption have been increasing. In 2013, China's total energy consumption was 2,852.4 million tonnes of oil equivalent. Within it, the largest part was coal consumption, accounting for 67 per cent. The second largest was oil which accounted for 18 per cent of the total consumption. Besides, hydroelectricity accounted for 7 per cent, natural gas accounted for 5 per cent, renewable energy accounted for 2 per cent, and nuclear energy accounted for 1 per cent (see Figure 3.16).[30]

According to the Chinese official White Paper China's Energy Conditions and Policies published in 2006, the structure of China's

[30] British Petroleum, *BP Statistical Review of World Energy June 2014.*

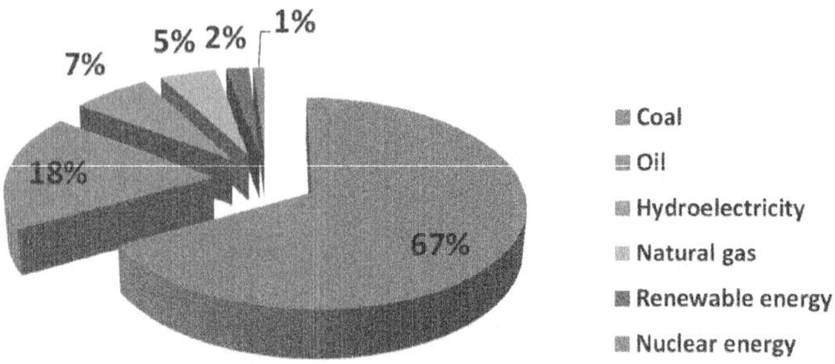

Figure 3.16: Composition of China's energy consumption, 2013 (Share of total)

Source: BP, *BP Statistical Review of World Energy June 2014.*

energy consumption has been changed and optimised since the 1980s. For example, the proportion of coal in its primary energy consumption dropped from 72.2 per cent in 1980 to 69.4 per cent in 2006.[31] This proportion continued to decline in the past few years to 67 per cent in 2013, according to the report published by BP. Meanwhile, the shares of oil and natural gas have risen, and more clean energy is being used in the people's daily life. During 1978 and 2005, the annual growth rate of China's oil consumption and natural gas consumption were 4.77 per cent and 4.6 per cent respectively, as the *White Paper* showed. While oil accounted for only 11.4 per cent of China's overall energy consumption on average during 1953 and 1978, this proportion has increased to 19.6 per cent on average during 1980 and 2005, and was 18 per cent in 2013. The share of natural gas in the country's total energy consumption was about 2 to 3 per cent in the first a few years of this century, and has reached 5 per cent in 2013.

Indeed, China has been making efforts to use more and more oil and natural gas, while reducing the proportion of coal in its total energy consumption. An important reason is that the government needs to address the serious air pollution caused by too much coal consumption. For instance, some Chinese cities have switched the

[31] The State Council Information Office, the People's Republic of China, *China's Energy Conditions and Policies*, 26 December 2007, http://www.china.org.cn/english/environment/236955.htm, accessed 27 December 2007.

energy used to generate power and heating from coal to natural gas; and the fuel for public transportation such as taxi and bus from oil to natural gas. Also, the fast development of transportation, petrochemical industry and agriculture contribute to the increase of oil consumption. These sectors are expected to continue developing in the next 20 years, promoting the country's oil and natural gas demand for a long time in the future.

In addition, China's domestic natural gas pipeline network is fragmented, while Chinese national oil companies (NOCs) have been investing in the expansion of the transmission system to connect more supplies to demand centres along the coast and in the southern regions as well as integrating local gas distribution networks.[32]

Particularly, two major pipeline projects will help raise the proportion of oil and natural gas in China's overall energy consumption. The first project is the West to East Gas Pipeline (*xi qi dong shu*), literally transporting natural gas from the western part to the eastern part of China. The project is constructed and operated by China National Petroleum Corporation (CNPC). It was built between 2002 and 2004, and started commercial operation at the end of 2004, with a capacity to transport more than 3.6 billion cubic metres of natural gas per year.[33]

The construction of the second route of this project began in 2009. Passing through 14 provinces or municipalities across China, this pipeline will be linked with the Central Asia Natural Gas Pipeline, and stretch from Tarim Basin in Xinjiang to major gas demand centres such as Guangzhou at the south and Shanghai at the east, with a total length of more than 9,000 km. The major sources of natural gas supply for the pipeline will come from Central Asian states like Kazakhstan and Turkmenistan. The whole project has been completed in 2011.

---

[32] U.S. Energy Information Administration, 'Country Analysis Brief: China', 4 February 2014, http://www.eia.gov/countries/analysisbriefs/China/china.pdf, accessed 14 January 2015.

[33] Xinhua, 'Xi qi dong shu: gaibian woguo nengyuan xiaofei jiegou' (West to East Natural Gas Project: Changing China's energy consumption structure), 7 September 2005, http://news.xinhuanet.com/fortune/2005-09/07/content_3457146.htm, accessed 27 January 2009.

To transport expanding gas import from Central Asia, CNPC has begun constructing the third route of the West to East Gas Pipeline, which is expected to become operational by 2015. This third pipeline will run partially parallel to the second route and end in the key markets in south–eastern part of China. This third route will transport natural gas from Central Asia and domestically produced gas from the Xinjiang Uygur Autonomous Region to Fujian Province and Guangdong Province. Additionally, the potential fourth and fifth routes of the West to East Gas Pipeline are under planning.[34]

The second project is the West to East Oil Pipeline (*xi you dong song*), literally transporting oil from the western part to the eastern part of China. It is also constructed and operated by CNPC and linked with China–Kazakhstan Oil Pipeline; and transport oil produced in western areas of China like Xinjiang and oil imported from Central Asia to eastern areas of China. It has the capacity to transport about 20 million tonnes of oil per year. The principal purpose of the two projects is to ease the energy shortage in China's eastern provinces which have been experiencing fast industrialisation for many years.

3.1.6 *The increasing use of transport facilities*

China's oil consumption has also been increasing with its growing use of transport facilities. The associated demand for fuel used to transport goods and to provide the growing fleets of private vehicles has kept upward pressure on oil consumption.[35] In the recent years, the vehicles sales in China have grown significantly and the country has been the world's largest vehicle market for five years. The annual growth rate of vehicle ownership in China is above 10 per cent, which

[34] U.S. Energy Information Administration, 'Country Analysis Brief: China'.

[35] Julie Jiang and Jonathan Sinton, *Overseas Investments by Chinese National Oil Companies: Assessing the Drivers and Impacts*, International Energy Agency, Information Paper, February 2011, http://www.iea.org/publications/freepublications/publication/overseas_china.pdf, accessed 24 August 2014.

is the fastest growth rate in the world. In January 2009, the monthly automobile sales volume in China reached 735.5 thousand, surpassing that of the United States (656.88 thousand), becoming the world's largest automobile market.[36] In the same year, China's annual automobile sales volume reached 13.64 million. Then, this volume nearly doubled in five years and expanded to 23.49 million in 2014, a 6.9 per cent increase from the previous year, keeping the country the number one automobile market for the sixth consecutive year.[37] It is estimated that the annual sales volume will surpass 30 million before 2020. Also, while in 2013 there were merely 137 million automobiles in China, the number is expected to reach 350 million by 2030.

Notably, three factors have contributed to the boom of China's automobile market. The first factor is the country's sound economic fundamentals, with annual GDP growth of about 10 per cent during the past two decades. The expanding rich and middle classes are willing to purchase vehicles, reflecting their high social and economic status. Second, the Chinese government has effectively stimulated domestic demand for automobile through the implementation of various policies to increase automobile production and boost domestic consumption. Third, the fast expansion of the highway system across China has helped promote the consumption of automobile domestically.[38]

---

[36] Souhu, '2009nian 1yue zhongguo qiche xiaoliang shouchao meiguo quanqiu diyi' ('In January 2009, the automobile sales volume in China overtook that of the United States for the first time, making the country the largest automobile market in the world'), 12 February 2009, http://auto.sohu.com/20090212/n262192899.shtml, accessed 2 October 2014.

[37] Xinhua, 'Zhongguo qiche chanxiaoliang lianxu liunian quanqiu diyi' ('China's automobile production and sales volumes are both the worlds' largest for the sixth consecutive year'), 13 January 2015, http://www.nmg.xinhuanet.com/xwzx/qctd/2015-01/13/c_1113975348.htm, accessed 19 January 2015.

[38] Yu Hong and Yang Mu, 'China as the world's largest automobile market', in Yang Mu and Yu Hong eds., *China's Industrial Development in the 21st Century*, (Singapore: World Scientific Publishing Co. Pte. Ltd., 2011), pp. 17–36.

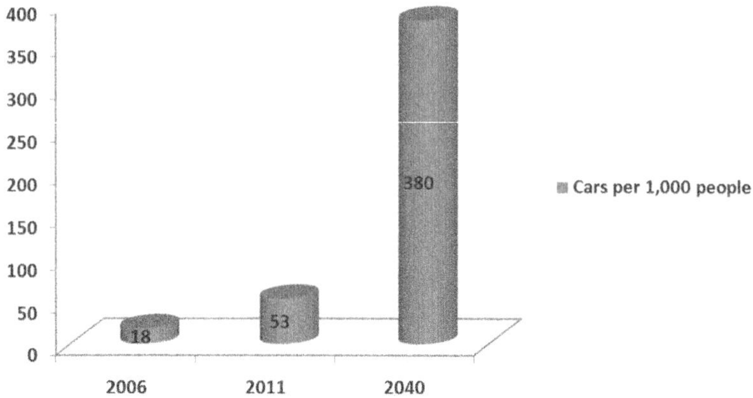

Figure 3.17: Passenger cars ownership in China, 2006–2040 (Cars per 1,000 people)

Source: OPEC, *World Oil Outlook 2008*, p. 56; and OPEC, *2014 World Oil Outlook*, p. 10.

Also, the growing Chinese economy has led to an increasing demand for commercial vehicles, as well as an expanding middle class with financial capability to purchase private cars. For example, according to statistics the OPEC, China's passenger car ownership has been increasing in the past decade and will continue growing in the next 20–30 years. The largest rise in passenger car volumes in the world between 2011 and 2040 will be in China. During this period, the passenger car volume in the country is expected to expand by more than 470 million. While China's passenger car volume was merely 18 cars per 1,000 people in 2006, it has risen to 53 cars per 1,000 in 2011, and is expected to reach over 380 cars per 1,000 in 2040 (Figure 3.17), which is a similar level to many OECD countries in the 1990s.[39]

Some authors pointed out that as for transport fuels, like all other countries, China has little choice but to continue relying on oil regardless of whether it is imported or produced domestically. There

[39] OPEC, *World Oil Outlook 2008*, p. 56; and OPEC, *World Oil Outlook 2014*, p. 10.

is no alternate fuel of the same density available at the same or better price. The use of bio fuels and electric/hybrid vehicles will not replace the use of petrol or diesel in the near future. Therefore, the role of oil in China's energy strategy is set to remain very significant over the next two decades.[40]

In addition, China has become the world's largest trading nation by 2014, and its fast expanding trading activities will continue to increase its oil demand in transportation facilities such as marine bunkers. Besides, based on experiences elsewhere, car sales are expected to grow faster than GDP growth in China, until the country's per capita income reach about USD 20,000.[41] In comparison, China's current per capita income is merely about USD 3,000. These factors are expected to continue boosting the country's use of transport facilities, and have significant implications for its energy demand in the future.

3.2 China's domestic energy supply

China is rich in energy resources, but the per capita share is small. In 2013, China's proved coal reserves was 114.5 billion tonnes, accounting for 12.8 per cent of the world's total reserves; but its proved oil reserves was 2.5 billion tonnes or 18.1 billion barrels, only accounting for 1.1 per cent of the world's total; and its proved natural gas reserves was 3.3 trillion cubic metres, only accounting for 1.8 per cent of the world's total. Therefore, just as the Chinese official White Paper described, China's energy resource endowment is not optimistic and the country's per capita share of coal, oil and natural gas is low. But China's energy consumption has been growing too quickly in the recent years, posing increasing challenge to the country's energy supply.[42]

---

[40] Elspeth Thomson and Augustin Boey, 'The role of oil and gas in China's energy strategy: An overview', *Asia Pacific Business Review* 21: 1, 2015, pp. 10–25.

[41] Rosen and Houser, 'China Energy', p. 15.

[42] The State Council Information Office, the People's Republic of China, *China's Energy Policy 2012*.

In specific, due to four characteristics of China's domestic energy resources, the country's energy outlook particularly in terms of oil and gas is not optimistic.[43]

First, China has abounding energy resources, but they are dominated by coal. Its oil and natural gas reserves are relatively small. Although its unconventional fossil energy resources like shale oil and coal-bed gas have much potential for exploitation, the prospects of these alternative energy resources are still unclear.

Second, due to China's large population, its per capita possession of energy resources is low by world standards. The country's per capita shares of coal, petroleum and natural gas account for merely 67 per cent, 5.4 per cent, and 7.5 per cent respectively of the world's averages.

Third, the distribution of energy reserves across the country is uneven. Its coal reserves are largely located in the north and the northwest; and oil and gas deposits in the eastern, central and western areas and along the coast. However, the consumption of energy resources is concentrated in the south–eastern areas, because the economy there is the most developed within China. The different location of major energy producing and consuming regions has resulted in long and costly energy flow routes within the country. Namely, coal and oil need transporting large-scale over long distances from the north to the south; and natural gas and electricity from the west to the east.

Fourth, the development of energy resources in the country is rather difficult. Compared with many other countries, China faces severe geological difficulties in exploring its coal resources; and its oil and gas reserves are mainly located in regions with complex geological conditions and at great depths, and so they need exploiting with advanced technologies and facilities. Although the country has been

[43] The description of China's domestic energy reserves in the following four paragraphs is mainly from the two Chinese official white papers released in 2007 and 2012, respectively. See the State Council Information Office, the PRC, *China's Energy Conditions and Policies*; and the State Council Information Office, the People's Republic of China, *China's Energy Policy 2012*.

improving its oil exploration and development technologies for years, but is still short of state of the art prospecting and tapping techniques and facilities.

With regard to domestic energy production, China became the world's largest energy producer in 2007,[44] and its energy production has increased substantially throughout the years. For instance, China's domestic oil production has doubled and natural gas production has expanded for more than eight times during the past three decades or so (see Figures 3.18 and 3.19).

In 2013, China's domestic oil output reached 208.1 million tonnes or 4,180 thousand bpd, a 0.6 per cent increase from the previous year,

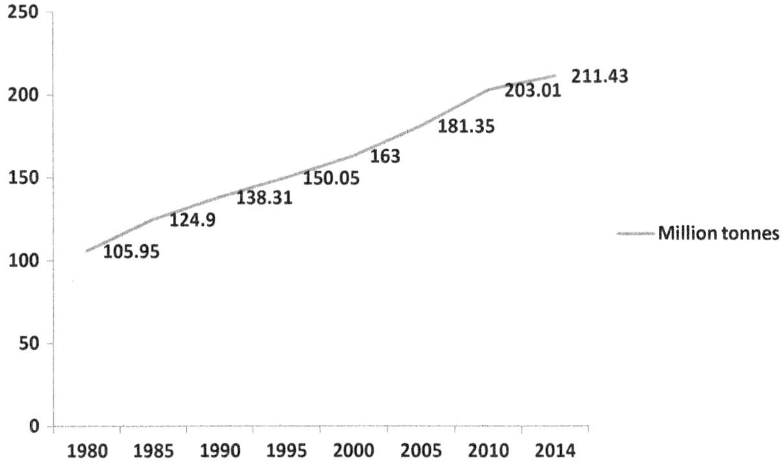

Figure 3.18: China's oil production, 1980–2014
(Million tonnes)

Source: Guojia tongjiju nengyuan tongjisi (Department for Energy Statistics, National Bureau of Statistics, the People's Republic of China), *Zhongguo nengyuan tongji nianjian 2013* (*China Energy Statistical Yearbook 2013*), pp. 118–119; and Guojia tongji ju (National Bureau of Statistics, the People's Republic of China), '2014 guomin jingji he shehui fazhan tongji gongbao' (Statistical Communiqué of the People's Republic of China on the 2014 National Economic and Social Development), 26 February 2015, http://www.stats.gov.cn/tjsj/zxfb/201502/t20150226_685799.html, accessed 8 March 2015.

[44] U.S. Energy Information Administration, 'Country Analysis Brief: China'.

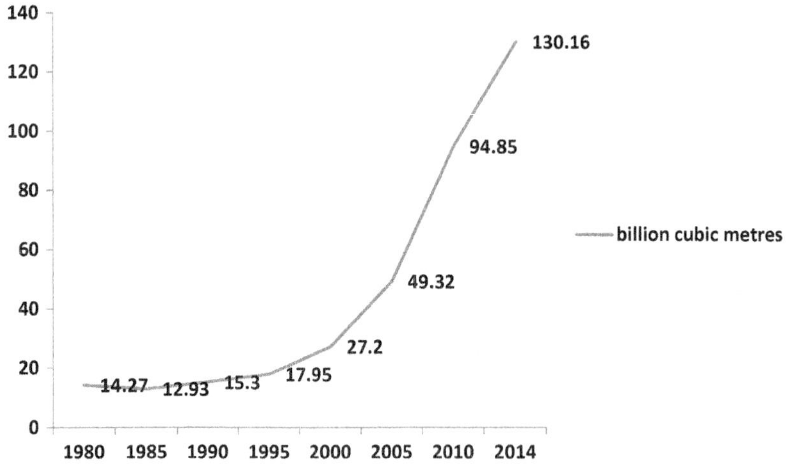

Figure 3.19: China's natural gas production, 1980–2014 (Billion cubic metres)

Source: Guojia tongjiju nengyuan tongjisi (Department for Energy Statistics, National Bureau of Statistics, the People's Republic of China), *Zhongguo nengyuan tongji nianjian 2013* (*China Energy Statistical Yearbook 2013*), pp. 128–129; and Guojia tongji ju (National Bureau of Statistics, the People's Republic of China), '2014 guomin jingji he shehui fazhan tongji gongbao' (Statistical Communiqué of the People's Republic of China on the 2014 National Economic and Social Development).

and accounted for 5 per cent of the world's total oil production, making the country the fourth largest oil producer in the world after Saudi Arabia, Russia and the United States (Figure 3.20). In the same year, China's domestic output of natural gas grew up to 117.1 billion cubic metres, a 9.5 per cent increase from the previous year, and accounted for 3.5 per cent of the world's overall natural gas production, becoming the sixth largest natural gas producer in the world after the United States, Russia, Iran, Qatar and Canada (Figure 3.21).[45]

China's oil industry was established in large scale in 1959 when Daqing oil field in North–eastern China's Heilongjiang Province was discovered and developed. Subsequently, the country's oil production rose significantly for about 20 consecutive years. But the increase rate dropped dramatically after the late 1970s. China has great difficulty

[45] British Petroleum, *BP Statistical Review of World Energy June 2014.*

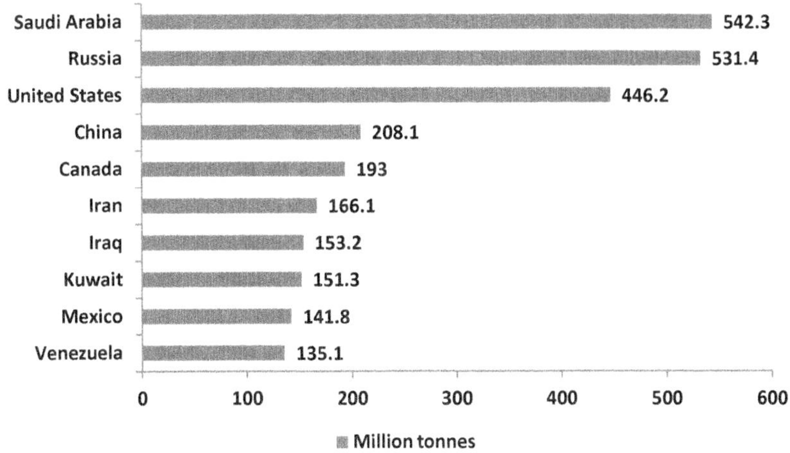

Figure 3.20: The top 10 oil producing countries in the world, 2013 (Million tonnes)

Source: BP, *BP Statistical Review of World Energy June 2014.*

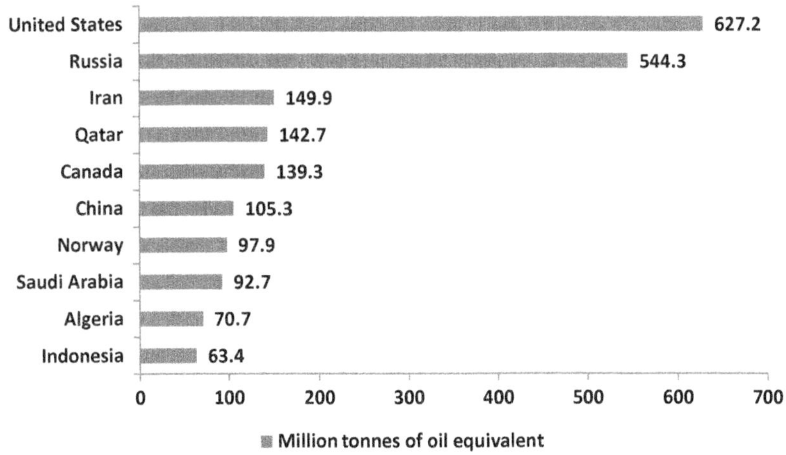

Figure 3.21: The top ten natural gas producing countries in the world, 2013 (Million tonnes of oil equivalent)

Source: BP, *BP Statistical Review of World Energy June 2014.*

in finding new big oil fields, and the costs to extract oil from mature oil fields are steadily increasing. Under such conditions, it is remarkable that its oil output has managed to maintain a positive growth rate. In the recent years, in order to maximise the output from existing oil fields, China has extensively adopted a number of enhanced oil recovery (EOR) techniques to promote its domestic oil production, and has been striving to optimise the development of the country's domestic resources, in order to mitigate the decline in mature or aging fields, while developing new capacity to offset these declines.

The older giant complexes such as Daqing, Shengli and Liaohe oil fields are the largest contributors to China's domestic oil supply. CNPC's Daqing oil field, the country's largest oil field, alone contributes about 20 per cent of China's overall domestic oil production. China Petroleum and Chemical Corporation's (Sinopec) Shengli oil field is the country's second largest oil field. But these mature oil fields' output has already started to decline. Although the application of EOR technologies in these fields has been able to slow the decline rates, Daqing, Shengli and other older fields have been heavily exploited since the 1960s, and their output is expected to decline within the next decade.[46]

Nowadays, much of the increased oil production comes from offshore oil fields that begun to offer significant output in the late 1990s; or from new oil fields in the North–western part of the country, in particular the Tarim Basin in Xinjiang. But development and transportation costs for these new oil fields are much higher than the other oil fields.[47]

---

[46] CNPC's application of various EOR technologies on Liaohe and Jilin oil fields in Northeast China, both of which are among China's oldest onshore oil fields, has helped prevent production declines in the recent years. U.S. Energy Information Administration, 'Country Analysis Brief: China'.

[47] Tatsu Kambara and Christopher Howe, *China and the Global Energy Crisis: Development and Prospects for China's Oil and Natural Gas* (Cheltenham: Edward Elgar Publishing Limited, 2008), pp. 12–67 and pp. 81–95; Armelle Guizot, *Chinese Energy Markets: Trading and Risk Management of Commodities and Renewables* (Basingstoke, UK: Palgrave Macmillan, 2007), pp. 1–3 and pp. 29–33; Naughton, *The Chinese Economy*, pp. 339–340; and OPEC, *2014 World Oil Outlook*, p. 152.

In fact, the country's inner areas, such as Junggar and Tarim basins in Northwest China's Xinjiang Uygur Autonomous Region, and Central China's Ordos Basin where locates the Changqing oil field, have achieved vigorous production growth in the recent years thanks to the application of improved drilling and advanced oil extraction techniques to tap into those geologically complex oil reserves. Especially, oil output at Changqing oil field which is China's third largest oil field and located in the north central Ordos basin, has expanded significantly in the past a few years.

Moreover, offshore exploration and production activities, mostly driven by China National Offshore Oil Corporation (CNOOC), have focused on the Bohai Bay Basin in the Yellow Sea, the Pearl River Mouth Basin in the South China Sea, and the East China Sea as well. Most of these fields are small and mature faster than the country's onshore oil fields, pushing CNOOC to explore deep water fields. In particular, the Bohai Bay Basin in north China is the oldest offshore oil producing area and holds the bulk of proven offshore reserves in China.[48]

In the recent years, China has made some new discoveries of oil reserves. The most exciting discovery in the past three decades was that in 2007, CNPC made a remarkable discovery of a giant oil field — Jidong Nanpu oil field — in the eastern part of Hebei Province and in the Bohai Bay Basin, with a 1.18 billion tonnes of oil reserves volume.[49] In the same year, CNPC initiated the first phase of the Jidong Nanpu oil field development project. A report published by the OPEC estimated that most of the increment of China's domestic crude oil production in the medium-term would come from Jidong Nanpu oil field and those oil fields in Xinjiang.[50] Figure 3.22 shows the major oil fields in China.

China's major onshore natural gas producing regions include Sichuan Basin in Southwest China, Tarim Basin (China's second largest gas producing area), Junggar Basin and Qaidam Basin in

[48] U.S. Energy Information Administration, 'Country Analysis Brief: China'.

[49] The 2007 Annual Report of CNPC, pp. 6, 9 and 33.

[50] OPEC, *2014 World Oil Outlook*, p. 152.

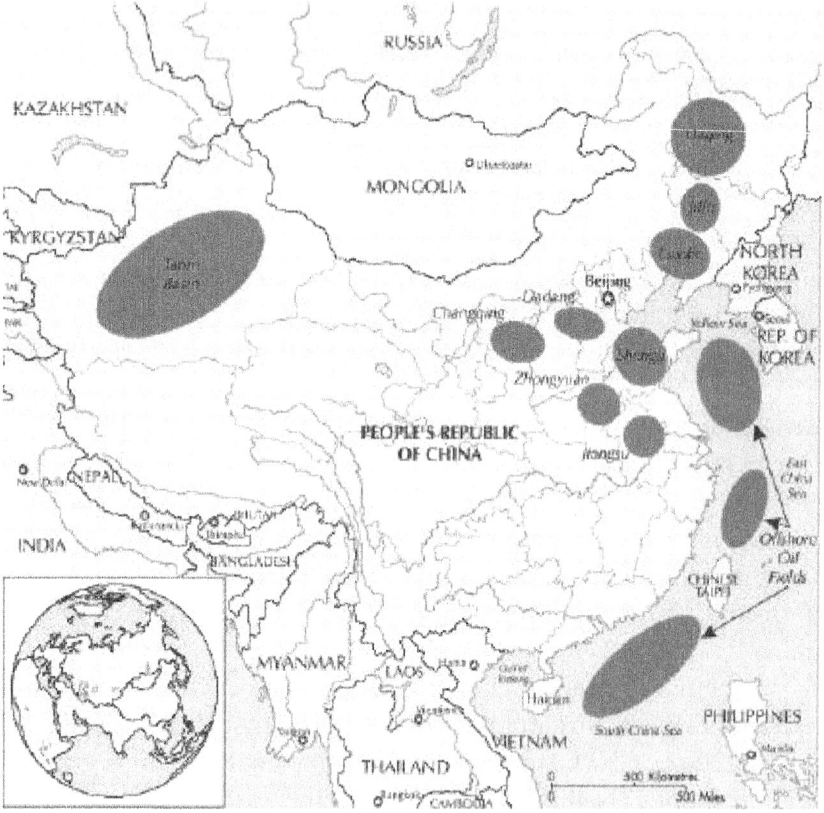

Figure 3.22: China's major oil fields

Source: U.S. Energy Information Administration, 'Country Analysis Brief: China', 4 February 2014, http://www.eia.gov/countries/analysisbriefs/China/china.pdf, accessed 14 January 2015.

Northwest China, Ordos Basin (China's largest gas producing area) in North China, and Songliao Basin in Northeast China. There are also some offshore natural gas fields located in the Bohai Basin and the Panyu complex of the Pearl River Mouth Basin. Chinese NOCs are also exploring more technically challenging areas such as deep water, coal bed methane and shale gas reserves with international oil companies (IOCs), such as Husky, Chevron, BG, BP, Anadarko and Eni. The western South China Sea is home to Yacheng 13-1 field,

China's largest offshore natural gas field. The eastern South China Sea is under intense exploration for gas, too.[51]

In addition, it is reported that according to the energy strategy and planning of China's 13th Five Year Plan (2016–2020), which is still under formulation, China's energy development strategy and domestic energy system will be further adjusted and optimised. Expanding the domestic energy production is among the seven major strategic tasks for China's energy development. With regard to oil production, China plans to increase the on shore oil production while exploring new off shore oil fields, and make great effort in exploring shale gas and off shore natural gas. Together with off shore energy resources, on shore oil and gas rich areas such as Tarim Baisin, Qaidam Basin, Changqing oil and gas field in Ordos Basin (the largest oil and gas field in China today), Sichuan and Chongqing will be the focus of China's future domestic energy exploration and production.[52]

3.3 The widening gap between demand and supply

Although China's energy especially oil and natural gas output has been increasing for years, it does not satisfy the country's fast expanding appetite. As a result, the recent years have witnessed a widening gap between China's oil demand and domestic supply, and a similar situation of the country's natural gas consumption and domestic production. In particular, while the self-sufficiency rate of China's overall energy consumption is about 90 per cent, its oil self-sufficiency rate is less than 40 per cent.

In the first place, the country's oil demand is growing much faster than its domestic supply. Since the 1990s, China's crude oil

---

[51] U.S. Energy Information Administration, 'Country Analysis Brief: China'.

[52] Wangyi, 'Shisanwu nengyuan luxiantu: zhongguo zhongyao nengyuan zhanlue ji yingxiang fenxi' (The 13th Five Year Plan energy development roadmap: An analysis of China's key energy strategy and its implications), 30 October 2014, http://money.163.com/14/1030/04/A9PEGUML00252G50.html, accessed 15 February 2015.

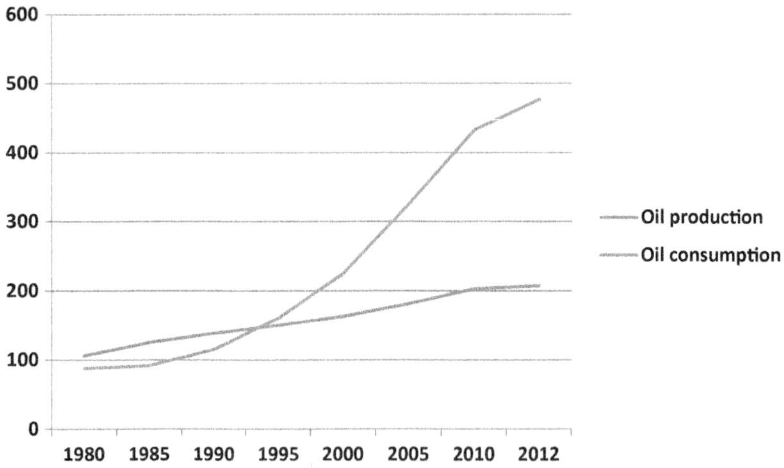

Figure 3.23: China's oil production and consumption, 1980–2012 (Million tonnes)

Source: Guojia tongjiju nengyuan tongjisi (Department for Energy Statistics, National Bureau of Statistics, the People's Republic of China), *Zhongguo nengyuan tongji nianjian 2013* (*China Energy Statistical Yearbook 2013*), pp. 118–119.

consumption expanded at 5.77 per cent annually, but the increase rate of its domestic crude oil output was just 1.67 per cent per year.[53] As an IEA report maintained, China's oil fields are aging, and their reserves-to-production ratios are low. Hence, the country's domestic oil production is nearing its peak. As a result, the country is almost entirely dependent on the international oil market to meet incremental oil demand.[54] Figure 3.23 shows that the gap between China's domestic oil supply and demand had been continuously widening in the past two decades.

In the late 1970s when China opened up and emerged into the world economy, it had significant oil surpluses and was an important oil exporter. It exported a peak of 27 million tonnes or about 500,000 bpd in 1985. But later, the country's oil surplus steadily

[53] Cui Minxuan, ed., *The Energy Development Report of China 2007*, p. 98.

[54] Jiang and Sinton, *Overseas Investments by Chinese National Oil Companies: Assessing the Drivers and Impacts.*

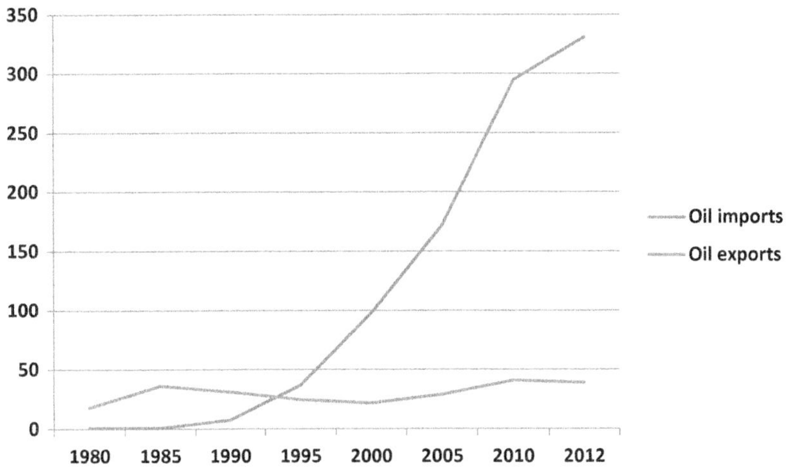

Figure 3.24: China's oil imports and exports, 1980–2012 (Million tonnes)

Source: Guojia tongjiju nengyuan tongjisi (Department for Energy Statistics, National Bureau of Statistics, the People's Republic of China), *Zhongguo nengyuan tongji nianjian 2013* (*China Energy Statistical Yearbook 2013*), pp. 118–119.

declined as its domestic demand substantially grew. It became a net oil importer in 1993, and since then its net oil import has been rapidly expanding (see Figure 3.24).[55]

Simultaneously, the dependency of China's oil consumption upon foreign supply has been growing. In 2009, China imported just under four million bpd of crude oil, increasing by 14 per cent from 2008, and the first time that China imported more than half of its crude oil consumption,[56] pushing its oil dependency on overseas supply to surpass 50 per cent. In the same year, China overtook Japan to become the world's second largest net oil import behind the United Sates. In 2013, according to the statistics from the BP, China's net oil import reached 311.5 million tonnes, the second largest only after the U.S. (326.9 million tons); meanwhile, its dependency on foreign oil

[55] Naughton, *The Chinese Economy*, p. 340.

[56] Jiang and Sinton, *Overseas Investments by Chinese National Oil Companies: Assessing the Drivers and Impacts.*

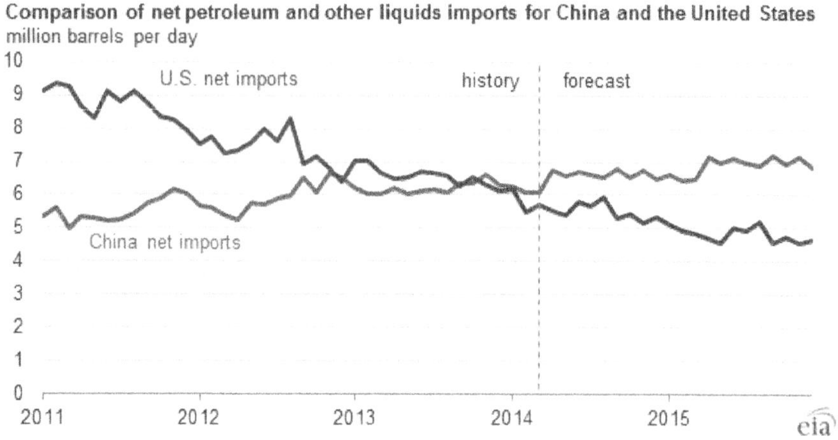

Figure 3.25: Comparison of net petroleum and other liquids imports for China and the United Sates, 2011–2015 (Million bpd)

Source: U.S. Energy Information Administration, 'China is now the world's largest net importer of petroleum and other liquid fuels', 24 March 2014, http://www.eia.gov/todayinenergy/detail.cfm?id=15531, accessed 14 January 2015.

reached 61.4 per cent.[57] The U.S. Energy Information Administration (EIA) estimated in February 2014 that China would surpass the U.S. as the largest net oil importer by 2014.[58] The EIA also declared a month later that in September 2013, China's net imports of petroleum and other liquids exceeded those of the United States on a monthly basis, making the country the largest net importer of crude oil and other liquids in the world (see Figure 3.25).[59]

The IEA projected that by around 2030, China will replace the United States as the world's largest oil consumer.[60] By that time, its

---

[57] Author's calculation based on the statistics from British Petroleum, *BP Statistical Review of World Energy June 2014.*

[58] U.S. Energy Information Administration, 'Country Analysis Brief: China'.

[59] U.S. Energy Information Administration, 'China is now the world's largest net importer of petroleum and other liquid fuels', 24 March 2014, http://www.eia.gov/todayinenergy/detail.cfm?id=15531, accessed 14 January 2015.

[60] International Energy Agency, *World Energy Outlook 2013 — Executive Summary*, OECD/IEA 2013, http://www.iea.org/publications/freepublications/publication/WEO2013_Executive_Summary_English.pdf, accessed 24 August 2014.

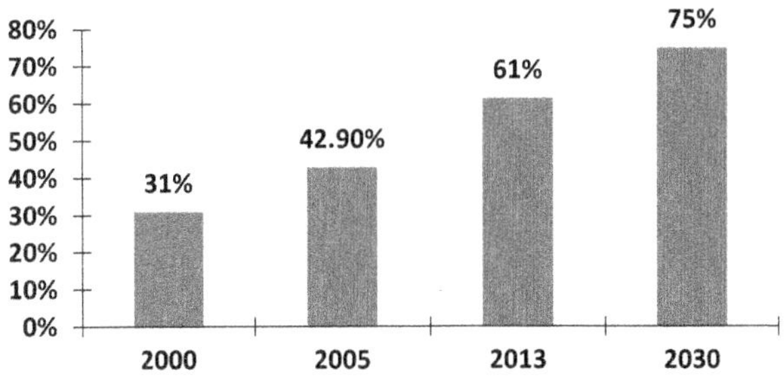

Figure 3.26: China's oil dependence on foreign supply, 2000–2030 (Percentage)

Source: Author's compilation.

dependency on foreign oil supply is expected to reach 75 per cent. Figure 3.26 shows the ratio of China's oil dependence on foreign supply from 2000 to 2030.

Nowadays, China imports oil and natural gas from various regions (the Middle East, Africa, Latin America, Russia, Central Asia, and the Asia Pacific) via different ways (shipment, pipeline and railway). Figure 3.27 shows the country's current and prospective major oil and gas imports routes.

Among them, the Middle East has been the largest oil supplier for China for a long time, although in the recent years, African countries in particular Angola have been contributing more to China's oil import.[61] In 2013, oil import from the region took up about half of China's overall oil import. For example, the statistics released by BP shows that in 2013, China imported 153.9 million tonnes of crude oil from the Middle East, accounting for 49.4 per cent of its overall oil import;[62] an EIA publication shows that 52 per cent of China's crude oil import in 2013 originated from the Middle East.[63] In the recent years, although China has been diversifying the sources of its

[61] U.S. Energy Information Administration, 'Country Analysis Brief: China'.

[62] Author's calculation based on British Petroleum, *BP Statistical Review of World Energy June 2014*.

[63] U.S. Energy Information Administration, 'Country Analysis Brief: China'.

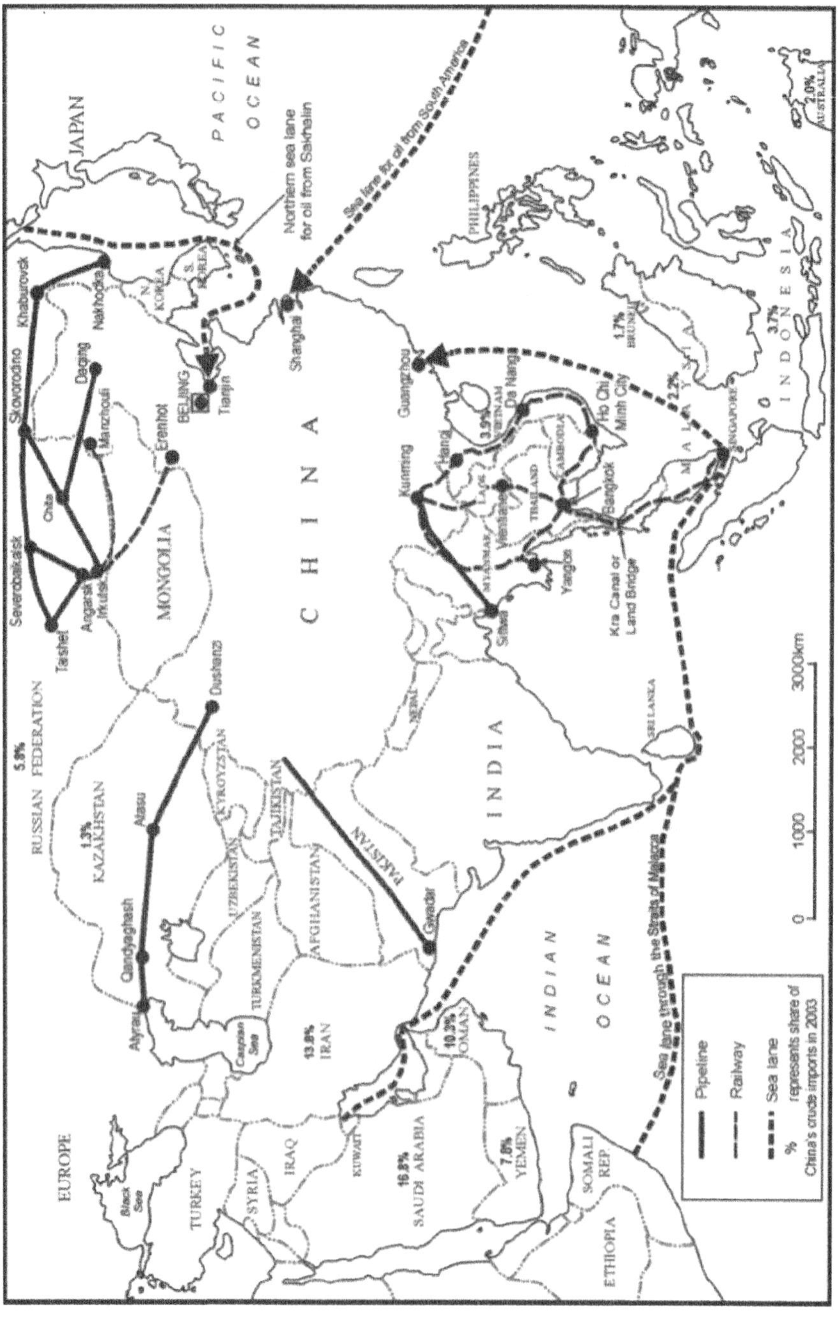

Figure 3.27: China's current and prospective oil and gas import routes

Source: Hongyi Harry Lai, 'China's Oil Diplomacy: Is It A Global Security Threat?', *Third World Quarterly* 28: 3, 2007, p. 524.

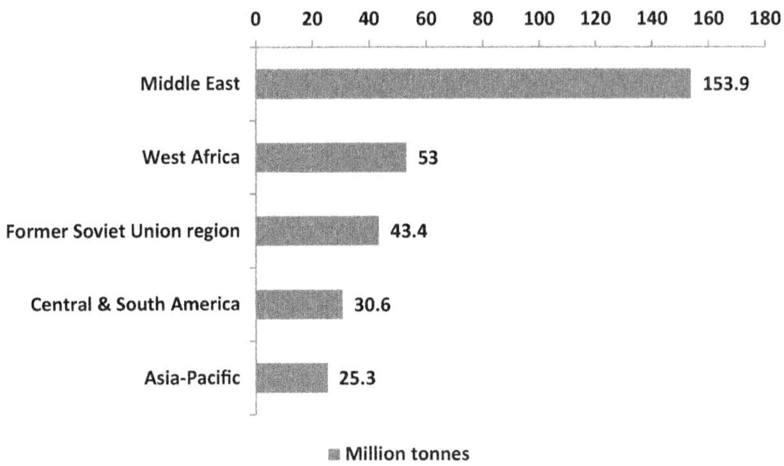

Figure 3.28: The volumes of China's oil import by major regions, 2013 (Million tonnes)

Source: Based on BP, *BP Statistical Review of World Energy June 2014*.

crude oil import due to its robust oil demand growth and the geopolitical uncertainties in some parts of the world, its heavy reliance on oil supply from the Middle East is unlikely to change in the foreseeable future. After the Middle East, West Africa, the former Soviet Union region, Latin America, etc. are also important sources of China's oil import,[64] as shown in Figure 3.28.

According to EIA, while the Middle East supplied 52 per cent (2.9 million bpd) of China's oil import in 2013, other regions exporting oil to China include Africa 23 per cent (1.3 million bpd), the Americas 10 per cent (562,000 bpd), the Asia–Pacific 2 per cent (129,000 bpd) and other countries 13 per cent (736,000 bpd),[65] as Figure 3.29 illustrated.

Among those countries exporting oil to China, Saudi Arabia continues to be the largest supplier of crude oil to China and provided 19 per cent of China's oil import in 2013. Since crude oil production in Iran, Libya, Sudan and South Sudan has dropped since 2011, China

[64] British Petroleum, *BP Statistical Review of World Energy June 2014*.

[65] U.S. Energy Information Administration, 'Country Analysis Brief: China'.

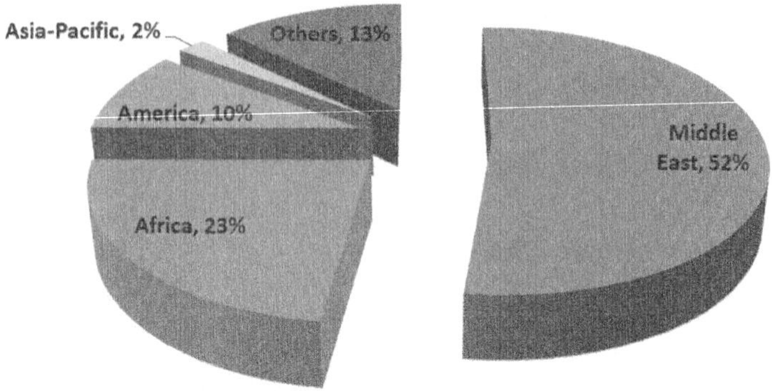

Figure 3.29: China's oil import by regions, 2013 (Percentage)

Source: U.S. Energy Information Administration, 'Country Analysis Brief: China', 4 February 2014, http://www.eia.gov/countries/analysisbriefs/China/china.pdf, accessed 14 January 2015.

replaced the lost shares of crude oil import from these countries with import from Oman, Iraq, the United Arab Emirates, Angola, Venezuela and Russia.[66] According to IEA, China's top ten crude oil suppliers in 2013, in order of import volume, were Saudi Arabia, Angola, Oman, Russia, Iraq, Iran, Venezuela, Kazakhstan, United Arab Emirates and Kuwait,[67] six of which are Middle Eastern countries.

Furthermore, since 2006, the expansion of China's natural gas consumption has been increasingly faster than the growth of the country's domestic production of natural gas, as Figure 3.30 demonstrates. As a result, although there has been a vigorous exploitation of China's domestic onshore and offshore natural gas resources, including unconventional gas, much of the country's natural gas demand must be met by imports.[68]

[66] U.S. Energy Information Administration, 'China is now the world's largest net importer of petroleum and other liquid fuels'.

[67] Julie Jiang and Chen Ding, *Updates on Overseas Investments by China's National Oil Companies: Achievements and Challenges since 2011*, International Energy Agency, Partner Country Series, 2014, http://www.iea.org/publications/freepublications/publication/PartnerCountrySeriesUpdateonOverseasInvestmentsbyChinasNationalOilCompanies.pdf, accessed 24 August 2014.

[68] Jiang and Sinton, *Overseas Investments by Chinese National Oil Companies: Assessing the Drivers and Impacts.*

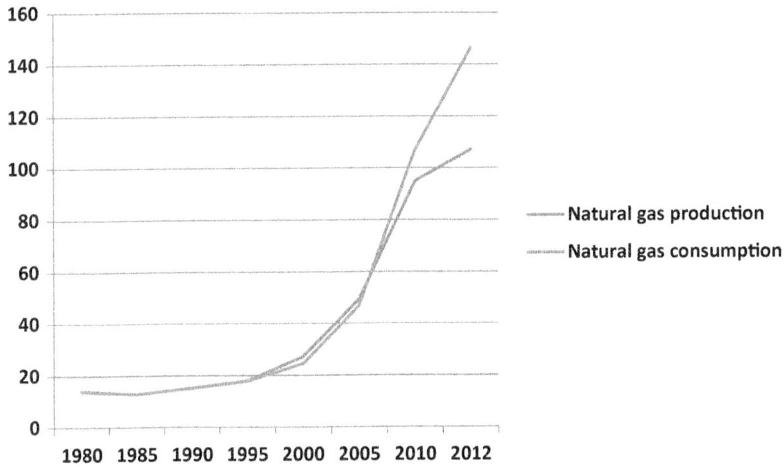

Figure 3.30: China's natural gas production and consumption, 1980–2012 (Billion cubic metres)

Source: Guojia tongjiju nengyuan tongjisi (Department for Energy Statistics, National Bureau of Statistics, the People's Republic of China), *Zhongguo nengyuan tongji nianjian 2013* (*China Energy Statistical Yearbook 2013*), pp. 128–129.

China begun to import natural gas in June 2006 with the opening of the Dapeng liquefied natural gas (LNG) import terminal in Guangdong Province, which is supplied mainly by Australia;[69] and started to import pipeline gas in early 2010 from Central Asia, mainly from Turkmenistan.[70] The past few years have witnessed a dramatic growth of China's natural gas import. In 2013, China's net imports of natural gas stood at 49.1 billion cubic meters, making the country the fifth largest net importer of natural gas in the world after Japan, Germany, Italy and South Korea (see Figure 3.31); and China's dependency on foreign natural gas was 30.4 per cent.[71] This depend-

[69] Eurasia Group, 'China's Overseas Investments in Oil and Gas Production', 2006, http://www.uscc.gov/researchpapers/2006/oil_gas.pdf, accessed 13 March 2009, p. 2.

[70] Jiang and Sinton, *Overseas Investments by Chinese National Oil Companies: Assessing the Drivers and Impacts.*

[71] Author's calculation on the statistics from British Petroleum, *BP Statistical Review of World Energy June 2014.*

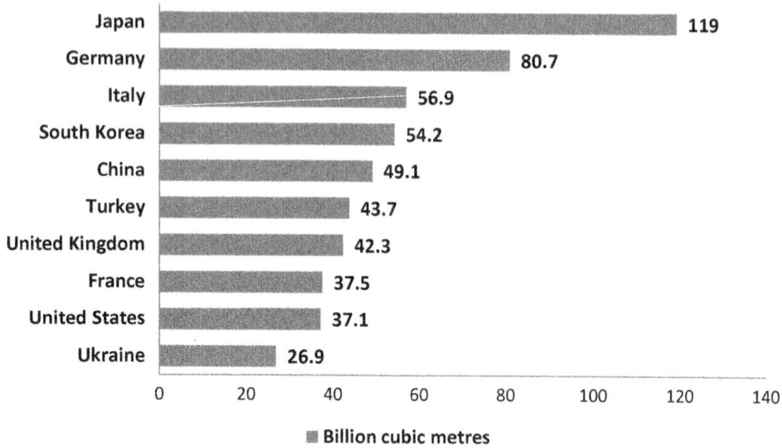

Figure 3.31: Major net natural gas importing countries in the world, 2013 (Billion cubic metres)

Source: BP, *Statistical Review of World Energy 2014.*

ency rate is reported to have risen to 32.2 per cent in 2014.[72] It is expected that within five years from then on, China's dependency on foreign natural gas will surpass 50 per cent.

Currently, China imports natural gas in two ways, namely by gas pipeline (from Central Asia and Myanmar) and via LNG terminals (from various exporters), with the former accounting for 52.79 per cent of China's total gas import in 2013 and the latter for the remainder. In 2013, China imported 27.4 billion cubic metres of natural gas through pipeline and 24.5 billion cubic metres in LNG. Among those countries exporting natural gas to China, Turkmenistan is China's major source of pipeline natural gas import, exporting 24.4 billion cubic metres to China alone in 2013; while Qatar, Australia, Malaysia, Indonesia and Yemen are major sources of China's LNG import (Figure 3.32).[73]

[72] Xinhua, 'Woguo tianranqi duiwai yicundu shengzhi 32.2 per cent' (China's natural gas dependency on foreign supply has rose to 32.2 per cent), 19 January 2015, http://news.xinhuanet.com/energy/2015-01/19/c_127398476.htm, accessed 8 February 2015.

[73] British Petroleum, *BP Statistical Review of World Energy June 2014.*

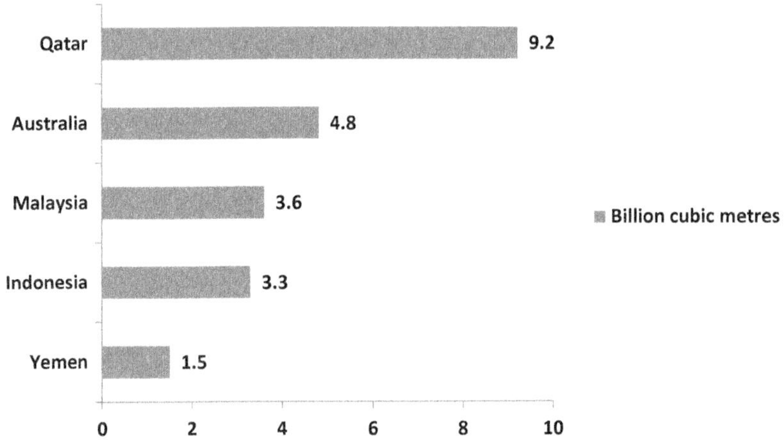

Figure 3.32: Major sources of China's LNG import, 2013 (Billion cubic metres)

Source: Based on BP, *BP Statistical Review of World Energy June 2014.*

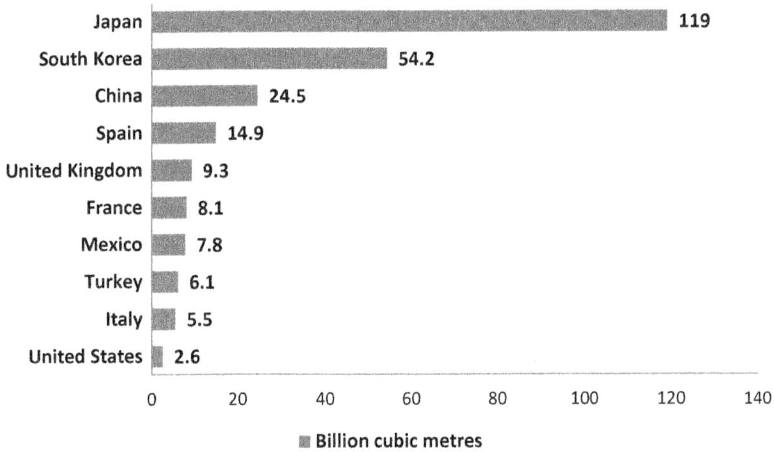

Figure 3.33: Major net LNG importing countries in the world, 2013 (Billion cubic metres)

Source: BP, *Statistical Review of World Energy 2014.*

China has surpassed Spain in 2012 to become the world's third largest LNG importer after Japan and South Korea (Figure 3.33), consuming above 6 per cent of the global LNG trade. CNOOC is the leading LNG player in China, operating six LNG regasification terminals. The company has taken the upper hand in China's LNG market compared to the other Chinese NOCs and continues to expand aggressively. CNPC recently entered the LNG market and has become the operator of three re-gasification terminals. Sinopec has also entered China's LNG market, operating one terminal.[74]

3.4 Conclusion

For more than three decades, China's energy consumption has been expanding with the country's socio–economic development. Its demand for oil and natural gas has increased much faster than the growth of its domestic production. The country is importing more and more energy from different regions and countries in the world, and this trend is likely to continue in the foreseeable future. This situation determines that the country has to secure more and more energy supply from abroad largely through energy diplomacy, which includes the Chinese government's energy diplomacy efforts and Chinese NOCs' overseas direct investment and transnational operation. The development of China's global search for energy resources will be discussed in the next chapter.

[74] U.S. Energy Information Administration, 'Country Analysis Brief: China'.

Chapter

4

Beijing's Energy Diplomacy and Chinese NOCs' Overseas Expansion

For the purpose of securing more overseas energy supply to satisfy China's growing appetite, China's energy diplomacy has been developing rapidly since the beginning of this century, seeking access to oil and natural gas resources around the world. On the one hand, the Chinese government has been actively engaging energy exporting countries in the world, exemplified by Chinese top leaders' frequent visits to those states and Beijing's intervention in some big energy deals. On the other hand, Chinese national oil companies (NOCs) have been fast expanding their outward direct investment (ODI) and transnational operation in many energy rich countries. This chapter is divided into two sections, demonstrating the recent development of these two aspects respectively.

4.1 The Chinese government's recent energy diplomacy

The recent years have witnessed the fast development of Beijing's energy diplomacy activities across major energy exporting regions and countries such as the Middle East, Africa, Central Asia, Russia, Southeast Asia, Latin America, Australia and Canada. In particular, China's top leaders' (President, Premier, top legislator and top political advisor, or other members of the Standing Committee of the CCP Central Committee's Political Bureau) have paid frequent visits abroad to engage energy exporting states. This section consists of six sub-sections, showing Beijing's energy diplomacy in six different regions around the world.

4.1.1 *The Middle East*

The Middle East has been the biggest source of China's crude oil import for a long time and is expected to remain so in the foreseeable future. In 2013, China imported 153.9 million tonnes of crude oil from the Middle East, accounting for 49.4 per cent of its overall oil import.[1] It is worth noting that this proportion has actually expanded during the past several years, although China has been making great efforts in geographically diversifying the country's oil import sources. For instance, the crude oil supply from the Middle East only took up 38.6 per cent of China's total oil import in 2007, more than 10 per cent lower than the current level.[2]

Especially, for Gulf countries particularly Saudi Arabia, their oil export to the United States may decrease because the United States is focusing to increase its domestic oil production, while China is a

[1] British Petroleum, *BP Statistical Review of World Energy June 2014*, http://www.bp.com/content/dam/bp/pdf/Energy-economics/statistical-review-2014/BP-statistical-review-of-world-energy-2014-full-report.pdf, accessed 24 August 2014.

[2] British Petroleum, *BP Statistical Review of World Energy June 2008*, http://www.bp.com/liveassets/bp_internet/globalbp/globalbp_uk_english/reports_and_publications/statistical_energy_review_2008/STAGING/local_assets/downloads/pdf/statistical_review_of_world_energy_full_review_2008.pdf, accessed 25 December 2008.

huge market.[3] In fact, China has replaced the United States as the largest buyer of oil from Saudi Arabia; while Saudi Arabia has maintained its position as the largest source of China's oil import for years, and is likely to remain so in the foreseeable future. It is an essential energy cooperation partner for China. Moreover, although a long-standing U.S. ally, Riyadh is pursuing a hedging strategy towards Washington, largely due to the anti-Saudi backlash in the United States after the September 11th.[4] Thus, Saudi Arabia is more willing to develop a robust economic and strategic relationship with China.

In the recent years, the mistrust between the United States and Saudi Arabia has increased and the differences and competition between the two countries have emerged against the background of the plunge in international oil prices in the second half of 2014.

The United States' mistrust of Saudi Arabia has swelled after the September 11th. Washington is increasingly concerned about the funding of terrorism by foundations from some Middle Eastern countries such as Saudi Arabia. Likewise, Saudi Arabia's mistrust of the United States has also grown. In 2011, the Obama administration was slow in its response to the Arab Spring; the Saudi government was disappointed to see that the United States only stood by or even forced the collapse of Hosni Mubarak's regime in Egypt and Muammar Kaddafi's regime in Libya.

Currently, Washington's non-involvement policy towards the Syrian civil war and its partial removal of sanctions against Iran are favourable to Iran and Syria, but not in the best interest of Saudi Arabia and Egypt. Saudi Arabia is particularly concerned that the United States may approach Iran, its powerful rival in the region. Currently, though it is still difficult for the West to reach an

---

[3] Joshy M. Paul, 'The role of energy security in China's foreign policy: A maritime perspective', *Maritime Affairs* 6: 2, 2010, pp. 49–71.

[4] Flynt Leverett and Jeffery Bader, 'Managing China–U.S. Energy Competition in the Middle East', *The Washington Quarterly* 29: 1, p.195; and Mikkal E. Herberg, 'Energy Security Survey 2007: The Rise of Asia's National Oil Companies', NBR Special Report, no. 14, December 2007, the National Bureau of Asian Research, http://www.nbr.org/publications/specialreport/pdf/SR14.pdf, accessed 13 March 2009, p. 18.

agreement with Iran over its nuclear programme in a short time, it is not impossible that the West may again remove some of the sanctions in exchange for Iran's further commitment.

More importantly, due to the opposition of member states like Saudi Arabia, OPEC oil ministers' meeting in November 2014 failed to reach an agreement to cut oil production to shore up the international oil prices. This move was generally seen as targeting North America's shale oil sector. In 1986, after Saudi Arabia dramatically increased its crude oil production, international oil prices plummeted by 69 per cent. As a result, many oil producers in Texas, the United States were driven out of business. In consequence, Saudi Arabia managed to maintain its share of the oil market. Therefore, this time, Riyadh intended to slow down or stop the growth of shale oil production in the United States by similar means, in order to keep or even expand U.S. reliance on Middle Eastern oil.[5]

Under such circumstances, economic and energy cooperation is more important for both China and Saudi Arabia. The bilateral energy cooperation has covered crude oil trade as well as mutual investment in upstream and downstream projects in both countries.

In March 2014, Chinese President Xi Jinping met in Beijing with Salman bin Abdulaziz Al Saud, who was Saudi Arabia's Crown Prince, Deputy Prime Minister and Defence Minister at that time. Salman was crowned as the new king of Saudi Arabia in January 2015 after the death of his half-brother, King Abdullah. During the meeting, Xi pointed out that it is China's firm and long-term policy to develop friendly cooperation with Saudi Arabia, and proposed to continuously promote the strategic relationship between the two countries. He maintained that energy cooperation is the pillar of China–Saudi partnership, encouraged Saudi Arabia to participate in China's strategy of building the Silk Road Economic Belt as well as the 21st Century Maritime Silk Road, and expressed Beijing's willingness to accelerate

---

[5] Zhang Chi, 'Changes in the global energy system and their implications for China', EAI Background Brief, No. 1006, East Asian Institute, National University of Singapore, 12 March 2015.

the negotiation to establish the free trade area (FTA)[6] between China and the Gulf Cooperation Council (GCC). On the other hand, Salman explicitly expressed that the purpose of his visit to Beijing was to reinforce Saudi Arabia's strategic relations with China.[7]

Also, as some authors noted, China has begun to engage the Middle East in a more coordinated manner and actively adjusted its policy to serve its growing commercial and energy interests. Beijing's diplomacy has succeeded in institutionalising China's relations with the region through mechanisms like the China–Arab Cooperation Forum (CACF) and the China–Gulf Cooperation Council (GCC) Framework agreement on economic, trade, investment and technological cooperation. Considering the existing tension between GCC countries and Iran, China has been playing balanced diplomacy by pulling Tehran into the Shanghai Cooperation Organisation (SCO) as an observer.[8]

Besides, Beijing has accelerated its engagement with the League of Arab States (LAS) that consists of 22 member states. Meanwhile, the CACF was established in 2004 and has become an important

---

[6] In July 2004, China and the Gulf Cooperation Council (GCC) which consists of Bahrain, Kuwait, Oman, Qatar, Saudi Arabia and the United Arab Emirates, announced that they would start talks to establish a free trade agreement (FTA). It is said that the FTA would promote China's energy import. Even those non-oil business agreements between China and the Gulf states were reached with a view of satisfying China's energy demand. Pak K Lee, 'China's quest for oil security: oil (wars) in the pipeline?', *Pacific Review* 18: 2, 2005, pp. 265–230. During former Chinese President Hu Jintao's visit to Saudi Arabia in February 2009, he pledged to seek an early FTA with the GCC, and proposed that the two sides should continue energy dialogue and make joint efforts to maintain stability in global energy markets. Xinhua, 'Chinese president continues visit in Saudi Arabia', 12 February 2009, http://news.xinhuanet.com/english/2009-02/12/content_10806379.htm, accessed 16 February 2009.

[7] Xinhua, 'Xi Jinping huijian shate wangchu jian fushouxiang, guofang dachen saleman' (President Xi Jinping met with Saudi Crown Prince, Deputy Prime Minister and Defence Minister Salman), 13 March 2014, http://news.xinhuanet.com/2014-03/13/c_119762148.htm, accessed 3 March 2015.

[8] Chen Gang and Ryan Clarke, 'China's diplomacy in the Middle East', in Wang Gungwu and Zheng Yongnian eds., *China: Development and Governance* (Singapore: World Scientific Publishing Co. Pte. Ltd., 2013), pp. 469–475.

platform for dialogue and cooperation between China and Arab countries. The Sixth CACF Ministerial Meeting was held in Beijing in June 2014, and Chinese President Xi Jinping delivered a speech at the opening ceremony, calling to deepen the cooperation between China and Arab countries.[9]

4.1.2 *Africa*

Africa is the second largest source of China's oil import. In 2013, China imported 64.9 million tonnes of crude oil from Africa (6 million tonnes from North Africa, 53 million tonnes from West Africa, and 5.9 million tonnes from East and Southern Africa), accounting for 20.8 per cent of its overall oil import.[10] The proportion has somewhat declined during the past few years, as Africa contributed to 26 per cent of China's total oil import in 2007.[11] In the recent years, China has been making efforts to diversify its oil import sources, alleviating its over-dependence on oil from the Middle East. Thus, Africa plays an important role in China's energy import.

According to a Chinese official policy document — *China's African Policy* — published in January 2006, Beijing would make efforts to promote economic cooperation between China and Africa, including cooperation on natural resources:

> 'The Chinese Government facilitates information sharing and cooperation with Africa in resources areas. It encourages and supports competent Chinese enterprises to cooperate with African nations in various ways on the basis of the principle of mutual benefit and common development, to develop and exploit rationally their resources, with a view of helping African countries to translate their

[9] People's Daily On-line, 'Xi Jinping chuxi zhonga hezuo luntan diliujie buzhangji huiyi kaimushi bing fabiao zhongyao jianghua' (President Xi Jinping attended the opening ceremony of the Sixth Ministerial-level Meeting of the China-Arab Cooperation Forum and delivered an important speech), 6 June 2014, http://politics.people.com.cn/n/2014/0606/c1024-25110595.html, accessed 3 March 2015.
[10] British Petroleum, *BP Statistical Review of World Energy June 2014.*
[11] British Petroleum, *BP Statistical Review of World Energy June 2008.*

advantages in resources to competitive strength, and realize sustainable development in their own countries and the continent as a whole'.[12]

With the rise of the Chinese economy in the recent years, China–Africa cooperation is gaining momentum, adding to the dynamics of China's energy diplomacy. The Forum on China–African Cooperation (FOCAC) was established in 2000 as a platform for promoting cooperation, and has 50 African member states. It was elevated to summit level in 2006. Highlighted by the FOCAC Beijing summit in November 2006, China's fast expanding engagement with Africa in the recent years has aroused global attention. The fifth FOCAC Ministerial Meeting was held in Beijing in July 2012. Notably, presidents of seven African states (South Africa, Benin, Equatorial Guinea, Djibouti, Niger, Cote d'Ivoire, Cape Verde and Kenya) as well as United Nations Secretary General Ban Ki-moon also attended the opening ceremony and met with then Chinese President Hu Jintao.[13] The meeting issued the Beijing Declaration[14] and Beijing Action Plan.[15] In particular, according to the Action Plan, the Chinese side and the African side would further promote their energy cooperation through various ways.

---

[12] See Xinhua, 'Full text: China's African Policy', 12 January 2006, http://news.xinhuanet.com/english/2006-01/12/content_4042521.htm, accessed 3 February 2009.

[13] Xinhua, 'Zhongfei hezuo luntan diwujie buzhangji huiyi longzhong kaimu' (The Fifth Ministerial Meeting of the Forum on China–African Cooperation opened), 19 July 2012, http://news.xinhuanet.com/world/2012-07/19/c_112481778.htm, accessed 5 March 2015.

[14] Xinhua, 'Zhongfei hezuo luntan diwujie buzhangji huiyi Beijing xuanyan — quanwen' (Beijing Declaration of the Fifth Ministerial Meeting of the Forum on China–African Cooperation — full text), 20 July 2012, http://news.xinhuanet.com/world/2012-07/20/c_112493071.htm, accessed 5 March 2015.

[15] Ministry of Foreign Affairs, the People's Republic of China, 'Zhongfei hezuo luntan diwujie buzhangji huiyi — Beijing xindong jihua, 2013 nian zhi 2015 nian' (The Fifth Ministerial Meeting of the Forum on China–African Cooperation — Beijing Action Plan, 2013–2015), 23 July 2012, http://www.fmprc.gov.cn/ce/cena/chn/xwdt/t957754.htm, accessed 5 March 2015.

Notably, the traditional China–Africa friendship promotes China's energy cooperation with some African states.[16] As Chris Alden wrote, China is in many ways Africa's most enduring partner. Beijing has pursued a foreign policy that has conformed to Africans' interests and needs more than any other external power.[17] Horace Campbell also noted that although the United States and World Bank claimed to be fighting poverty in Africa, the conditions of the poor people in Africa have worsened after two decades of structural adjustment. 'Chinese investment potentially provides an alternative for African leaders and entrepreneurs, while providing long-term potential for the development of African economies.' Also, Beijing's diplomacy 'provided space for manoeuvre for Africans by laying the basis for an alternative international system in the 21st century'.[18] Moreover, the Chinese government held that Africa is on the verge of a developmental take-off, and it is time for China to expand its role in the region. This idea is well received among African countries.[19] Besides, China is considered by more and more Africans as an alternative political and economic model to the Washington Consensus.[20]

In August 2013, the Chinese government issued an official white paper *China–Africa Economic and Trade Cooperation* (2013), declaring that Chinese companies have helped transform the resource advantages in some African countries into their economic development

[16]For detailed discussion and analysis, see Zhang Chi, 'China's oil diplomacy in Africa: The convergence of national and corporate interests', in Christopher M. Dent eds., *China and Africa Development Relations* (London and New York: Routledge, 2011).

[17]Chris Alden, *China in Africa* (London and New York: Zed Books, 2007), p. 126.

[18]Horace Campbell, 'China in Africa: Challenging US Global Hegemony', in Firoze Manji and Stephen Marks eds., *African Perspectives on China in Africa* (Cape Town, Nairobi and Oxford: Fahamu — Networks for Social Justice, 2007), pp. 119–137.

[19]Bates Gill, Chin-hao Huang and J. Stephen Morrison, 'Assessing China's Growing Influence in Africa', *China Security* 3: 3, 2007, p. 8.

[20]See, for example, Ndubisi Obiorah, 'Who's Afraid of China in Africa? Towards an African Civil Society Perspective on China–Africa Relations', in Firoze Manji and Stephen Marks eds., *African Perspectives on China in Africa* (Cape Town, Nairobi and Oxford: Fahamu — Networks for Social Justice, 2007), p. 44.

opportunities.[21] In the past two years, four members of the new generation of China's top leadership,[22] that formally assumed duty in March 2013, have paid visits to 10 African states (see Table 4.1).

For instance, in May 2014, Chinese Premier Li Keqiang visited Nigeria. When talking with Nigerian President Goodluck Jonathan, Li proposed to expand and deepen cooperation with Nigeria on

Table 4.1: Chinese top leaders' visits to Africa, 2013 to 2014

Time	Visitors	Countries
March 2013	President Xi Jinping	Tanzania, South Africa, Congo-Brazzaville
September 2013	Chairman Zhang Dejiang[23]	Uganda, Nigeria,
May 2014	Premier Li Keqiang	Ethiopia, Nigeria, Angola, Kenya,
November 2014	Chairman Yu Zhengsheng[24]	Algeria, Morocco

[21] The State Council Information Office, the People's Republic of China, *China–Africa Economic and Trade Cooperation (2013)*, white paper, August 2013, http://english.gov.cn/archive/white_paper/2014/08/23/content_281474982986536.htm, accessed 3 March 2015.

[22] In this book, the new generation of China's top leadership refers to the seven members in the Standing Committee of the Political Bureau of the 18th Chinese Communist Party's (CCP) Central Committee, namely Xi Jinping, Li Keqiang, Zhang Dejiang, Yu Zhengsheng, Wang Qishan, Liu Yunshan and Zhang Gaoli. They became members of the Standing Committee — the top CCP leadership — during the 18th CCP National Congress in November 2012, and then Xi Jinping, Li Keqiang, Zhang Dejiang, Yu Zhengsheng and Zhang Gaoli formally took official positions as President, Premier, Chairman of the National People's Congress (top legislator), Chairman of the Chinese People's Political Consultative Conference (top political advisor) and Executive Vice-premier, respectively.

[23] Zhang Dejiang is China's top legislator. He is member of the Standing Committee of the Political Bureau of the 18th CCP Central Committee and Chairman of the 12th Standing Committee of China's National People's Congress.

[24] Yu Zhengsheng is China's top political advisor. He is member of the Standing Committee of the Political Bureau of the 18th CCP Central Committee and Chairman of the 12th Chinese People's Political Consultative Conference.

infrastructure construction, agriculture, energy, etc. President Jonathan expressed his willingness to expand the bilateral cooperation in several areas including energy, and welcomed the expansion of Chinese investment in Nigeria.[25]

After Nigeria, Premier Li went to Angola, an increasingly important oil supplier to China. Angola was Africa's top oil exporter to China and the world's second largest oil supplier to China in 2013, only behind Saudi Arabia.[26] During his talks with Angolan President Jose Eduardo Dos Santos, Li advocated the promotion of China–Angola cooperation in a series of areas such as energy, infrastructure facilities, agriculture, etc. President Dos Santos welcomed the expansion of Chinese investment in Angola and participation in Angola's socio–economic development.[27]

4.1.3 *Central Asia*

Central Asia is a critical energy supplying region for China, largely because it borders China and is located in the inner area of the Euro–Asia continent. China's energy import from Central Asia is mainly transported via pipelines, avoiding the Straits of Malacca that have caused Beijing's concern over energy security.[28] In the past few years, crude oil and natural gas pipelines such as Kazakhstan–China Oil Pipeline and Central Asia–China Gas Pipeline have been built or have new routes and sections under construction. China's energy cooperation with countries in the region, such as Kazakhstan, Uzbekistan, Turkmenistan, Tajikistan, etc. has been expanding rapidly in the recent years. With more and

---

[25] Xinhua, 'Li Keqiang wu niriliya zongtong: jiang zhanlue huoban guanxi tisheng dao xinshuiping'(Premier Li Keqiang met with Nigerian President: promoting the level of the strategic partnership to a new height), 8 May 2014, http://www.sn.xinhuanet.com/2014-05/08/c_1110591264.htm, accessed 4 March 2015.

[26] Julie Jiang and Chen Ding, *Updates on Overseas Investments by China's National Oil Companies: Achievements and Challenges since 2011*, International Energy Agency, Partner Country Series, 2014, http://www.iea.org/publications/freepublications/publication/Partner Country Series Updateon Overseas Investments by Chinas National Oil Companies.pdf, accessed 24 August 2014.

[27] Xinhua, 'Li Keqiang tong angela zongtong duosi sangtuosi juxing huitan'(Li Keqiang held talks with Angolan President Jose Eduardo Dos Santos), 9 May 2014, http://news.xinhuanet.com/world/2014-05/09/c_1110623607.htm, accessed 4 March 2015.

[28] For details of the Straits of Malacca and China's energy security, see the next chapter.

more crude oil and natural gas flowing to China through the pipelines, the region is expected to play an increasingly important role for China's energy import as well as energy security.

Moreover, the SCO serves as a platform for further energy and strategic cooperation between China and Central Asian states. This organisation is made up of China, Russia, Kazakhstan, Kyrgyzstan, Tajikistan and Uzbekistan; India, Pakistan and Iran participate as observers. The SCO groups both important energy producers like Russia, Kazakhstan and Iran, and large energy consumers like China and India. Its member states control about a quarter of the world's oil supplies and are building several pipelines across the region. The SCO has established a working group to study cooperation projects, cooperation orientation and the implementation of cooperation projects in energy sector.[29] In the past a couple of years, China's top leaders have paid a series of visits to Central Asia (see Table 4.2).

Table 4.2: Chinese top leaders' visits to Central Asia, 2013 to 2014

Time	Visitors	Countries
September 2013	President Xi Jinping	Turkmenistan, Kazakhstan, Uzbekistan, Kyrgyzstan
November 2013	Premier Li Keqiang	Uzbekistan
August 2014	Vice-premier Zhang Gaoli[30]	Turkmenistan
September 2014	President Xi Jinping	Tajikistan
December 2014	Premier Li Keqiang	Kazakhstan

[29] Beijing Review, 'Image Improvement', http://www.bjreview.com.cn/06-24-e/w-3.htm, 24 June 2006, accessed 25 July 2006.

[30] Zhang Gaoli is member of the Standing Committee of the Political Bureau of the 18th CCP Central Committee and Executive Vice-premier of the State Council of the People's Republic of China. He is in charge of China's energy affairs and the National Development and Reform Commission (NDRC) which is China's major energy management administration.

Currently, Kazakhstan is China's most important energy cooperation partner in Central Asia. In September 2013, Chinese President Xi Jinping visited Kazakhstan, and met with his Kazakh counterpart President Nursultan Nazarbayev. The two leaders agreed on deepening the China–Kazakhstan Comprehensive Strategic Relationship, and emphasised that the bilateral energy cooperation enjoys complementary advantages and is a win–win situation; the two countries are long-term, stable and reliable partners for energy cooperation.[31] During his visit, President Xi delivered a speech at Nazarbayev University. In the speech, he proposed for the first time for the construction of the Silk Road Economic Belt with regional countries through innovative cooperation.[32] Now the Silk Road Economic Belt strategy has become a pillar of China's international strategy under Xi.

A year later, in September 2014, Chinese President Xi Jinping went to Tajikistan to attend the 14th Meeting of the Council of Heads of State of the SCO. During this visit, President Xi Jinping and President Emomali Rahmon of Tajikistan attended the commencement ceremonies of the cooperation project of China–Tajikistan power plant and Tajik section of Line D of Central Asia–China Gas Pipeline. This 1,000 km long pipeline will transmit natural gas from Turkmenistan through Uzbekistan, Tajikistan and Kyrgyzstan to China. The project is scheduled to complete and start operation in 2016, with a designated capacity of 30 billion cubic metres. National Petroleum Corporation's (CNPC) will cooperate with its Central Asian partners to construct this pipeline. During this visit, President Xi also proposed to uphold the so-called Spirit of the Silk Road for

---

[31] Xinhua, 'Xi Jinping tong hasakesitan zongtong nazhaerbayefu juxing huitan' (President Xi Jinping held talks with Kazakh President Nursultan Nazarbayev), 7 September 2013, http://news.xinhuanet.com/world/2013-09/07/c_117272640.htm, accessed 3 March 2015.

[32] Xinhua, 'Xi Jinping zai nazhaerbayefu daxue de yanjiang, quanwen' (President Xi Jinping's speech at Nazarbayev University, full text), 8 September 2013, http://news.xinhuanet.com/world/2013-09/08/c_117273079_2.htm, accessed 3 March 2015.

the further cooperation between China and Tajikistan.[33] He suggested that the two countries take the opportunity of the building of the Silk Road Economic Belt to promote the bilateral cooperation in several areas including oil and gas, electric power, economic and trade, transportation facilities, etc., and raise the bilateral trade volume to USD 3 billion.[34]

Moreover, Turkmenistan is an increasingly important energy cooperation partner for China. The country owns the world's fourth largest proved reserves of natural gas.[35] In 2007, CNPC inked a production sharing agreement (PSA) for reserves on the right bank of Amu Darya River in eastern Turkmenistan and a natural gas purchase agreement for 30 to 40 billion cubic metres per year for 30 years, as well as a gas pipeline from Turkmenistan to China. In the same year, China signed a transit agreement with Uzbekistan and Kazakhstan for this pipeline. The USD 4 billion loan from China Development Bank, a Chinese state-owned policy bank, to Turkmengaz for the development of the South Yoloten gas field (Block A and B) further enhanced the cooperation agreement. Notably, Chinese NOCs' success in securing the pipelines and oil and gas supplies from Central Asia should be credited in part to the Chinese government's long-term lobbying efforts in the region. In 2013, during Chinese President Xi Jinping's visit to Turkmenistan, the presidents of the two states launched the construction of the second phase of Galkynysh gas field and signed a *Joint Declaration*

---

[33] People's Daily On-line, 'Xijinping he lahemeng zongtong gongtong chuxi zhongta dianli he zhongguo-zhongya tianranqi guandao hezuo xiangmu kaigong yishi' (President Xi Jinping and President Emomali Rahmon attended the commencement ceremonies of the cooperation project of China–Tajikistan power plant and Central Asia–China Gas Pipeline), 14 September 2014, http://politics.people.com.cn/n/2014/0914/c1024-25655846.html, accessed 2 March 2015.

[34] People's Daily On-line, 'Xijinping tong tajikesitan zongtong lahemeng huitan' (President Xi Jinping held talks with Tajik President Emomali Rahmon), 14 September 2014, http://politics.people.com.cn/n/2014/0914/c1024-25655845.html, accessed 2 March 2015.

[35] British Petroleum, *BP Statistical Review of World Energy June 2014.*

on Establishing a Strategic Partnership between Turkmenistan and China. Galkynysh is the second largest gas field in the world.[36] Just as Chinese Vice-premier Zhang Gaoli pointed out during his visit to Turkmenistan in August 2014, the natural gas cooperation between China and Turkmenistan is the key focus of the bilateral economic cooperation, which has been promoted by the heads of state of both countries.[37]

4.1.4 *Russia*

As the world's second largest producer of oil and gas[38] and China's neighbour, Russia is a critical energy cooperation partner for China. The strategic cooperation between Beijing and Moscow in the international politics also boosts the bilateral energy cooperation. It is worth noting that since coming into power, Chinese President Xi Jinping have visited Russia for three times during the past two years, apart from meeting his Russian counterpart, President Vladimir Putin, at some multilateral meetings. In particular, from 2013 to 2014, the two presidents met 7 times at various occasions and venues. Other Chinese top leaders such as Premier Keqiang and Vice-premier Zhang Gaoli have also visited Russia (see Table 4.3).

Take the three visits to Russia by Chinese President, Premier and Vice-premier respectively in 2014 for example. In February 2014, during his visit to Russia to attend the opening ceremony of the Sochi Winter Olympics, Chinese President Xi Jinping met with Russian President Vladimir Putin. Xi proposed to push forward China–Russia

[36] Jiang and Ding, *Updates on Overseas Investments by China's National Oil Companies: Achievements and Challenges since 2011.*

[37] Xinhua, 'Zhang Gaoli yu tukumansitan fuzongli gongtong zhuchi zhongtu hezuo weiyuanh di sanci huiyi' (Vice-premier Zhang Gaoli chaired together with Turkmenistan Deputy Prime Minister the 3rd meeting of China–Turkmenistan Cooperation Commission), 27 August 2014, http://news.xinhuanet.com/politics/2014-08/27/c_1112255872.htm, accessed 2 March 2015.

[38] British Petroleum, *BP Statistical Review of World Energy June 2014.*

Table 4.3: Chinese top leaders' visits to Russia, 2013–2014

Time	Visitors
March 2013	President Xi Jinping
June 2013	Vice-premier Zhang Gaoli
September 2013	President Xi Jinping
September 2013	Chairman Zhang Dejiang
February 2014	President Xi Jinping
August to September 2014	Vice-premier Zhang Gaoli
October 2014	Premier Li Keqiang

cooperation on some strategic projects such as the scheduled increase of Russia's oil and gas supply to China, the expansion of China–Russia Oil Pipeline, the construction of joint venture oil refinery, etc. Putin also expressed his willingness to promote the bilateral cooperation on several areas including the exploration of oil and gas.[39]

On 30 August 2014, Chinese Vice-premier Zhang Gaoli and Russian Deputy Prime Minister Arkady Dvorkovich co-chaired the 11th meeting of China–Russia Energy Cooperation Committee in Moscow. On 1 September 2014, Zhang met with Russian President Vladimir Putin in Yakutsk. They attended the commencement ceremony for the construction of the Russian section of the eastern route of Russia–China Gas Pipeline. At the ceremony, Zhang explicitly declared that this pipeline is of strategic importance, and the bilateral cooperation on this significant project is directly promoted by the heads of state of China and Russia in person. He also proposed that the two countries push forward their energy cooperation covering oil, natural gas, nuclear energy, coal, electric power and new energies, simultaneously construct the eastern and the western

[39] People's Daily On-line, 'Xi Jinping huijian eluosi zongtnog pujing' (President Xi Jinping met with Russian President Vladimir Putin), 6 February 2014, http://politics.people.com.cn/n/2014/0206/c1024-24285729.html, accessed 2 March 2015.

routes of the pipelines, and enhance both upstream and downstream cooperation.[40]

In October 2014, Chinese Premier Li Keqiang visited Russia and attended the 19th regular meeting between Chinese Premier and Russian Prime Minister. During his meeting with Russian Prime Minister Dmitry Medvedev, Premier Li proposed to promote the bilateral cooperation especially energy cooperation.[41]

4.1.5 *Southeast Asia*

China's engagement with Southeast Asia developed rapidly since the beginning of this century. For Beijing, Southeast Asia is not only important in its energy resources, but also in its geographical position, as it is an unavoidable vital energy import route for China. Some countries in the region, such as Malaysia and Indonesia, are important liquefied natural gas (LNG) supplier to China. In 2013, Malaysia and Indonesia were the third and the fourth largest LNG exporter to China respectively.[42]

Notably, all of the shipping or transportation routes for China's energy import from the Middle East and Africa pass through Southeast Asia, either via South China Sea or via China–Myanmar Oil Pipeline. There are at least a dozen strategic straits and water areas in the region including Malacca Strait, Sunda Strait, Gaspar

---

[40] Xinhua, 'Zhang Gaoli fu eluosi juxing zhonge nengyuan hezuo weiyuanhui di shiyici huiyi huijian eluosi zongtong pujing bing gongtong chuxi zhonge dongxian tianranqi guandao ejingneiduan kaigong yishi' (Vice-premier Zhang Gaoli visited Russia to attend the 11th meeting of China–Russia Energy Cooperation Commission, met with Russia President Vladimir Putin, and attended together with President Putin the commencement ceremony of the Russia section of the eastern route of China–Russia Gas Pipeline), 2 September 2014, http://news.xinhuanet.com/politics/2014-09/02/c_1112316107.htm, accessed 2 March 2015.

[41] People's Daily On-line, 'Li Keqiang tong eluosi zongli meideweijiefu gongtong zhuchi zhonge zongli di shijiuci dingqi huiwu'(Premier Li Keqiang and Russian Prime Minister Dmitry Medvedev together chaired the 19th regular meeting between Chinese Premier and Russian Prime Minister), 13 October 2014, http://politics.people.com.cn/n/2014/1013/c1024-25826219.html, accessed 2 March 2015.

[42] British Petroleum, *BP Statistical Review of World Energy June 2014.*

Strait, Lombok Strait, Makassar Strait, Maluku Strait and the South China Sea.[43]

Moreover, the South China Sea has abundant oil and natural gas reserves. But China, Vietnam, the Philippines, Malaysia, Indonesia and Brunei have maritime territorial disputes over the Xisha (Paracel) and the Nansha (Spratly) islands. In 1995, Beijing formally agreed to discuss the dispute on a collective basis with the Association of Southeast Asian Nations (ASEAN), and begun to actively engage Southeast Asia.[44] Particularly, to ease Southeast Asian countries' concern about China's potential use of military force to settle the disputes, and to improve China's relationship with the regional countries, Beijing signed *the Declaration on the Conduct of Parties in the South China Sea* in 2002 and *the Treaty of Amity and Cooperation in Southeast Asia* in 2003,[45] committing itself to peaceful resolution of disputes and providing opportunities for joint energy exploitations in disputed sea areas. Notably, the South China Sea is considered to have more gas than oil potential. Most of the hydrocarbon fields explored in the South China Sea areas of China, Brunei, Indonesia, Malaysia, the Philippines and Vietnam, contain gas instead of oil.[46]

---

[43] There are at least four sea lanes of communication (SLOC) available for China in Southeast Asia. The first is the route of the Middle East/Africa-Malacca Strait-South China Sea-China, which is for ships less than 100,000 tonnes. The second is the route of the Middle East/Africa-Sunda Strait-Gaspar Strait-South China Sea-China, which is for ships over 100,000 tonnes. The third is the route of the Latin America-South/North Pacific-Philippine Sea-Balintang/Bashi Strait-South China Sea-China. The fourth is the alternative route of the Middle East/Africa–Lombok Strait–Makassar Strait-Maluku Strait-Philippine Sea-West Pacific-China, which is a Japanese transportation route. See Zhang Xuegang, 'Southeast Asia and Energy: Gateway to Stability', *China Security* 3: 2, 2007, pp. 19–24.

[44] For details, see Michael Yahuda, *The International Politics of the Asia–Pacific* (London and New York: RoutledgeCurzon, 2004), pp. 298–302.

[45] For more information, see David Shambaugh, 'China Engages Asia: Reshaping the Regional Order', *International Security* 29: 3, pp. 64–99.

[46] Michael Richardson, 'A Southward Thrust for China's Energy Diplomacy in the South China Sea', China Brief 8: 21, the Jamestown Foundation, 2008, http://www.jamestown.org/programs/chinabrief/single/?tx_ttnews%5Btt_news%5D=34140&tx_ttnews%5BbackPid%5D=168&no_cache=1, accessed 13 March 2009.

Also, Myanmar's geographical position is significant for China. It lies between China and the Indian Ocean seaways to the Persian Gulf. The China–Myanmar Oil Pipeline transport part of China's crude oil import from the Middle East and Africa, circumventing the Straits of Malacca. In addition, the Lancang Jiang-Mekong River route, which passes through China, Thailand, Laos, Myanmar, Cambodia and Vietnam, is a potential supplementary energy import transportation route for China.

In the past two years, China's new generation of leadership have frequently travelled to Southeast Asia (see Table 4.4), visiting 8 countries among the 10 ASEAN members.

In October 2013, Chinese President Xi Jinping attended the 21st APEC Summit in Indonesia and delivered a speech at the Indonesian Parliament. In the speech, Xi proposed for the first time to build the 21st Century

Table 4.4: Chinese top leaders' visits to Southeast Asia, 2013–2014

Time	Visitors	Countries
September 2013	Mr Liu Yushan[47]	Cambodia
October 2013	President Xi Jinping	Indonesia, Malaysia
October 2013	Premier Li Keqiang	Brunei, Thailand, Vietnam
October 2013	Vice-premier Zhang Gaoli	Singapore
November 2014	Premier Li Keqiang	Myanmar
December 2014	Premier Li Keqiang	Thailand
December 2014	Chairman Yu Zhengsheng	Vietnam

[47] Liu Yunshan is member of the Standing Committee of the Political Bureau of the 18th CCP Central Committee, and President of the CCP Central Party School. He is in charge of China's publicity, ideological and cultural affairs as well as the CCP's party affairs.

Maritime Silk Road through cooperation with ASEAN countries, and to realise common development and prosperity.[48] In November 2014, when attending the 17th China–ASEAN (10+1) Summit, the 17th ASEAN-China, Japan and South Korea Summit, and the 9th East Asian Summit in Naypyitaw, Myanmar, Chinese Premier Li Keqiang paid an official visit to Myanmar and met with Myanmar President Thein Sein. He proposed to deepen the cooperation between China and Myanmar in several areas including energy and agriculture.[49]

4.1.6 *Latin America*

Latin America is also an energy rich region and important energy supplier for China. For example, Venezuela was China's seventh largest oil supplier in 2013.[50] There is significant potential for energy cooperation between China and Latin America. In November 2008, the Chinese government published *China's Policy Paper on Latin America and the Caribbean*, articulating its policy towards the region of Latin America and the Caribbean, including cooperation in resources and energy:

> 'The Chinese Government encourages and supports qualified Chinese companies with good reputation in investing in manufacturing, agriculture, forestry, fishing, energy, mineral resources, infrastructure, and service sector in Latin America and the Caribbean to promote

---

[48] Xinhua, 'Xi Jinping zai yinni guohui fabiao yanjiang: Xieshou jianshe zhongguo-dongmeng mingyun gongtongti' (Xi Jinping delivered a speech at Indonesian parliament: Jointly building the China–ASEAN community of common destiny), 13 October 2013, http://news.xinhuanet.com/world/2013-10/03/c_117591652_3.htm, accessed 4 March 2015.

[49] People's Daily On-line, 'Li Keqiang zongli tong miandian zongtong wudengsheng juxing huitanshi qiangdiao, quanmian tisheng zhongmian zhanlue hezuo shuiping, liangguo yongyuan zuo haolingju, haopengyou, haohuoban' (When meeting with Myanmar President Thein Sein, Premier Li Keqiang proposed to comprehensively promote China–Myanmar cooperation, and stressed that the two countries should remain good neighbours, friends and partners forever), 14 November 2014, http://politics.people.com.cn/n/2014/1114/c1024-26026867.html, accessed 2 March 2015.

[50] Jiang and Ding, *Updates on Overseas Investments by China's National Oil Companies: Achievements and Challenges since 2011.*

> the economic and social development of both sides...The Chinese side wishes to expand and deepen mutually beneficial cooperation with Latin American and Caribbean countries in resources and energy within bilateral cooperation frameworks'.[51]

China's engagement with Latin America began to accelerate at the beginning of this century. In the past two years, Chinese President Xi Jinping visited Latin America each year (see Table 4.5).

In July 2014, China-Community of Latin American and Caribbean States (CELAC) Forum was established in Brazil, during Chinese President Xi Jinping's state visit to the country.[52] In January 2015, the first China-CELAC[53] Forum Ministerial Meeting was held in Beijing. Notably, the presidents Venezuela and Ecuador were also in Beijing to have talks with Chinese President Xi Jinping. The two-day meeting, co-chaired by Chinese Foreign Minister Wang Yi and his Costa Rican counterpart, Manuel Gonzalez Sanz, passed a five-year cooperation

Table 4.5: Chinese top leaders' visits to Latin America, 2013–2014

Time	Visitors	Countries
June 2013	President Xi Jinping	Trinidad and Tobago, Costa Rica, Mexico
July 2014	President Xi Jinping	Brazil, Argentina, Venezuela, Cuba
November 2014	Chairman Zhang Dejiang	Peru, Columbia, Mexico

[51] Xinhua, 'Full text: China's Policy Paper on Latin America and the Caribbean', 5 November 2008, http://news.xinhuanet.com/english/2008-11/05/content_10308117.htm, accessed 3 February 2009.

[52] Xinhua, 'China-CELAC Forum established in Brazil', 18 July 2014, http://news.xinhuanet.com/english/china/2014-07/18/c_133493520.htm, accessed 4 March 2015.

[53] The Community of Latin American and Caribbean States (CELAC) is a regional organisation established in December 2011 in Caracas, Venezuela. It includes 33 member states — all the sovereign states in Latin America and Caribbean except the United States and Canada.

plan, regulations for the forum and the Beijing Declaration.[54] The forum's next meeting will be held in Chile in 2018.

In particular, China-CELAC Cooperation Plan (2015–2019) was released during the meeting. The plan formulated measures for the comprehensive cooperation between China and Latin America, covering many aspects including energy. Actually, energy cooperation was mentioned several times in the plan. For example, Article Four of the plan declared to promote the energy infrastructure construction; Article Five explicitly stressed to strengthen the energy cooperation between China and Latin America, and proposed to hold the China–Latin America forum on energy and mining; Article Seven touched upon furthering China-CELAC cooperation on petro–chemical industry.[55]

When meeting with Argentinean President Cristina Fernández de Kirchner in Buenos Aires in July 2014, Chinese President Xi Jinping proposed to actively carry out China–Argentina cooperation on oil and gas, mining, agriculture, etc. During the visit, the two presidents promoted the relations between China and Argentina to comprehensive strategic partners.[56] When meeting with Venezuelan President Nicolas Maduro, President Xi maintained that energy cooperation is the pillar of China–Venezuela cooperation. The two presidents also promoted China–Venezuela relations to comprehensive strategic partners.[57]

---

[54] Xinhua, 'China, CELAC deepens cooperation with major documents', 9 January 2015, http://news.xinhuanet.com/english/china/2015-01/09/c_133908824.htm, accessed 4 March 2015.

[55] Xinhua, 'Zhongguo yu lamei he jialebi guojia hezuo guihua 2015–2019' (China-CELAC Cooperation Plan 2015–2019), 9 January 2015, http://news.xinhuanet.com/world/2015-01/09/c_1113944648.htm, accessed 4 March 2015.

[56] People's Daily On-line, 'Xi Jinping tong agenting zongtong juxing huitan' (President Xi Jinping held talks with Argentinean President), 19 July 2014, http://politics.people.com.cn/n/2014/0719/c1024-25302513.html, accessed 2 March 2015.

[57] People's Daily On-line, 'Xi Jinping tong weineiruila zongtong maduluo juxing huitan' (President Xi Jinping held talks with Venezuelan President Nicolas Maduro), 21 July 2014, http://politics.people.com.cn/n/2014/0721/c1024-25312095.html, accessed 2 March 2015.

4.2 Chinese NOCs' overseas investment

When Chinese leaders actively engage energy exporting states, Chinese NOCs also spare no effort in expanding their transnational operation, especially acquisitions of oil and natural gas assets abroad. This section shows the three leading Chinese NOCs' overseas operation particularly investment, focusing on some of their major acquisition deals and service contracts reached in the recent years. It is divided into two sub-sections. The first part provides brief background discussion about the three major NOCs' business, especially their overseas operations. The second part demonstrates the major development of their overseas investment across different parts of the world.

4.2.1 *The three major Chinese NOCs*

There are three major Chinese NOCs: China National Petroleum Corporation (CNPC), China Petroleum and Chemical Corporation (Sinopec) and China National Offshore Oil Corporation (CNOOC).[58] The parent companies — CNPC Group, Sinopec Group and CNOOC General Corporation respectively — are full SOEs; and their respective three subsidiaries — PetroChina Ltd., Sinopec Ltd. and CNOOC Ltd. — are listed companies on the domestic and international stock exchanges with integrated upstream, midstream and downstream businesses. Meanwhile, CNPC Group, Sinopec Group and CNOOC General Corporation are respectively controlling shareholders of their listed subsidiaries PetroChina Ltd., Sinopec Ltd. and CNOOC Ltd. Also, CNPC Group and Sinopec Group carry out some of their ODI activities through their respective fully invested

[58] Apart from these three leading NOCs, there is a relatively minor Chinese NOCs — Sinochem Corporation that has been carrying out transnational oil investment since 2002; and its overseas investment subsidiary is Sinochem Oil Exploration and Development Co. Ltd. Besides, a couple of central SOEs dominated in other sectors like CITIC Group have also been involved in overseas oil investments. Moreover, there are some provincial SOEs such as Shaanxi Yanchang Petroleum (Group) Co. Ltd that has been carrying out overseas investments for years.

overseas investment subsidiaries — China National Oil and Gas Exploration and Development Corporation (CNODC) and Sinopec International Oil Exploration and Development Company Ltd. (SIOEDC). But in the case of CNOOC General Corporation, all of its overseas direct investment projects are done by its listed subsidiary CNOOC Ltd.

Chinese NOCs began to pursue overseas investment and transnational operations in 1993. After venturing overseas for more than 20 years, Chinese NOCs have become emerging important global players in merger and acquisition (M&A) of oil and gas assets. In the recent years, they have greatly increased their overseas production, mainly through upstream M&A.

The International Energy Agency estimated that by the end of 2013, the combined overseas oil and gas production by Chinese companies (including the NOCs and other firms) has totalled 2.5 million barrels of oil equivalent per day. Within it, the oil production reaching 2.1 million bpd, which is equivalent to about the half of China's domestic oil production in 2013, or is equivalent to Brazil's overall oil production in the same year. Notably, Chinese NOCs' overseas portfolios have become more and more diversified. Currently, about half of the Chinese overseas oil production is located in the Middle East and Africa (see Figure 4.1), while an increasing percentage comes from Organisation for Economic Cooperation and Development (OECD) countries such as North America and OECD Europe.

In 2013, Chinese companies signed M&A deals valued at USD 30 billion largely to acquire offshore, LNG and unconventional oil and gas assets, the majority of which were in the OECD member countries. By the end of 2013, there have been 10 Chinese companies spearheaded by the three major NOCs — CNPC, Sinopec and CNOOC — have oil investment and production in 42 countries. Among these countries, Iraq, Kazakhstan and Nigeria are the top three host countries contributing to Chinese production abroad (Figure 4.2).[59]

[59] Jiang and Ding, *Updates on Overseas Investments by China's National Oil Companies: Achievements and Challenges since 2011.*

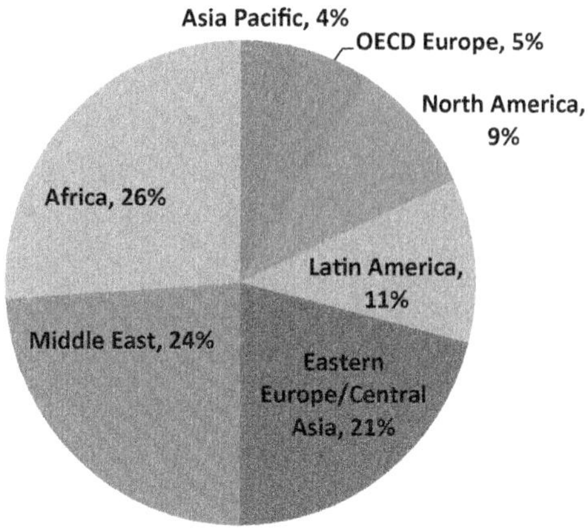

Figure 4.1: Chinese overseas oil and gas production by region, 2013 (Percentage)

Source: Julie Jiang and Chen Ding, *Updates on Overseas Investments by China's National Oil Companies: Achievements and Challenges since 2011*, International Energy Agency, Partner Country Series, 2014, http://www.iea.org/publications/freepublications/publication/Partner Country Series Updateon Overseas Investments by Chinas National Oil Companies.pdf, accessed 24 August 2014.

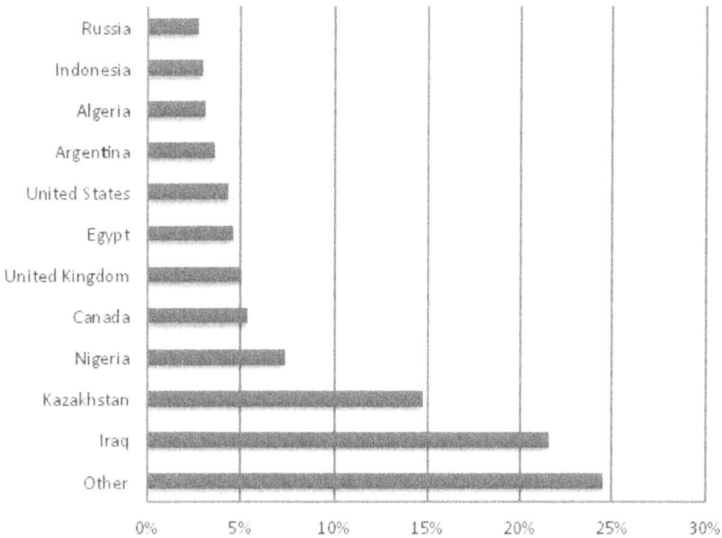

Figure 4.2: Chinese overseas oil and gas production by major countries, 2013 (Percentage)

Source: *Ibid.*

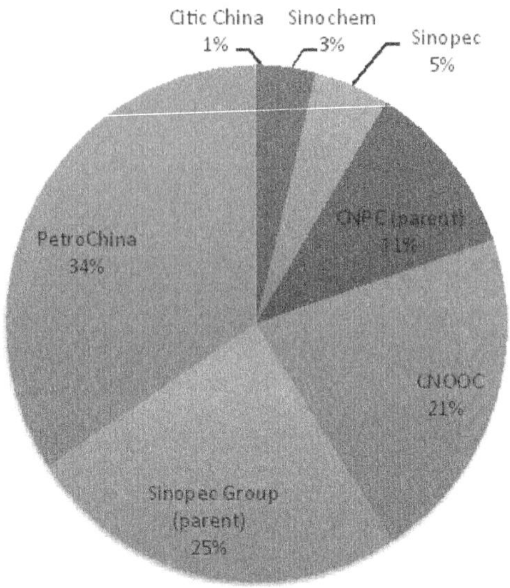

Figure 4.3: Chinese overseas oil and gas production by company, 2013 (Percentage)
Source: *Ibid*.

Notably, the overseas oil and gas production by the three major NOCs together accounts for 96 per cent of the total overseas oil and gas production by all Chinese companies, with CNPC/PetroChina taking up 45 per cent, Sinopec accounting for 30 per cent and CNOOC taking up 21 per cent (see Figure 4.3). The following paragraphs provide a brief introduction to these three NOCs.

CNPC

China National Petroleum Corporation (CNPC Group) is the flagship Chinese NOC venturing overseas, with worldwide businesses covering both upstream and downstream oil and natural gas operations, oil field services, pipeline construction and operation, engineering and construction, petroleum materials and equipment manufacturing and supply, and capital management, finance and insurance services as well. Thus, the narrative of Chinese NOCs' transnational operation and overseas investment in this section will focus on CNPC's activities.

CNPC was the first Chinese NOCs to expand overseas. In the early 1990s, CNPC started to invest in Peru, Sudan and Kazakhstan. In 2013 CNPC's oil output was 112.60 million tonnes within China (54 per cent of China's domestic oil production) and 105.86 million tons overseas; its natural gas output was 88.84 billion cubic meters in China (75 per cent of China's domestic natural gas production) and 21.7 billion cubic meters overseas. Meanwhile, CNPC has oil and gas assets and interests in 37 countries across Africa, Central Asia and Russia, North America, the Middle East, the Asia Pacific, etc., and provides oil field services and engineering construction in 67 countries around the world.[60] Notably, many overseas investment and operation of CNPC are carried out by its overseas investment arm and subsidiary — CNODC, which has been actively engaged in overseas oil and natural gas business since the early 1990s.

Moreover, some of CNPC's overseas investment and operation are carried out by its subsidiary PetroChina Company Limited (PetroChina). This firm is the most active and prominent Chinese player in overseas oil and natural gas investment. PetroChina was established in 1999 as a subsidiary of CNPC. CNPC is the sole sponsor and controlling shareholder of PetroChina. As the largest oil and natural gas producer and distributor in China and one of the largest oil companies in the world, PetroChina engages in wide range of activities such as 'exploration, development, production and marketing of crude oil and natural gas; refining, transportation, storage and marketing of crude oil and oil products; the production and marketing of primary petrochemical products, derivative chemicals and other chemicals; transportation of natural gas, crude oil and refined oil, and marketing of natural gas.'[61] It aims to become an international energy company with strong competitiveness and one of the major producers and distributors of petroleum and petrochemical products in the world.

---

[60] CNPC website, http://www.cnpc.com.cn/en/index.shtml, accessed 23 February 2015.

[61] PetroChina website, http://www.petrochina.com.cn/ptr/gsjj/gsjs_common.shtml, accessed 23 February 2015.

According to its website, PetroChina is dedicated to implementing three strategies of resources, markets and internationalisation. First, as to resources, the strategy of PetroChina includes attaching equal importance to oil and gas, enhancing the exploration of domestic resources, increasing the acquisition of overseas assets, expanding the exploration of offshore resources, increasing strategic reserves and developing alternative energies. Second, with regard to market, PetroChina's strategy includes taking the advantages of is integrated operation from upstream to downstream sectors, enhancing operation in mature markets, developing the international market, promoting its competitiveness in both domestic and international markets, etc. Third, as to internationalisation, the company's strategy includes attaching more importance to overseas oil and gas exploration and development, developing mid-stream and downstream businesses in prudently and effectively, diversifying the sources of energy import, and expanding the scale of international oil and gas trade.[62]

Sinopec

China Petrochemical Corporation (Sinopec Group) ranked the fourth in the Fortune Global 500 in 2013. The company's key business activities include the exploration, production, storage and transportation and marketing of oil and natural gas; oil refining; the wholesale of oil products; the production, marketing, storage, transportation of petrochemicals and other chemical products; etc.[63] Currently, Sinopec Group has overseas investment in Africa, Central Asia, the Americas, the Middle East and the Asia Pacific. Moreover, SIOEDC is a subsidiary of Sinopec Group. It was established in 2001 and is responsible for the management and operation of Sinopec Group's overseas oil and natural gas investment. Sinopec has operated in 58 countries across the world.[64]

---

[62] PetroChina website, http://www.petrochina.com.cn/ptr/fzzl/gsjs_common.shtml, accessed 23 February 2015.

[63] Sinopec Group website, http://www.sinopecgroup.com/group/en/companyprofile/AboutSinopecGroup/, accessed 23 February 2015.

[64] Sinopec Group website, http://www.sinopecgroup.com/group/scjy/gjhjy/, accessed 24 February 2015.

CNOOC

China National Offshore Oil Corporation (CNOOC General Corporation) was founded in 1982. Its business covers oil and gas exploration and development, engineering and technical services, refining and marketing, etc. In 2013, the company's overseas business ran across more than 40 countries,[65] with overseas assets accounted for 40 per cent of the company's total assets. In the same year, CNOOC produced 27.46 million tonnes of crude oil and 8.9 billion cubic metres of natural gas abroad.[66]

Its subsidiary CNOOC Ltd. was established in 1999 and is China's largest producer of offshore crude oil and natural gas and one of the largest oil and gas exploration and production companies in the world. Its business covers oil and natural gas exploration, development, production and sales.[67] In 2013, the company had overseas oil and gas assets in 20 countries around the world (Canada, the United States, Trinidad and Tobago, Columbia, Brazil, Argentina, Iceland, the United Kingdom, Algeria, Nigeria, Equatorial Guinea, Republic of Congo, Uganda, Kenya, Iraq, Qatar, Myanmar, Cambodia, Indonesia and Australia.).[68]

4.2.2 *Major approaches of Chinese NOCs' overseas investment*

During the past two decades of ODI and transnational operation, Chinese NOCs have learnt much from their early years of business abroad, and gradually become more experienced in securing foreign energy supply and overseas equities. Notably, they have been venturing overseas mainly through six approaches, including direct mergers

---

[65] CNOOC website, http://www.cnooc.com.cn/col/col6141/index.html, accessed 24 February 2015.

[66] CNOOC website, http://www.cnooc.com.cn/col/col5581/index.html, accessed 24 February 2015.

[67] CNOOC Ltd. website, http://www.cnoocltd.com/col/col7261/index.html, accessed 24 February 2015.

[68] CNOOC Ltd. website, http://www.cnoocltd.com/col/col7321/index.html, accessed 24 February 2015.

and acquisitions (M&A) of oil and gas assets, building partnership with international oil companies (IOCs), concluding loan-for-energy deals with the support from the Chinese government and policy banks, accessing to some oil and gas resources overseas through market-for-resources approach, reaching service contracts with some energy-rich countries, and building and operating transnational and overseas oil and gas pipelines. This sub-section demonstrates the six approaches one by one.

4.2.2.1 *Mergers and acquisitions*

After more than 20 years of operation and investment overseas, Chinese NOCs have been emerging as important global players in M&A of oil and natural gas assets, and sparing no effort in expanding and diversifying their reserves and assets across the world.

The U.S. Energy Information Administration (EIA) pointed out that Chinese NOCs are taking advantage of the world economic downturn to increase their global M&A; and they benefit from the country's huge foreign exchange reserves in purchasing equity in overseas projects or acquiring stakes in foreign energy companies.[69] Indeed, the 2008 global financial crisis provided opportunities for Chinese NOCs to purchase quality assets abroad from stricken companies and to secure long-term supply deals by extending loans to energy-rich countries in need of capital. Particularly in 2009, at a time when most IOCs cut back on their investment spending, Chinese NOCs, along with other Chinese companies, invested in 10 overseas M&A projects for a total of USD 18.2 billion, accounting for 13 per cent of total global acquisitions (USD 144 billion) and 61 per cent of all acquisitions by national oil companies (USD 30 billion).[70]

---

[69] U.S. Energy Information Administration, 'Country Analysis Brief: China', 4 February 2014, http://www.eia.gov/countries/analysisbriefs/China/china.pdf, accessed 14 January 2015.

[70] Julie Jiang and Jonathan Sinton, *Overseas Investments by Chinese National Oil Companies: Assessing the Drivers and Impacts*, International Energy Agency, February 2011, http://www.iea.org/publications/freepublications/publication/overseas_china.pdf, accessed 24 August 2014.

In 2013, Chinese companies, including the three major NOCs and other firms, invested USD 38 billion in M&A of upstream oil and gas assets abroad. In this way, all of the three main Chinese NOCs have realised obvious improvement of their overseas production portfolio, especially in OECD member countries, while increasingly focusing focus on unconventional oil and natural gas production, deep-water exploration and LNG projects.

It is worth noting that the recent development Chinese NOCs' oil and gas M&A has indicated that their continuously expanding ODI has been switching from those 'leftover' assets deserted by Western oil companies, as the cases often be in the 1990s, and some politically risky and unstable areas in the world, to assets with much better quality and located in more politically and economically stable places. For example, in the past few years, Chinese NOCs' ODI and M&A initiatives have achieved some success in OECD countries across North America, West Europe and Australia. This comes at a time when Chinese NOCs are facing increasingly complicated and uncertain geopolitical situations in the Middle East and Africa.[71] The details of the development of Chinese NOCs' M&A and ODI around the world in the past two decades will be provided in the next section.

4.2.2.2 *Partnership with international oil companies*

In the recent years, Chinese NOCs have begun to increasingly cooperate with IOCs to build partnerships in their overseas investment and operation. This helps diversify the risks is their transnational operation, and is beneficial for Chinese NOCs to gain advanced technologies and operational and managerial experiences in unfamiliar environments.[72]

On the one hand, after more than two decade of operation abroad, Chinese NOCs have realised that successful transnational

---

[71] Jiang and Ding, *Updates on Overseas Investments by China's National Oil Companies: Achievements and Challenges since 2011.*

[72] Jiang and Sinton, *Overseas Investments by Chinese National Oil Companies: Assessing the Drivers and Impacts.*

operation requires them to operate more like IOCs, and they are learning and evolving from their early overseas experiences. Instead of working alone, as in their early years of overseas expansion, today Chinese NOCs often establish partnerships with IOCs. Cooperation with IOCs has proven crucial for Chinese NOCs' overseas investment and operation. So far, forming alliances with IOCs has brought Chinese NOCs four benefits.

Firstly, cooperation with IOCs gives Chinese NOCs access to state of the art technologies and helps them improve their managerial capacity. Currently, Chinese NOCs still lack some of the up-to-date energy technologies especially those in deepwater drilling and unconventional oil and gas exploration. Cooperation and partnership with IOCs is the best approach for Chinese NOCs to acquire such technical expertise.

For instance, CNOOC is working with TOTAL in Nigeria's Akpo and Egina deepwater oil fields, which allows CNOOC to gain the necessary knowhow and techniques for the exploration of China's domestic deepwater oil and gas reserves. Also, CNOOC's acquiring of 36.5 per cent stake of the Golden Eagle project in 2014 through its subsidiary Nexen, 70 km northeast of Aberdeen, Scotland, gave it opportunity to gain offshore operational and technical expertise. In 2011, CNPC concluded a USD 1 billion agreement with INEOS to establish trading and refining joint ventures related to the refining operations in Grangemouth in the United Kingdom and Lavéra in France. Meanwhile, the two companies reached a strategic cooperation agreement to share refining and petrochemical technologies. In 2013, CNPC and CNOOC jointly won a 35-year production service contract in a consortium with Petrobras, Shell and Total to develop a pre-salt discovery in Brazil's Libra oil field. This is another case that Chinese NOCs have been seeking to gain deepwater technical and managerial knowhow by participating as minority shareholders in a large project to work side by side with experienced IOCs.[73]

[73] Jiang and Ding, *Updates on Overseas Investments by China's National Oil Companies: Achievements and Challenges since 2011.*

Also, Chinese NOCs are making efforts in gaining technologies and experiences in operating LNG projects, so as to better satisfy the surging natural gas demand in China. Their acquiring of stakes in LNG projects overseas such as those in Indonesia and Australia helps them gain expertise of the LNG supply chain.

Moreover, Chinese NOCs are interested in unconventional resources including shale gas and oil, coal bed methane and sands projects, both in China and abroad. In particular, as China's demand for natural gas grows rises, there is an urgent need to explore its vast domestic reserves of unconventional gas. Due to their lack of experiences in shale gas and oil production, Chinese NOCs are interested in investing abroad in the form of partnership and joint venture.

A report by the United Nations Conference on Trade and Development (UNCTAD) maintained that transnational corporations and state-owned enterprises (SOEs) from countries rich in shale resources, such as China, are strongly motivated to establish partnerships, typically in the form of joint ventures, with US companies to acquire the technical expertise needed to lead the shale gas development in their countries and replicate the success of the United States shale revolution in their home countries.[74]

For example, Sinopec's establishment of a joint venture with U.S. energy company Devon Energy which is the largest oil and gas exploration and production company in the world and the largest LNG supplier in North America, not only provides it with access to US upstream oil and gas business, but also allow it to benefit from Devon's advanced technical and managerial knowhow for developing unconventional oil and gas.

Secondly, cooperating with IOCs reduces Chinese NOCs' risks in overseas ventures. For example, bidding in partnership with foreign firms diversified the risk for each partner in those highly risky and politically unstable host countries. This was particularly the case in 2009 when

---

[74] United Nations Conference on Trade and Development (UNCTAD), *World Investment Report 2014: Investing in the SDGs: An Action Plan* (New York and Geneva: United Nations, 2014), http://unctad.org/en/PublicationsLibrary/wir2014_en.pdf, accessed 11 March 2015.

Chinese NOCs cooperated with IOCs to participate in bidding rounds in Iraq. Chinese NOCs have won three contract bids and gained rights to develop the Rumaila, Halfaya and Missan oil fields with IOCs such as British Petroleum (BP), TOTAL, Turkish Petroleum and Petronas.

Also, by partnering with IOCs, Chinese NOCs can reduce the risks of operating in unfamiliar cultures. This kind of partnership could help the NOCs avoid political risks at a time of rising resource nationalism in some resource-rich countries and accusations of Chinese NOCs blocking resources to others. The NOCs have become more aware of the political sensitivities as they have gained experiences in transnational operation. Meanwhile, the partnership with IOCs helps CNPC mitigate negative international attention, and decrease demands for greater transparency.

Thirdly, partnering with IOCs gives Chinese NOCs access to the IOCs' established local connections and networks. For instance, CNPC carried out a joint acquisition of Arrow Energy in Australia with Shell, and joined Shell's share in Syria, which gave it the opportunity to take advantage of Shell's established local connections, instead of having to build its own network.

Lastly, the partnership with IOCs allows Chinese NOCs to leverage the IOCs' cross-cultural knowledge and experiences in transnational operation, which Chinese NOCs lack and would need years to build up. It has been proved that successful overseas acquisitions do not automatically bring successful operation. Chinese NOCs have begun to cultivate cross-cultural awareness among their work forces, and even started to hire non-Chinese employees to facilitate this task.

On the other hand, from the perspective of many IOCs, they are interested in forming partnership with Chinese NOCs. One reason is that this kind of partnership also diversifies and lowers the risks and costs for the IOCs' transnational operation, such as in Iraq. Another key reason is that such partnership with Chinese NOCs would help the IOCs enter China's domestic resources. The Chinese government has explicitly expressed its willingness to see more IOCs with advanced techniques cooperating with Chinese NOCs in oil and gas exploration and production within China, as shown in the Chinese official white paper *China's Energy Policy 2012*.

> 'China upholds a policy of opening to the rest of the world in the field of energy. To provide a favourable environment for foreign investment and protect the legitimate rights and interests of investors, it has promulgated a series of laws and regulations in succession, like the Law on Sino-foreign Equity Joint Ventures, Law on Sino-foreign Cooperative Joint Ventures and Law on Foreign Investment Enterprises, and framed such policy documents as the Catalogue of Industries for Guiding Foreign Investment and the Catalogue of Advantageous Industries for Foreign Investment in the Central and Western Regions. The Chinese government encourages foreign investment to engage in the exploration and development of oil, natural gas and unconventional oil and gas resources, such as shale gas and coal-bed gas, by way of cooperation ...'[75]

In the recent years, some cases of assets swap have taken place between Chinese NOCs and IOCs. For example, in March 2013, CNPC reached a deal with Italy's Eni for USD 4.1 billion to acquire a 29 per cent stake of its offshore Area 4 in Mozambique. This is a mutually beneficial deal, as CNPC became the first Chinese NOCs to tap into Mozambique's offshore area, while Eni was looking to reduce its funding commitment and finding a potential buyer for the LNG exports from East Africa. But there is another layer to this deal. Both CNPC and Eni announced a cooperation agreement for shale gas exploration in the Rongchang block in China's Sichuan province. Also in 2013, CNPC/PetroChina acquired a 25 per cent share in the West Qurna-1 oil project in Iraq from ExxonMobil. This deal was the result of several years of negotiation. ExxonMobil is working with PetroChina in China's Ordos Basin to conduct a joint study on the Changdong block. This type of asset swap has become an unspoken rule for foreign companies to work with Chinese majors.[76]

---

[75] The State Council Information Office, the People's Republic of China, *China's Energy Policy 2012*, October 2012, http://english.gov.cn/archive/white_paper/2014/09/09/content_281474986284499.htm, accessed 14 January 2015.

[76] Jiang and Ding, *Updates on Overseas Investments by China's National Oil Companies: Achievements and Challenges since 2011.*

To some extent, much of the onshore oil production in China is limited to Chinese NOCs. Thus, IOCs have been often granted greater access to offshore oil blocks and technically challenging gas fields in China, mainly through reaching production sharing contracts (PSCs) and establishing joint ventures with their Chinese counterparts. For instance, CNOOC typically uses PSCs with IOCs wanting to jointly develop upstream offshore projects in China and has the right to acquire up to a 51 per cent working interest in all offshore discoveries once the IOC recovers its development costs. Nowadays, those IOCs engaged in offshore exploration and production activities in China include ConocoPhillips, Shell, Chevron, BP, BG, Husky, Anadarko, Eni, etc. In those PSCs, Chinese NOCs hold the majority working interest and can become the operator once development costs have been recovered.[77]

4.2.2.3 *Loan-for-energy deals*

In the past a few years, Chinese NOCs have often utilised strong financial resources provided by China's policy banks and government policy support to secure long-term loan-for-energy deals. It is apparent that the 2008 global financial crisis brought Chinese NOCs important opportunities to expand overseas through bilateral loan-for-energy approach, particularly in 2009, because some resource-rich countries were more eager to secure money but reluctant to sell assets. By the end of 2012, Chinese NOCs have signed loan-for-oil and loan-for-gas deals with atleast nine countries including Angola, Bolivia, Brazil, Ecuador, Ghana, Kazakhstan, Russia, Turkmenistan and Venezuela.[78]

While Chinese policy banks providing essential financial support for the NOCs, the Chinese government also played an indispensable and more active role in facilitating these deals than before. In some cases, such as those deals with Kazakhstan and Russia, the Chinese government was even directly involved in finalising the deals.

[77] U.S. Energy Information Administration, 'Country Analysis Brief: China'.
[78] *ibid.*

To some extent, Chinese NOCs enjoyed a competitive advantage because of their access to the country's huge foreign exchange reserves. In some cases, China's state-owned policy banks are indispensable players in Chinese NOCs' loan-for-energy deal with foreign companies. The China Development Bank (CDB) and the China Export–Import Bank (CEIB) are the two major banks that provided critical funding for Chinese NOCs' long-term loan-for-oil or loan-for-gas deals. For instance, the CNPC's package deal with Rosneft in 2009 provides a good example to examine the roles of various players in the deal.

In January 2009, Russia's state-owned oil company Rosneft had debt of USD 13 billion that needs repaying by the summer of 2009. Since the Russian government suffered from the 2008 financial crisis, it was looking for new source of revenue. Moreover, funds were needed to develop Russia's Eastern Siberian oil and gas fields in order to supply the Asian market. Thus, the Russian government and Rosneft approached the Chinese government and CNPC. In February 2009, after 15 years of negotiation between the Chinese side and the Russian side, it was announced that CDB would lend Russia's Rosneft and Transneft USD 15 billion and USD 10 billion respectively. Consequently, CNPC became the beneficiary as it reached a long-term oil supply deal with the Russians and finally secured the oil pipeline it had long desired into the Northeastern part of China.

According to the loan-for-oil deal, CDB agreed to provide the financing that the Russian side needed, with an interest rate of merely 5.69 per cent, a very favourable rate given that few commercial banks could lend at that time. As this was a bundled package deal, CNPC would obtain the right to purchase 300,000 bpd of crude oil at market price for 20 years; and CNPC would deposit the payment for the oil import into a designated account at CDB so that CDB can be guaranteed to receive payments from Rosneft. The USD 10 billion deal with Transneft works the same way. Besides, the oil pipeline connects Russia's East Siberia Pacific Pipeline System at Skovorodino to China's Daqing refinery in Heilongjiang Province.[79]

---

[79] Jiang and Sinton, *Overseas Investments by Chinese National Oil Companies: Assessing the Drivers and Impacts.*

Moreover, in February 2009, China and Venezuela reached an agreement to establish a USD 4 billion joint development fund. According to the EIA, by 2013, China has signed with Venezuela several loan-for-oil deals for about USD 40 billion in exchange for 600,000 bpd of crude oil and oil products supply. Also, China has extended USD 2 billion to another Latin American country, Ecuador, in 2013, becoming Ecuador's major crude oil buyer.[80]

Indeed, China's policy banks have become essential strategic partners for Chinese NOCs to secure more energy supplies from abroad. Notably, in September 2010, CNPC and Sinopec established strategic alliances with CDB; and CDB agreed to provide USD 30 billion at low rates over the next five years to support CNPC's overseas expansion.[81] Thanks to such effective coalition between China's policy banks and the NOCs, the following years have witnessed a series of significant energy deals between China and some energy-rich countries.

For instance, in 2011, China reached a deal with Turkmenistan to secure an additional 25 billion cubic metres of natural gas to feed into the Central Asia–China Gas Pipeline, with CDB providing USD 4.1 billion loan. In 2012, China reached a deal with Venezuela to develop the Orinoco oil field, with CDB providing USD 4 billion loan that is to be repaid by 230 thousand bpd of crude oil for three years. In 2013, China reached a deal with Venezuela again to promote the production of the Orinoco oil field, with CDB offering another USD 4 billion loan which needed repaying with 310 thousand bpd of oil for four years. Also in 2013, China reached a deal with Russia to double its oil import from Russia. CDB provided USD 25 billion of loan. A part of the loan (USD 15 billion) was lent to Rosneft to secure a further 300 thousand bpd of crude oil supply from Rosneft to CNPC for 25 years; and the rest of the loan (USD 10 billion) was lent to Transneft for pipeline construction. This was the second loan-for-oil deal reached between China and Russia.

---

[80] U.S. Energy Information Administration, 'Country Analysis Brief: China'.
[81] Jiang and Sinton, *Overseas Investments by Chinese National Oil Companies: Assessing the Drivers and Impacts*.

4.2.2.4 *Market-for-resources cooperation*

For some years, Chinese NOCs have been using the market-for-resource approach to expand overseas partnership and secure more energy supply from abroad. In general, this approach exchanges IOCs' access to China's domestic market for Chinese NOCs' access to overseas energy resources. To some extent, China's huge domestic unconventional resources and energy market is the vital attraction for IOCs and NOCs from other countries to form partnerships with Chinese NOCs. For example, some leading IOCs such as BP, Shell, SK and TOTAL have worked together with Chinese NOCs to build fuel filling stations in China; and ExxonMobil, BP, Shell, TOTAL and BASF have invested in refineries in China.

First of all, the market-for-resource approach is especially useful for building partnership with NOCs from some energy-rich countries. The cooperation between Sinopec and Saudi Aramco is a typical case. For instance, Saudi Aramco has reached cooperation agreements with Sinopec to build 750 filling stations and a refinery in Fujian Province and a crude stockpile facility in Hainan Province. Saudi Aramco has also signed a memorandum of understanding (MOU) with CNPC for cooperation on a refinery in Yunnan. Due to the operation of China–Myanmar Oil Pipeline, part of China's crude oil import from the Middle East and Africa would be transported through the pipeline to this refinery to supply the local markets in Southwest China.

On the other hand, in January 2012, Sinopec entered the refining industry in Saudi Arabia by investing USD 4.5 billion to build a joint venture with Saudi Aramco on the Yasref refinery at Yanbu on the Red Sea, with Sinopec holding a 37.5 per cent share. This refinery is scheduled to begin to operate at the end of 2014 and will be able to process 400,000 bpd of Arabian heavy crude oil. This project marked Sinopec's first international downstream investment, indicating the effectiveness and significance of Chinese NOCs' market-for-resources approach that is mutually beneficial for Chinese NOCs and NOCs from energy-rich countries. Indeed, this strategic partnership between the world's top oil exporter and top oil importer has led to a win–win situation.

Therefore, through offering a piece of China's vast domestic market, Chinese NOCs are able to leverage the partnership and credibility

they have built, enjoy preferential treatment for cooperation in these energy-rich countries, or simply increase their opportunities to purchase more energy resources from these countries.

Moreover, China's significant potential for shale gas development and the vast domestic oil and gas markets have increasingly encouraged IOCs to partner with Chinese NOCs. Such cooperation has increased significantly in the recent years. International IOCs including Shell, ConocoPhillips, Eni and TOTAL have all reached cooperation agreements with their Chinese counterparts to conduct seismic surveys, exploration and joint research to explore some Chinese domestic shale oil and gas blocks. Hess was the first IOC to sign a production sharing contract (PSC) with Chinese NOCs.

It is worth noting that the cooperation and partnership between Chinese NOCs and these IOCs actually began outside China, when Chinese NOCs formed partnership with these IOCs to venture overseas for investment and operation. Such partnership has allowed Chinese NOCs to leverage China's domestic energy resources and market in exchange for greater access to IOC assets abroad. For example, since the establishment of CNOOC in 1982, the company has signed 200 PSAs with 78 companies from over 20 countries to develop China's offshore reserves. Recently, BP and Canada's largest energy company, Husky, reached an agreement with CNOOC to develop China's deepwater blocks. But actually their partnership started outside China when CNOOC purchased equities in Husky's Indonesia project in 2008, and BP's projects in Australia in 2003 and in Latin America in 2010.[82]

4.2.2.5 *Reaching service contracts*

Apart from ODI through partnership with IOCs or by themselves, Chinese NOCs have reached some significant service contracts with some energy-rich countries, further expanding their upstream operation overseas.

---

[82] Jiang and Ding, *Updates on Overseas Investments by China's National Oil Companies: Achievements and Challenges since 2011.*

Some energy-rich countries, especially those in the Middle East, only offer service contracts to foreign oil companies. The typical example is Iraq that owns the world's second largest proven oil reserves. In 2009, Iraq opened its oil fields to foreign companies in the form of service contracts; and Chinese NOCs won three contracts in collaboration with IOCs. Specifically, CNPC jointly bid with BP to reach a service contract of 20 years in Iraq to develop the country's largest oil field, Rumaila, and with TOTAL and Petronas to develop the Halfaya oil field, and Iraqi South Oil holds a 25 per cent stake in both of these bids; CNOOC won the contract with Turkish Petroleum (TPAO) to develop Iraq's Missan oil field, and Iraq Drilling Company is the Iraqi partner and holds a 25 per cent share. Just like many IOCs, Chinese NOCs consider Iraq as a key investment destination, due to the lack of other high quality oil assets for overseas investment.

Actually, reaching service contracts has become a key approach of Chinese NOCs' cooperation in energy-rich countries which are reluctant to sell assets. Chinese NOCs have recognised this situation and are willing to bid for these contracts with or without the partnership with IOCs. At the same time, they have to face political risks and uncertainties for accessing the energy resources in countries like Iraq.[83]

4.2.2.6 *Construction of transnational and overseas pipelines*

Apart from carrying out ODI and securing service contracts to expand their upstream exploration and production abroad, Chinese NOCs have been investing in transnational and foreign oil and gas pipelines in Central Asia (Kazakhstan, Turkmenistan and Uzbekistan), Southeast Asia (Myanmar), the Middle East (Iraq) and Canada, mainly for the purpose of securing more oil and gas supply from more diversified sources. For this part, most of the pipeline construction has been carried out by CNPC.

---

[83] Jiang and Sinton, *Overseas Investments by Chinese National Oil Companies: Assessing the Drivers and Impacts.*

4.2.2.6.1 Central Asia

In June 2003, during then Chinese President Hu Jintao's visit to Kazakhstan, the two countries agreed to reactivate the oil pipeline construction plan that was shelved in 1999.[84] When Kazak Prime Minister Daniel Akhmetov visited China in September 2003, the two governments discussed the possibilities of the project. In May 2004, during Kazak President Nursultan Nazarbayev's trip to Beijing, the presidents of the two states signed a joint declaration on the construction of the first phase of the China–Kazakhstan Crude Oil Pipeline, which would become China's first transnational oil pipeline.[85] Notably, the construction of the pipeline also fits Kazakhstan's interests in diversifying its energy exports to reduce heavy reliance on Russian pipelines to export energy to Western markets.[86]

In May 2006, CNPC inaugurated the Kazakhstan–China Oil Pipeline, and began to import crude oil from Kazakhstan and Russian through this pipeline. The pipeline, with a designated capacity of 240,000 bpd, spans 2,200 km and connects Atyrau in Kazakhstan's Caspian shore with Alashankou on the Chinese border in Xinjiang Uygur Autonomous Region. It was developed by the Sino–Kazakh Pipeline Company, a joint venture between CNPC and Kazakhstan's KazMunaiGaz. The expansion work is underway on the Atasuto-Alashankou section of the pipeline to nearly double the capacity to 400,000 bpd in 2014.[87] Kazakhstan–China Oil Pipeline brings oil

---

[84] It is said that in the mid-1990s, Kazakhstan was interested in building a cross-border oil pipeline to export oil to China; but at that time, Chinese leaders were not very interested in the project. The reason is unclear. It may include some factors such as the high construction costs, the relatively low international oil prices at that time, and China was not so heavily dependent on overseas oil supply. As a result, the proposal was shelved.

[85] This pipeline is Kazakhstan's first pipeline that directly links this energy exporting country with the consuming market without passing through a third country. The 2007 Annual Report of CNPC, p. 11; and People's Daily On-line, 'Kazakhstan oil piped into China', 25 May 2006, http://english.people.com.cn/200605/25/eng20060525_268539.html, accessed 25 May 2006.

[86] Herberg, 'Energy Security Survey 2007', p. 18.

[87] U.S. Energy Information Administration, 'Country Analysis Brief: China'.

from the fields in central Kazakhstan and Russia's West Siberian region to China, and is expected to account for six per cent of China's overall crude oil imports by 2015.

As to the natural gas pipeline, the Central Asia–China Gas Pipeline, starts in Turkmenistan and spans across Uzbekistan and Kazakhstan before reaching China. This 1,840 km-long gas pipeline, with the designated capacity to transport 25 billion cubic metres per year, will cost an estimated USD 2.2 billion. It runs parallel to Lines A and B, and is scheduled to be finished by 2014 and reach its designed capacity by the end of 2015.

The Central Asia–China Gas Pipeline is connected to gas fields in each of these three countries, and enters China in the Xinjiang Uygur Autonomous Region, where it is connected to the Phase 2 of China's domestic West to East Gas Pipeline. The entire pipeline extends 7,000 km across four countries with a total cost estimated at USD 7.31 billion. CNPC demonstrated strong capability in completing this longest pipeline in the world with record speed. The Central Asia–China Gas Pipeline was inaugurated in December 2009, and the natural gas started to flow into China in January 2010.

The Lines A and B of the Central Asia–China Gas Pipeline have a combined capacity of 30 to 40 billion cubic metres per year. At the end of 2011, construction of the Uzbek section (529 km long) of Line C started in Bukhara Oblast, Uzbekistan. The planned Line D of the project will run along a new route and the construction is expected to commence in 2014. It will pass through three Central Asian states including Uzbekistan (205 km), Tajikistan (415 km) and Kyrgyzstan (225 km), with a designated capacity of 30 billion cubic metres per year. It is scheduled to be put into service by 2020, and will transport natural gas from the Galkynysh gas field in Turkmenistan to China. Uzbekistan has started to supply natural gas to China since April 2012. In addition, Tajikistan is a potential new natural gas source in Central Asia for the supply to China. The country might be able to export gas to China via Line D in the future. In March 2014, CDB agreed to loan Kazakhstan's KazTransGas USD 700 million for 15 years. The loan will be used to complete the second, 311 km portion of a gas pipeline from southern Kazakhstan and is expected to

transport additional volumes of natural gas to China through the existing Kazakhstan–China gas pipeline.[88]

4.2.2.6.2 Russia

In 1996, China and Russia signed energy cooperation agreements that included an oil pipeline from Russia's East Siberia to Daqing in China's Heilongjiang Province. After several more years of negotiation, CNPC and the private Russian oil company Yukos signed in 2003 an agreement to jointly construct the Angarsk–Daqing pipeline. Whereas, the political battle between Russian President Putin and Mikhail Khodorkovsky, the head of Yukos, ended the deal.

Finally, in October 2008, CNPC and Transneft signed an agreement in principle on the construction and operation of a crude oil pipeline from Russia's Skovorodino to China's Mohe at the border. According to the agreement, the two countries would jointly construct and operate the Russia–China Oil Pipeline based on the Phase I project of Russia's Far East Pipeline.

In February 2009, CNPC signed with Transneft 'The Contract on Construction and Operation of the Crude Oil Pipeline from Skovorodino to the Russia–China Border', as well as agreements with Rosneft and Transneft respectively for long-term bilateral crude oil trade for USD 25 billion. Under the agreements, Russia would supply 15 million tonnes of crude oil to China annually for the 20 years of the contract term. Construction of the Russian section and the Chinese section (from Mohe to Daqing) of the Russia–China Oil Pipeline commenced in April and May 2009, respectively. This 1,030 km pipeline, which connects Russia's East Siberia–Pacific Oil Pipeline to China's Daqing refinery complex via Skovorodino, has a capacity of 300,000 bpd. Transneft received a USD 10 billion loan from China to build the 65 km long section of this branch within Russia. Most of the length of the pipeline, 965 km, is located within China. CNPC completed this section in June 2010, and oil started to flow

[88] Jiang and Ding, *Updates on Overseas Investments by China's National Oil Companies: Achievements and Challenges since 2011.*

through the pipeline on 1 January 2011. By 2015, this pipeline could transport 9 per cent of China's oil import.[89]

4.2.2.6.3 Myanmar

CNPC has invested in and built the parallel crude oil and natural gas pipelines running through Myanmar to Southwest China. The China–Myanmar Oil and Gas Pipelines not only transport Myanmar's natural gas resources to China, but also tranship part of China's crude oil import from the Middle East and Africa, therefore short-circuiting the long maritime transportation route, avoiding the passage through the chokepoint of the Strait of Malacca, and bringing more energy supply to China's landlocked southwest region The Strait of Malacca the vital channel linking the Indian and Pacific Oceans, and is the major route for the shipping of oil and LNG from the Persian Gulf and Africa to East Asian markets.

Before the pipelines were built, all of China's oil and LNG imports from the Middle East and Africa had to pass through the Strait of Malacca, which is quite busy and narrow (less than 3 km at its narrowest point). Alternative sea lanes of communication (SLOC) do exist, but require more travel time, enjoy less protection and are more costly. Moreover, the rampant pirate activities, the risk of oil spills and shipping accidents may cause the disruption or even blockage of the shipping through the strait. Also, there has been increasing concern in China for years on the country's heavy reliance on the vulnerable Strait of Malacca, as well as on the dominating military presence in the region by the United States. In this sense, Chinese NOCs' investment and construction of the China–Myanmar Oil and Gas Pipelines could provide an alternative energy transportation route to lower China's reliance on the Strait of Malacca and help diversify its gas imports.

In June 2009, CNPC and Myanmar's Ministry of Energy signed an MOU to construct, operate and manage the parallel China–Myanmar

---

[89] CNPC website, http://www.cnpc.com.cn/en/Russia/country_index.shtml, accessed 16 February 2015.

Oil and Gas Pipelines. The construction of both pipelines commenced in June 2010. The oil pipeline with a capacity of 440,000 bpd requires an estimated USD 1.5 billion investment; and the gas pipeline with a designated capacity of 12 billion cubic metres capacity requires USD 1 billion investments. The oil pipeline, starting at the Kyaukpyu port of Arakan coast in the Bay of Bengal, has been completed and put into service. This 1,100 km oil pipeline links the Indian Ocean with China's south-western Yunnan Province. Compared with the traditional SLOC through the Strait of Malacca, the pipeline route could save about 1,200 km of maritime shipping distance for China's oil import from the Middle East and Africa. Meanwhile, the parallel gas pipeline will transport gas from Blocks A1 and A3 in Myanmar to China. The parallel pipelines are expected to transport 7 per cent of China's crude oil and natural gas import by 2015.[90]

In addition, the oil pipeline opens a critical oil import gateway for China's inner southwest region, which traditionally had to receive oil products from refineries on China's eastern coast. Nowadays, the direct import of crude oil to the region could boost the newly established refinery sector there. CNPC has built refineries in Kunming and Chongqing.

4.2.2.6.4 Iraq

In July 2013, CNPC won the engineering, procurement and construction contract to construct the 272 km pipeline project to connect Iraq's Halfaya, Buzurkan, Fuka and Abu Ghareb oil fields to the Al-Fao export terminal on the Persian Gulf. Also, CNPC has been shortlisted to build Iraq's USD 18 billion oil export pipeline to Jordan, the country's first such pipeline in decades. Besides, CNPC is the operator of the Halfaya project, and the Halfaya export pipeline will be completed in 2014 and the Phase 2 of Halfaya Central Processing Facility will be put into service in 2015.

---

[90] Jiang and Sinton, *Overseas Investments by Chinese National Oil Companies: Assessing the Drivers and Impacts.*

4.2.2.6.5 Canada

CNPC has expanded its downstream operation by partnering with TransCanada to build a pipeline, and with Shell and Asian partners to build LNG exporting facilities in Canada for the purpose of exporting to Asian markets.

4.2.2.6.6 Tanzania

CNPC has been building natural gas processing centres and pipelines in Tanzania since 2012 and gained a foothold in East Africa's gas boom.[91]

4.2.3 *The development of Chinese NOCs' overseas investment*

Since the 1990s, CNPC, Sinopec and CNOOC have been engaging in transnational operations mainly through ODI activities, and expanding their investment and equity in oil and natural gas assets across various parts of the world, hence continuously promoting their overseas production of oil and natural gas. Among them, CNPC's performance is particularly outstanding. This sub-section demonstrates the development of Chinese NOCs' overseas investment and operation particularly M&A activities across various regions and countries of the world during the past two decades.

4.2.3.1 *The Middle East*

As the region with the largest volume of proved oil reserves in the world,[92] the Middle East is always the most important investment destination and operation area for oil companies including Chinese

[91] Jiang and Ding, *Updates on Overseas Investments by China's National Oil Companies: Achievements and Challenges since 2011.*

[92] In 2013, the proved oil reserves in the Middle East were 109.4 billion tonnes, accounting for 47.9 per cent of the world's total. See British Petroleum, *BP Statistical Review of World Energy June 2014.*

NOCs. In 2013, the Middle East contributed to 24 per cent of Chinese companies' oil and gas production abroad.

4.2.3.1.1 Iran

CNPC has been present in Iran since 2004 and is engaged in oil and gas operations and oil field services in the country. In May 2004, CNPC acquired the MIS Oil field in Iran, holding 75 per cent shares of the project. In August 2007, the company received a notice from National Iranian Oil Company (NIOC) that approved the Supplementary Agreement of the Iran MIS project contract, marking the start of the MIS project. This marked the beginning of the term of the MIS project. Drilling of the project's first exploration well started in November 2007.

In May 2005, CNPC won the tender of Block-3 — an integrated exploration and development project with a buy-back contract mode. In June that year, the project was formally launched. In 2007, the Block-3 exploration project saw a test oil output of 1,250 bpd from the first exploration well BAB-1. In 2008, the oil and gas output appeared promising in multiple layers in key risk exploration well DBE-1. In 2009, the overall evaluation plan of Block-3 in BAB oil field has been approved by NIOC and has successfully entered the evaluation period. Major breakthroughs were made in 3-D seismic acquisition, processing and interpretation.[93]

In 2009, CNPC signed a USD 4.7 billion contract with the NIOC to develop the North Azadegan oil field and the Phase 11 of South Pars gas field.[94]

Apart from CNPC, Sinopec and CNOOC have also engaged in investment and operation in Iran. In October 2004, Sinopec signed an MOU with the Iranian government to acquire 51 per cent equity in the huge Yadavaran oil field, which industry reports suggest could

---

[93] CNPC website, http://www.cnpc.com.cn/en/Iran/country_index.shtml, accessed 16 February 2015.

[94] Jiang and Sinton, *Overseas Investments by Chinese National Oil Companies: Assessing the Drivers and Impacts.*

produce 300,000 bpd. In April 2006, Sinopec and Saudi Arabia's Aramco signed a framework accord, under which the two companies will enhance cooperation in oil and natural gas exploration. Sinopec was already drilling for natural gas in the Saudi desert. Also, Aramco promised 1 million bpd crude oil transfer to China before 2010.[95]

In December 2006, CNOOC and its Iranian counterpart signed an MOU on the development of the North Pars natural gas field in Iran. According to the preliminary agreement, natural gas produced from the field will be liquefied and shared equally between the two companies.[96]

4.2.3.1.2 Iraq

In 2013, Chinese companies had a combined 553,000 bpd of oil production in Iraq, accounting for 26 per cent of total Chinese overseas oil production in the same year. Chinese companies have been active in Iraq's oil sector since jointly winning tenders to service contracts in partnership with IOCs in 2009. CNPC/PetroChina is a partner in the consortium development of the Rumaila oil field and the operator of the Halfaya and Al-Ahdab oil fields, while CNOOC is operating the Missan oil fields, and Sinopec producing oil in the Iraq's Kurdistan regional government ruled area. For Iraq, Chinese investments offered the much needed capital to promote Iraq's oil production, which recently reached a record high.[97]

In June 1997, Al-Waha Petroleum Co., Ltd, a joint venture of CNPC and China North Industries Corporation, signed an agreement with then Iraqi Saddam Hussein regime to develop the Al-Ahdab oil field, which was postponed by the UN sanctions on Iraq

---

[95] Ding Ying, 'Seeking More Than Oil', *Beijing Review*, 19 June 2006, http://www.bjreview.com.cn/06-19-e/w-1.htm, accessed 25 July 2006.

[96] People's Daily On-line, 'CNOOC to develop Iranian gas field', 22 December 2006, http://english.peopledaily.com.cn/200612/22/eng20061222_335088.html, accessed 13 February 2009.

[97] Jiang and Ding, *Updates on Overseas Investments by China's National Oil Companies: Achievements and Challenges since 2011.*

and the following U.S. led invasion into the country. In November 2008, CNPC signed a Development Service Contract of Al-Ahdab Oil field with Iraq's Ministry of Oil, reviving the contract it had negotiated under the former Saddam Hussein regime.

In November 2009, CNPC partnered with BP in winning the bidding for the operating rights of Rumaila oil field, and with Total and Petronas in winning the contract for Halfaya oil field. In 2010, the Rumaila Field Operating Organization (ROO), a consortium by CNPC, BP and Iraq's South Oil Company, took over the operation management and began to operate the oil field. Rumaila oil field is the largest oil field in Iraq and the sixth in the world, with proven oil reserves of about 17 billion barrels. Currently, the production of Rumaila oil field accounts for 44 per cent of Iraq's overall oil output.

Also in 2009, CNPC partnered with Total and Petronas in winning the contract for Halfaya oil field. In January 2010, the service contract was formally signed for the Halfaya oil field, which is operated by CNPC in consortium with Total, Petronas and Iraq's Missan Oil. In June 2012, CNPC's 5 Mt/a capacity building project in Halfaya oil field, in partnership with Total, started crude oil production ahead of schedule.[98]

Recently, China has become the top foreign player in Iraq's oil sector. CNPC/PetroChina, benefited from its rich upstream experiences and early-entry into Iraq, taking up the leading position among the three major Chinese NOCs. It is reported that in 2013, CNPC/PetroChina joined ExxonMobil by purchasing 25 per cent of its share in Iraq's giant West Qurna oil field, which is 50 km northwest of the southern oil hub of Basra. This would add synergies to CNPC/PetroChina's production capacity in Iraq, making the company the biggest single foreign investor in Iraq's oil sector.[99]

---

[98] CNPC website, http://www.cnpc.com.cn/en/Iraq/country_index.shtml, accessed 16 February 2015.

[99] Reuters, 'Exclusive: PetroChina to join Exxon at giant Iraqi oil field', 9 August 2013, http://www.reuters.com/article/2013/08/09/us-petrochina-iraq-idUSBRE9780D520130809, accessed 20 February 2015.

4.2.3.1.3 Saudi Arabia

In April 2006, Sinopec and Saudi Arabia's Aramco signed a framework accord, under which the two companies will enhance cooperation in oil and natural gas exploration. Sinopec was already drilling for natural gas in the Saudi desert.[100]

4.2.3.1.4 Syria

CNPC started its oil and natural gas operation in Syria in 2002. It has oil and gas assets and equity while providing oil field services there. In December 2002, CNPC won a bid to operate an EOR project in Syria's Gbeibe oil field. In March 2003, CNPC signed a development and production contract on this oil field with Syria's Ministry of Petroleum and Mineral Resources and the Syrian Petroleum Corporation.

In December 2005, CNPC and the Indian oil company Oil and Natural Gas Corporation (ONGC) made a joint purchase of PetroCanada's 38 per cent stake in Syria's Al-Furat Petroleum Company (AFPC), a block with an area of 2,027 square km and covering 39 oil and natural gas fields, three oil processing plants, a natural gas processing plant and oil and natural gas pipelines. In 2010, CNPC concluded an agreement with Shell to acquire shares in Al Furat Petroleum Company. Under the agreement, CNPC acquired a 35 per cent interest in Syria Shell Petroleum Development (SSPD), which had been wholly owned by Shell. SSPD has 31.25 per cent equity in three production licenses — Deir-Ez-Zor, Fourth Annex and Ash Sham. [101]

Besides, in 2008, Sinopec purchased 100 percent stake of Tanganyika for assets in Syria, spending USD 1.8 billion.[102]

---

[100] Ding Ying, 'Seeking More Than Oil', *Beijing Review*, 19 June 2006, http://www.bjreview.com.cn/06-19-e/w-1.htm, accessed 25 July 2006.

[101] CNPC website, http://www.cnpc.com.cn/en/Syria/country_index.shtml, accessed 16 February 2015.

[102] Jiang and Sinton, *Overseas Investments by Chinese National Oil Companies: Assessing the Drivers and Impacts.*

4.2.3.1.5 Oman

CNPC has been operating in Oman since 2002, owning oil and gas assets while providing oil field services. It provides a wide range of oil field services covering geophysical prospecting, well drilling, logging, perforation and formation testing. The company also won the tender of a 3D prospecting project from Petroleum Development Oman (PDO) and reached a drilling service contract with PDO on the Lekhwair project.

In April 2002, CNPC signed a PSC for 50 per cent equity in Block 5 in Oman. The operator of this oil and natural gas exploration and development project is a joint operation company established by CNODC, CNPC (Hong Kong) Ltd. and Oman MB Group.

In June 2005, CNPC and Oman Oil Company S.A.O.C (OOC) signed an MOU on the investment and cooperation in oil and gas development and downstream operations.

In early 2006, we launched the drilling service project to provide engineering and technical services to 232 wells of PDO for three years as the master contractor, including drilling, logging, directional drilling, casing running, cementing and acid pickling.

In 2008, CNPC completed U.S. firm OXY's Phase One acquisition of the 4D/3C time-shift seismic prospecting contract in Oman. In 2010, CNPC's PDO acquisition project set a record with an average of 20,000 shots per day with Distance Simultaneous Separated Sweep (DSSS) technology.

In 2011, CNPC won the bidding of PDO's 3D project. In 2012, by employing a pile mark-free navigation system, low-frequency scanning and DS3 acquisition technologies, CNPC's 3D seismic project for Oman's PDO registered the largest number of 20,651 shots in a single day and 19,000 shots per day on average.[103]

4.2.3.2 *Africa*

Apart from the Middle East, Africa is another key area for Chinese NOCs' transnational operation and overseas investment. In 2013,

[103] CNPC website, http://www.cnpc.com.cn/en/Oman/country_index.shtml, accessed 16 February 2015.

Africa contributed to 26 per cent of Chinese overseas oil and gas production.

4.2.3.2.1 Sudan and South Sudan

CNPC' has been present in Sudan and South Sudan since 1996. Its oil and natural gas business there mainly includes four upstream projects — Block 1/2/4, Block 3/7, Block 6 and Block 15, three downstream projects — Khartoum Refinery, Khartoum Petrochemical Co., Ltd. and petrochemical trading, and crude pipelines in Block 1/2/4, Block 3/7 and Block 6. In September 1995, CNPC acquired the development rights of Block 6. It has 95 per cent equity of the block and is also the operator, while the former Sudan's Sudapet owning the remaining 5 per cent equity. In March 1997, CNPC won the tender for Block 1/2/4. In June that year, it established a joint venture — the Greater Nile Petroleum Operating Company — with Sudapet, Malaysia's Petronas and India's ONGC. CNPC has 40 per cent shares of the joint venture. In November 2000, CNPC won the tender for Block 3/7 and with a 41 per cent ownership. The other shareholders are Petronas, Sudapet, Sinopec and ENOC. In August 2005, CNPC, Petronas, Nigerian Express, Sudapet and the Hi-tech Group signed a PSC with the Sudanese government to explore and develop oil and natural gas in Block 15. CNPC has a 35 per cent ownership in the block. Also, CNPC and the former Sudanese Ministry of Energy and Mining jointly invested in and constructed Khartoum Refinery. CNPC has a 50 per cent share in the project. The refinery started operations on 16 May 2000 and its processing capacity is 5 million tonnes per year.[104]

4.2.3.2.2 Algeria

CNPC began to operate in Algeria in 2003, and has acquired the exploration licences for Block 102a /112, Block 350 and Block 438b in the country, and operates the Adrar upstream and downstream integrated project.

[104] CNPC website, http://www.cnpc.com.cn/en/cnpcworldwide/sudan/, accessed 12 February 2009.

In May 2003, CNPC won the bidding for the Adrar upstream and downstream integrated project with a contract term of 23 years. The project includes oil field appraisal and development, construction of a refinery and marketing of refined products. In July 2003, CNPC signed with Algeria State Oil Company (SONATRACH) the Shareholders Agreement of the Refining Company of Adrar project. CNPC and SONATRACH respectively have 70 per cent and 30 per cent of ownership in the project.

In November 2003, CNPC won the bidding for two risk exploration blocks in Algeria: Block 102a/112 and 350. In December 2003, CNPC and SONATRACH signed two agreements on an exploration licence for Block 102a /112 and Block 350 respectively, owning 75 per cent equity in both blocks.

In February 2004, CNPC and SONATRACH signed a protocol on cooperation in oil sector. In July 2004, CNPC won the bid for Block 438b with 100 per cent equity. In May 2005, CNPC and SONATRACH signed a contract for the construction of the Skikda Gas Condensate Refinery.[105]

4.2.3.2.3 Niger

CNPC started its business in Niger in 2003. On 23 November 2003, CNPC signed an agreement with the Nigerien government on the exploration and development licences for two blocks — Block Bilma and Block Tenere, owning 100 per cent shares in Block Bilma and an 80 per cent stake in Block Tenere.

In June 2008, CNPC and Niger signed integrated upstream and downstream deals in Agadem block, involving oil field exploration and development, construction and operation of a long-distance pipeline and a joint venture refinery. Under the contract, within three years CNPC would complete the first phase of construction and bring the oil field, pipeline, and refinery into operation. In November 2011, Phase-I of Agadem upstream and downstream integrated

---

[105] CNPC website, http://www.cnpc.com.cn/en/Algeria/country_index.shtml, accessed 16 February 2015.

project was completed and began operation, as the joint venture, Zinder Refinery, became operational.[106]

4.2.3.2.4 Chad

In Chad, CNPC operates an upstream and downstream integrated project, with the full equity of Block H and a 60 per cent share of N'Djamena JV refinery, while providing oil field services.

In December 2003, CNPC signed an agreement with the Swiss company Cliveden to purchase shares of Block H in Chad for the risk exploration. The block covers the whole or partial area of seven depositional basins — Lake Chad, Madiago, Bongor, Doba, Doseo, Salamat and Erdis. In 2006, CNPC obtained all the equity of Block H, and began to produce crude oil for the first time.

In September 2007, CNPC and Chad's Ministry of Petroleum signed an agreement to establish a joint venture refinery. In 2011, the upstream and downstream integrated project was completed and put into service. In July 2011, it began to deliver diesel to the local market.[107]

4.2.3.2.5 Mozambique

In March 2013, CNPC became the first Chinese NOC to enter Mozambique's huge offshore natural gas field, after reaching a deal with Italy's Eni for USD 4.1 billion to acquire a 29 per cent stake of its offshore Area 4.[108]

4.2.3.2.6 Mauritania

In 2004, CNPC began its oil and natural gas operations in Mauritania. It operated four exploration projects in the country, namely Block Ta13, Block Ta21, Block 12 and Block 20.

---

[106] CNPC website, http://www.cnpc.com.cn/en/Niger/country_index.shtml, accessed 16 February 2015.

[107] CNPC website, http://www.cnpc.com.cn/en/Chad/country_index.shtml, accessed 16 February 2015.

[108] Jiang and Ding, *Updates on Overseas Investments by China's National Oil Companies: Achievements and Challenges since 2011.*

4.2.3.2.7 Tunisia

CNPC has been present in Tunisia since 2002. On 12 October 2004, in Tunisia, CNPC purchased the NK exploration block and a 50 per cent ownership in the SLK oil field from Kuwait Foreign Petroleum Exploration Company.[109]

4.2.3.2.8 Libya

Currently, CNPC is engaged in a risk exploration project in Libya, while providing oil field services and engineering and construction services in the country

In 2002, China Petroleum Pipeline Bureau of CNPC won the bidding for the construction of a pipeline in western Libya. The 1,050 km long pipeline, jointly invested by Libya's National Oil Corporation and Italy's Agip, starts at the WAFA oil field in the inner area of desert and ends at Mellitah on the Mediterranean coast. There are two parallel lines paved in different channels with a diameter of 32 inches for the natural gas pipeline and a diameter of 16 inches for the crude oil pipeline. This project was completed and put into service in July 2004.

In December 2005, CNPC and the National Oil Corporation of Libya signed a risk exploration contract for Block 17-4 in Libya. The contract is an exploration and PSA covering five years of exploration and 25 years of production.

In 2008, CNPC concluded a contract to deliver Carbon/Oxygen Ratio data processing and interpretation services for 24 development wells of WAHA Oil, and made a breakthrough in high-end marine service by providing a mud logging service for DP3 and DP4 offshore rigs of Libya's ENI OIL.[110]

---

[109] CNPC website, http://www.cnpc.com.cn/en/Tunisia/country_index.shtml, accessed 16 February 2015.

[110] CNPC website, http://www.cnpc.com.cn/en/Libya/country_index.shtml, accessed 16 February 2015.

4.2.3.2.9 Equatorial Guinea

In May 2006, CNPC signed a Simplified Purchase Agreement for Block M with Fruitex in Equatorial Guinea. On 21 July 2006, an E&P contract on Block M was approved by the Equatorial Guinea government. CNPC owns 70 per cent shares in the block and is implementing the new PSC as operator.

4.2.3.2.10 Nigeria

CNPC started to cooperate with the Nigerian government in 2006. In 2006, CNPC reached an oil and natural gas cooperation agreement with the Nigerian government and won the bidding for four blocks — OPL298, OPL471, OPL721 and OPL732. These oil fields are located in the Niger Delta, with one onshore and the others offshore.[111]

In January 2006, CNOOC spent USD 2.3 billion in purchasing from South Atlantic Petroleum Ltd. a 45 per cent stake in the licence covering the OML 130 field in Nigeria. The field is an offshore oil and natural gas field in deep water near the Niger Delta.[112]

4.2.3.2.11 Gabon

In February 2004, Sinopec signed a technical evaluation deal with the Gabonese Oil Ministry for three onshore oil fields before taking a decision on whether to take up an exploration and PSCs.[113]

---

[111] CNPC website, http://www.cnpc.com.cn/en/Nigeria/country_index.shtml, accessed 16 February 2015.

[112] BBC News, 'China oil firms buys into Nigeria', 9 January 2006, ehttp://news.bbc.co.uk/2/hi/business/4594058.stm, accessed 9 January 2006; and Jiang and Sinton, *Overseas Investments by Chinese National Oil Companies: Assessing the Drivers and Impacts*.

[113] Wenran Jiang, 'Fueling the Dragon: China's Quest for Energy Security and Canada's Opportunities', Asia Pacific Foundation of Canada, April 2005, http://www.asiapacific.ca/analysis/pubs/pdfs/can_in_asia/cia_fueling_dragon.pdf, accessed 28 June 2006.

4.2.3.2.12 Angola

In 2004, CNPC purchased Block 18 from Angolan government with USD 2 billion, when Shell existing the country, and acquiring 50 per cent of the equity.

In 2006, Angola–China Petrochemical International Company, which is a joint venture of Sinopec Group (with 75 per cent of the company's stock ownership) and Angola State Oil Company, won the biddings with high prices to purchase a 26.75 per cent stake in Block 17, a 40 per cent stake in Block 18 and a 20 per cent stake in Block 15 in Angola.[114]

In 2009, CNPC and Sinopec jointly purchased 20 per cent stake of Block 32 in Angola from Marathon Oil, spending USD 1.3 billion.

4.2.3.2.13 Uganda

In October 2010, CNOOC and Total jointly purchased two third of Tullow Oil's stake in three blocks in Uganda.[115]

4.2.3.3 *Central Asia*

Central Asia has been an increasingly important area for Chinese NOCs' investment and operation for years. Chinese NOCs' business in Central Asia helps the improvement and diversification of their investment portfolio.

---

[114] Xinhua, 'Kuozhan haiwai chuliang, zhongshihua gaojia jinghuo angela san gaochan youtian' (Sinopec won bids for three highly productive oil fields in Angola with high prices, expanding its overseas oil reserves), 13 June 2006, http://news.xinhuanet.com/fortune/2006-06/13/content_4687360.htm, accessed 27 January 2009.

[115] Jiang and Sinton, *Overseas Investments by Chinese National Oil Companies: Assessing the Drivers and Impacts.*

4.2.3.3.1 Kazakhstan

In 2013, Chinese companies have combined production of 315,000 bpd of oil in Kazakhstan, the second largest after Iraq.[116]

CNPC's business in Kazakhstan is a typical example of Chinese NOCs' investment and operation in the region. The company has been present in Kazakhstan since 1997, and operates five oil field development projects (CNPC AktobeMunaiGas, North Buzachi, PK, KAM and ADM), two exploration projects, and jointly operates four oil and gas pipelines (Kazakhstan–China Oil Pipeline, Kazakhstan–China Gas Pipeline, Kenkiyak–Atyrau Pipeline and Zhanazhol-KC13 Gas Pipeline), while providing oil field services and engineering and construction services in the country.

In June 1997, CNPC acquired a 60.3 per cent stake in AktobeMunaiGas, and obtained the production licence for the Zhanazhol, Kenkiyak Oversalt and Kenkiyak Subsalt oil fields and a contract for an exploration block, owning an 85.42 per cent share in AktobeMunaiGas.

In June 2002, CNPC signed a risk exploration contract with Kazakhstan's Ministry of Energy and Mineral Resources on the central block at the eastern edge of the Precaspian Basin.

In 2003, CNPC purchased 35 per cent and 65 per cent shares of Chevron Texaco North Buzachi Inc. from Nimir and Texaco respectively, owning 100 per cent shares in the North Buzachi oil field. CNPC and Russia's Lukoil each have 50 per cent stakes in the North Buzachi Oil field after the stock rights changed hands several times. The oil field is jointly operated by the two companies.

In November 2004, CNPC purchased 50 per cent shares in Konys and Bektas oil fields (KAM Project), owning a 25 per cent stakes in both of the oil fields.

In January 2005, CNPC and Ay-Dan signed an agreement to purchase shares for ADM and acquired 100 per cent shares in it. Ay-Dan has exploration licences for the Aryss and Blinov blocks.

In October 2005, CNPC acquired PetroKazakhstan. According to its agreement with Kazakhstan's Ministry of Energy and Mineral

[116] Jiang and Ding, *Updates on Overseas Investments by China's National Oil Companies: Achievements and Challenges since 2011.*

Resources, CNPC transferred 33 per cent of its shares in Petro-Kazakhstan to KazMunaiGaz in July 2006, keeping 67 per cent stake in the company.

In 2011, CNPC and KazMunaiGas concluded an in-principle cooperation agreement to jointly explore and develop Urikhtau Gas Field, which would be a natural gas source for Phase-II of Kazakhstan–China Gas Pipeline.[117]

Besides, in 2004, Sinopec purchased oil assets from First International Oil Corporation of Kazakhstan with USD 153 million, acquiring 100 per cent of the equity.

In April 2009, CNPC and KMG jointly purchased 100 per cent of the MMG Ltd. assets in Kazakhstan, with USD 3.3 billion in all, while CNPC spending USD 1.7 billion alone.[118]

In 2013, CNPC purchased from ConocoPhillips an 8.33 per cent share for USD 5 billion in the giant Kashagan oil project in Kazakhstan's Caspian Sea; and spent an additional USD 3 billion in financing the second phase of the Kashagan project. The sale and purchase agreement was signed by the heads of Kazakh state oil and gas company KazMunaiGas and CNPC in the presence of Chinese President Xi Jinping and Kazakh President Nursultan Nazarbayev.[119]

4.2.3.3.2 Uzbekistan

CNPC has been operating in Uzbekistan since 2006. In June 2006, CNPC signed an oil and natural gas exploration agreement with Uzbekneftegaz. In August 2006, the Aral Sea Oil and Gas Development Consortium of Investors, a consortium consists of CNPC, Uzbekistan's Uzbekneftegaz, Russia's Lukoil, Malaysia's Petronas and South

[117] CNPC website, http://www.cnpc.com.cn/en/Kazakhstan/country_index.shtml, accessed 16 February 2015.

[118] Jiang and Sinton, *Overseas Investments by Chinese National Oil Companies: Assessing the Drivers and Impacts.*

[119] Reuters, 'China buys into giant Kazakh oilfield for USD 5 billion', 7 September 2013, http://www.reuters.com/article/2013/09/07/us-oil-kashagan-china-idUSBRE98606620130907, accessed 21 February 2015.

Korea's Korea National Oil Corporation (KNOC), signed a PSC with Uzbekistan government to explore and develop oil and natural gas projects in the Uzbek section of the Aral Sea.

In June 2008, the construction of the Uzbek section of the Central Asia–China Gas Pipeline began in Bukhara, a city in southern Uzbekistan. ATG, a natural gas pipeline joint venture between CNPC and Uzbekneftegaz, was in charge of the construction and operation of the Central Asia–China Gas Pipeline on the Uzbek territory.

In October 2008, CNPC and Uzbekneftegaz signed a cooperation agreement to create a joint venture to develop Mingbulak oil field, which is located at the northern edge of the Fergana Basin.

In December 2009, the Central Asia–China Gas Pipeline became operational. In September 2011, CNPC and Uzbekneftegaz signed an agreement on the construction and operation of Line C of the Central Asia–China Gas Pipeline.

In June 2010, CNPC signed a framework agreement on purchase and sale of natural gas with Uzbekneftegaz, whereby Uzbekistan would supply 10 billion cubic meters of natural gas to China annually.

In September 2013, CNPC and the Uzbek Government signed an MOU for feasibility study on oil and gas exploration and development in two blocks in Uzbekistan, and an agreement on the principles of establishing a joint venture company for oil and gas exploration and development in the Karakul block. Under the agreement, CNPC and Uzbekneftegaz would establish a joint venture to develop three gas fields and other potential oil and gas resources in the Karakul block.[120]

4.2.3.3.3 Turkmenistan

CNPC has been involved in business in Turkmenistan since 2002. In January 2002, CNPC signed an EOR contract for Gumda Oil field with Turkmenneft and acquired a 100 per cent share for five years. In April 2006, CNPC reached a basic agreement on a joint natural gas

[120] CNPC website, http://www.cnpc.com.cn/en/Uzbekistan/country_index.shtml, accessed 16 February 2015.

project with the Ministry of Oil and Gas Industry and Mineral Resources of Turkmenistan.

In July 2007, CNPC signed a PSA to explore and develop natural gas fields on the right bank of the Amu Darya River in Turkmenistan with the Turkmen State Agency for Management and Use of Hydrocarbon Resources, and a natural gas sale-and-purchase agreement with Turkmengaz. According to the agreements, Turkmenistan would export 30 billion cubic metres of natural gas to China per year for 30 years.

In December 2009, the Central Asia–China Gas Pipeline began to operate. In June 2012, CNPC signed a cooperation agreement with Turkmengaz to increase natural gas supply from Turkmenistan to China via the Central Asia–China Gas Pipeline.[121]

In 2011, a long-term loan-for-gas deal was reached between China and Turkmenistan to supply an additional 25 billion cubic metres of natural gas supply to China.[122]

4.2.3.3.4 Azerbaijan

In January 2002, CNPC signed a PSC on K&K (Kursangi and Karabagli) oil field in Azerbaijan and acquired a 50 per cent share of the field. In January 2003, CNPC purchased 62.83 per cent shares in CGL (Commonwealth Gobustan Limited) and thereby acquired a 50.26 per cent holding in the Gobustan oil field.

4.2.3.4 *Russia*

CNPC entered Russia in 2003. In December 2003, CNPC and Sakhalin Energy signed a frame agreement on exploration and development in Russia's Sakhalin oil field. In July 2005, CNPC signed a

[121] CNPC website, http://www.cnpc.com.cn/en/Turkmenistan/country_index.shtml, accessed 16 February 2015.

[122] Jiang and Ding, *Updates on Overseas Investments by China's National Oil Companies: Achievements and Challenges since 2011.*

long-term cooperation agreement with Rosneft. In March 2006, CNPC signed with Gazprom, Rosneft and Transneft respectively, an MOU on supplying natural gas from Russia to China, an Agreement on the Basic Principles for the Establishment of Joint Ventures in China and Russia for Strengthening Oil Cooperation, and a Minute on CNPC–Transneft Meeting. In September 2006, CNPC signed a strategic cooperation agreement with Lukoil Company.

In October 2006, CNPC and Rosneft established a joint venture Vostok Energy Ltd. CNPC holds a 49 per cent share in the company. In August 2007, this company won a bid for licences to explore oil and natural gas in two eastern Siberian blocks (Verkhneichersky and West Chonsky) in North Irkutsk Province.

In October 2008, CNPC and Transneft signed an agreement in principle to build the Russia–China Oil Pipeline in February 2009, CNPC signed agreements with Rosneft and Transneft respectively for long-term oil trade. In 2010, the Russia–China Oil Pipeline began to operate.

Moreover, CNPC inked a general agreement with Transneft over the operation of the Russia–China Oil Pipeline, and a framework agreement with Gazprom to import natural gas from Russia, an agreement with Rosneft on increasing oil supply to the Russia–China Oil Pipeline, and an agreement with Lukoil on expanding strategic cooperation.

In October 2009, CNPC signed with Gazprom the framework agreement on Russia exporting natural gas to China, and an MOU on upstream–downstream cooperation with Rosneft.

On 27 September 2010, then Chinese President Hu Jintao and then Russian President Dmitry Medvedev took part in the ceremony to mark the completion of the Russia–China Crude Pipeline. In January 2011, the Russia–China Crude Pipeline was put into service. On the same day, CNPC inked a general agreement with Transneft over the operation of the Russia–China Crude Pipeline that stretches from Russia's Skovorodino station to China's Mohe station, a framework agreement with Gazprom to import natural gas to China, an agreement with Rosneft on extending oil supply to the Russia–China Oil Pipeline, and an agreement with Lukoil on expanding strategic

cooperation. In addition, CNPC's joint venture refinery with Rosneft was inaugurated in Tianjin Binhai New District in September 2010.[123]

2013 was a breakthrough year for the energy cooperation between China and Russia, as Chinese NOCs reached multiple deals with their Russia counterparts.

In early 2013, CNPC joined ExxonMobil, Statoil and Eni in entering Russia's immense offshore Arctic area via a joint venture with Russia's Rosneft. Later in the year, CNPC bought a 20 per cent stake for approximately USD 27 billion in the Yamal LNG project in the Arctic from Novatek, a Russian gas producer.[124]

In June 2013, CNPC reached 25 year long-term loan-for-oil deal with Rosneft to double Russia's crude oil delivery to China to 600,000 bpd through pipelines.

In May 2014, CNPC finally signed with Gazprom a historical natural gas supply contract to supply 38 billion cubic metres per year from Russia to China via the eastern pipeline route for 30 years.[125]

Apart from CNPC, in June 2006, Sinopec spent USD 3.5 billion to purchase a 97 per cent ownership in Udmurtneft, a mid-sized unit of BP's Russia vehicle TNKBP.

4.2.3.5 *Southeast Asia*

4.2.3.5.1 Thailand

CNPC has been engaged in business in Thailand since 1993, and owns rights and interests in the Banya development block, BYW-NS development block and L21/43 risk exploration block. In 1993, CNPC purchased a 95.67 per cent interest of Banya Block. In July 2003 CNPC signed with the Ministry of Energy of Thailand a concession agreement on the L21/43 Block. In June 2006 CNPC

---

[123] CNPC website, http://www.cnpc.com.cn/en/Russia/country_index.shtml, accessed 16 February 2015.

[124] Jiang and Ding, *Updates on Overseas Investments by China's National Oil Companies: Achievements and Challenges since 2011.*

[125] CNPC website, http://www.cnpc.com.cn/en/Russia/country_index.shtml, accessed 16 February 2015.

acquired the equity of the BYW-NS development block with the approval of the Ministry of Energy of Thailand. In November 2011, CNPC's subsidiary China Petroleum Pipeline Bureau (CPPB) and Thailand's PTTEPI signed the EPCIC general contracting contract on the Myanmar–Thailand Gas Pipeline. In March 2012, CPPB began to construct the Myanmar–Thailand Gas Pipeline. In 2013, CPPB won the contract for the Nakhon Sawan Gas Pipeline Project.[126]

4.2.3.5.2 Myanmar

CNPC has been operating in Myanmar since 2001 and currently runs two onshore oil and gas development and production projects, one deep-water exploration and development project. In November 2001, CNPC bought the Bagan project, including Block IOR-3, TSF-2 and RSF-3, from TG World, acquiring a 100 per cent interest of the three blocks. These blocks are implementing the PSC within 25 years. In December 2001, CNPC reached with Myanmar's Ministry of Energy a 20 years contract of improving oil recovery with Myanmar Oil and Gas Enterprise (MOGE) on Block IOR-4, which situated in southern Myanmar. In January 2007, CNPC entered a PSC with MOGE and acquired oil and natural gas exploration and exploitation licences for three deep-water blocks (AD-1, AD-6 and AD-8).

In June 2008, CNPC, the Myanmar government and Daewoo Combo signed an MOU on sale and transportation of natural gas from Myanmar offshore blocks A-1 and A-3. In December 2008, CNPC signed a 30 year purchase and sale agreement with South Korean conglomerate Daewoo International on importing natural gas from offshore blocks A-1 and A-3 in Myanmar. In June 2009, CNPC signed an MOU with Myanmar's Ministry of Energy on the construction, operation and management of the Myanmar–China Crude Pipeline. In June 2010, CNPC signed with MOGE the shareholder

---

[126] CNPC website, http://www.cnpc.com.cn/en/Thailand/country_index.shtml, accessed 19 February 2015.

agreement of Southeast Asia Crude Pipeline Limited, and that of Southeast Asia Natural Gas Pipeline Limited.[127]

4.2.3.5.3 Indonesia

CNOOC entered Indonesia in 2003 when it purchased a 16.93 per cent interest of Tangguh LNG project with USD 340 million from BP and then sold 3.06 per cent to Talisman. In 2005, CNOOC YPF Repsol's oil field interests in Indonesia, making itself the largest operator in the offshore oil sector in the country.[128] In 2008, CNOOC spent a USD 125 million in purchasing 50 per cent stake in Husky (Madura) Energy's assets in Indonesia.

Moreover, CNPC has stakes in eight oil and gas blocks in Indonesia, namely Jabung, Tuban, Salawati Basin, Salawati Island, Bangko, SP and South Jambi B and Madura. The company is the operator of all the blocks except SP. In 2012, CNPC acquired all of Devon Energy's assets and activities in Indonesia's six blocks and operated eight E&P projects in the country.[129]

In December 2010, Sinopec spent USD 680 million in purchasing 18 per cent of Chevron's Gendalo–Gehem deep water gas project in Indonesia.[130]

4.2.3.5.4 Singapore

In 2009, CNPC spent USD 2 billion in acquiring 96 per cent stake in Singapore Petroleum Company, which has strong downstream assets in the Asia–Pacific. This deal will promote CNPC's international oil-trading position with refining capacity, product storage,

---

[127] CNPC website, http://www.cnpc.com.cn/en/Myanmar/country_index.shtml, accessed 16 February 2015.

[128] US Energy Information Administration, 'China', p. 5.

[129] CNPC website, http://www.cnpc.com.cn/en/Indonesia/country_index.shtml, accessed 16 February 2015.

[130] Jiang and Sinton, *Overseas Investments by Chinese National Oil Companies: Assessing the Drivers and Impacts.*

pipelines and other logistic assets in Singapore, a major oil-trading hub in the Asia–Pacific.[131]

4.2.3.6 *Other countries in the Asia–Pacific*

4.2.3.6.1 Afghanistan

CNPC also ventured into Afghanistan in 2011, winning that country's first bidding round for the oil extraction project in the Amu Darya river basin in the northern part of Afghanistan.[132]

4.2.3.6.2 Australia

In 2003, CNOOC bought a 5.56 per cent share of production in Australia's Northwest Shelf Project for delivery of natural gas to China's Guangdong Province, and a 12.5 per cent stake in the Gorgon gas field off the Australian western coast, with options for further development.[133] CNOOC also owns a 50 per cent stake in the Queensland Curtis LNG project, which is scheduled to start to supply LNG to China by 2015.[134] In 2007, CNPC signed a contract with an Australian company to purchase two million tonnes of LNG every year for 15 years, starting from a time between 2013 and 2015.[135] Moreover, in 2008, Sinopec spent USD 561 million in purchasing 60 per cent of Australia's AED oil for assets in Australia.[136] Sinopec also owns 25 per cent of Australia Pacific LNG project, which is expected

---

[131] *Ibid.*

[132] Jiang and Ding, *Updates on Overseas Investments by China's National Oil Companies: Achievements and Challenges since 2011.*

[133] Department of Foreign Affairs and Trade, the Australian Government, 'PRC Country brief — May 2006', http://www.dfat.gov.au/geo/china/cb_index.html#betr, accessed 29 June 2006.

[134] U.S. Energy Information Administration, 'Country Analysis Brief: China'.

[135] The 2007 Annual Report of CNPC, p. 39.

[136] Jiang and Sinton, *Overseas Investments by Chinese National Oil Companies: Assessing the Drivers and Impacts.*

to begin operation and supply LNG to China by 2015.[137] Currently, about 17 per cent of Australia's LNG export goes to China.[138]

4.2.3.6.3 Mongolia

In April 2005, CNPC purchased from SOCO International PLC (UK) a 94.4 per cent stake in three oil blocks, namely Block 19, 21 and 22, which are Located in the Tamsag Basin, Dornod Province, Mongolia. In August 2011, CNPC signed an MOU with Mongolian Ministry of Mineral Resources and Energy to expand the company's downstream operations in Mongolia.[139]

4.2.3.7 *Latin America*

4.2.3.7.1 Peru

CNPC has been present in Peru since 1994. It owns the risk exploration Block 111/113 and two production blocks 1-AB/8 and 6/7 in Talara oil field located in north-western Peru. Block 6/7 was the first overseas oil field development project operated by CNPC. The company took over the block in January 1994 and October 1995 respectively. In July 2003, CNPC signed with PLUSPETROL a cooperation agreement on Block 1-AB/8, holding 45 per cent shares. In December 2005, CNPC signed with Peru's Ministry of Energy and Mining risk exploration contracts covering Block 111 and Block 113 in the MDD basin. In 2013, CNPC acquired the entire shares of Petrobras Energia Peru S.A.[140]

---

[137] U.S. Energy Information Administration, 'Country Analysis Brief: China'.

[138] Jiang and Ding, *Updates on Overseas Investments by China's National Oil Companies: Achievements and Challenges since 2011.*

[139] CNPC website, http://www.cnpc.com.cn/en/Mongolia/country_index.shtml, accessed 16 February 2015.

[140] CNPC website, http://www.cnpc.com.cn/en/Peru/country_index.shtml, accessed 19 February 2015.

4.2.3.7.2 Venezuela

CNPC has been operating in Venezuela since 1997 and owns oil and gas assets and equity in the country. The company is involved in the development project of Caracoles and Intercampo oil fields, a joint-venture Orimulsion development project, the development and joint-venture operation of the Zumano oil field, and the development and joint-venture operation of Junin 4 block in the Orinoco heavy oil belt.

In June 1997, CNPC won biddings for the Intercampo Oil field and the Caracoles oil field,[141] located at Maracaibo Lake and the East Venezuela Basin respectively. In December 2004, CNPC and PDVSA signed a cooperation agreement on Zumano oil field. In August 2006, CNPC and PDVSA signed a joint venture agreement, with CNPC holding a 40 per cent stake. In 2007, then Venezuelan President Hugo Chavez signed a Presidential Order to transfer CNPC's rights in exploration and development of Zumano oil field to the joint venture Petrozumano.

In April 2001, CNPC and PDVSA established a joint venture to operate the Orimulsion Project that includes the MPE-3 Oil field and an emulsification plant, with CNPC owning a 70 per cent stake. The project was completed and put into service in November 2006.

In August 2006, CNPC and PDVSA signed an agreement for the joint exploration of the Orinoco Heavy Oil Belt's Junin-4 Block which was located in south–eastern part of Venezuela. In May 2008, CNPC reached two agreements with PDVSA. One agreement has led to the establishment of a joint venture to explore the super-heavy oil at Junin-4 Block in the Orinoco heavy oil belt, with a target annual

[141] The Intercampo and Caracoles oil fields are two marginal oil fields that had been explored for more than 50 years when taken over by CNPC, with many faults, a complicated geologic structure, and a small scale. In less than three years, the production peak of these oil fields has increased from 700 tonnes per day to 5,500 tonnes per day after CNPC's takeover. This was largely due to CNPC's technologies in releasing the potential of mature oil fields. This example was described as 'the CNPC model' by local oil companies in Venezuela. See CNPC website, http://www.cnpc.com.cn/en/Venezuela/country_index.shtml, accessed 19 February 2015.

production of 20 million tonnes; CNPC has a 40 per cent stake in the joint venture. The other agreement would lead to the establishment of a joint venture refinery in China, in which CNPC would have a 60 per cent stake. In December 2010, CNPC signed with the Venezuelan Ministry of Energy and Petroleum an agreement on the joint venture operation of the Junin 4 project. In September 2013, a framework agreement on the joint development of the Junin 10 Block was signed with the Ministry of Oil and Mining of Venezuela.[142]

4.2.3.7.3 Ecuador

CNPC operates two projects of Andes and Amazon in Ecuador. In August 2003, CNPC and Petroecuador signed a management rights transfer agreement for Block 11 in Ecuador. In October that year, it acquired and took over Block 11, and began oil and natural gas exploration and development in the block. In September 2005, CNPC and Sinopec made a joint acquisition of oil and natural gas assets and development rights and interests in five blocks owned by Encana, and established Andes Petroleum Ecuador Ltd., with CNPC owning 55 per cent shares. In February 2006, Andes Petroleum took over the rights to explore and develop five blocks in the east of the Oriente Basin. In 2010, Andes Petroleum converted its PSC into service contracts.[143]

4.2.3.7.4 Colombia

In 2006, Sinopec cooperate with the ONGC and jointly purchased a 50 per cent holding in the Colombian oil company Omimex de Colombia with USD 800 million. Sinopec and ONGC respectively have 25 per cent shares in it.

---

[142] CNPC website, http://www.cnpc.com.cn/en/Venezuela/country_index.shtml, accessed 19 February 2015.

[143] CNPC website, http://www.cnpc.com.cn/en/Ecuador/country_index.shtml, accessed 16 February 2015.

4.2.3.7.5 Argentina

In March 2010, CNOOC spent USD 3.1 billion in purchasing 50 per cent stake in the Argentinean oil company, Bridas Corporation, which has oil and gas exploitation operation in Argentina, Bolivia and Chile. In November 2010, CNOOC spent USD 2.47 billion in acquiring 60 per cent of the Argentinean oil and gas producer Pan American Energy from BP under Bridas.[144]

4.2.3.7.6 Brazil

In October 2010, Sinopec spent USD 7.1 billion in purchasing 40 per cent stake in Brazilian subsidiary of Spanish oil company Repsol.[145] In 2013, CNPC and CNOOC jointly won a 35-year production service contract in a consortium with Petrobras, Shell and Total to develop a pre-salt discovery in Brazil's Libra oil field.[146]

4.2.3.8 *North America and West Europe*

4.2.3.8.1 United States

In October 2009, CNOOC spent USD 1 million in purchasing partial share of Statoil's U.S. assets in deepwater areas of Gulf of Mexico. In November 2010, CNOOC spent USD 2.16 billion in acquiring 33.3 per cent interest in Chesapeake's 600,000 net acres in the Eagle Ford Shale.[147] In January 2012, Sinopec made its first entrance into the U.S. upstream oil and gas business, with the conclusion of a USD 2.5 billion joint venture with U.S.-based Devon Energy for a one-third share of five shale blocks. In February 2013, Sinopec reached a USD 1.02 billion joint venture deal with Chesapeake Energy for a 50 per

---

[144] Jiang and Sinton, *Overseas Investments by Chinese National Oil Companies: Assessing the Drivers and Impacts.*
[145] *Ibid.*
[146] Jiang and Ding, *Updates on Overseas Investments by China's National Oil Companies: Achievements and Challenges since 2011.*
[147] Jiang and Sinton, *Overseas Investments by Chinese National Oil Companies: Assessing the Drivers and Impacts.*

cent interest in the Mississippi Line asset with nearly 2.2 million acres.[148]

4.2.3.8.2 Canada

CNPC has been present in the country since 1993. It has interests in eight oil and natural gas blocks in the country and operates three of them. In June 1993, CNPC purchased a 15.8865 per cent operating interest in North Twining oil field and 11.477 per cent equity in the natural gas processing plant in Alberta province. In August 2009, CNPC acquired 60 per cent equity of MacKay River and Dover oil sands assets of Athabasca Oil Sands Corporation, which was a breakthrough for CNPC's unconventional energy sector overseas. In December 2009, the deal was approved by the Investment Review Department of Canada's Federal Ministry of Industry. In February 2010, CNPC signed documents with Canada's Athabasca Oil Sands Corporation to complete the handover of the MacKay River and Dover oil sands projects in Alberta. In June 2010, CNPC and Canadian Encana agreed to form a joint venture to develop Encana's shale gas assets in British Columbia.

In May 2012, PetroChina, Shell Canada Ltd., Korea Gas Corporation (KOGAS) and Mitsubishi Corporation announced they would jointly develop a proposed LNG export facility near Kitimat, British Columbia. Shell holds a 40 per cent working interest, with KOGAS, Mitsubishi and PetroChina each holding a 20 per cent working interest. In February 2012, an agreement was signed and the deal was completed between PetroChina and Royal Dutch Shell Plc, with PetroChina buying a 20 per cent stake in Shell's 100 per cent-owned Groundbirch assets in north–eastern British Columbia.[149]

Moreover, Sinopec has acquired 40 per cent equity in Synenco Energy's USD 4.5 billion Northern Lights oil sands project, which

[148] Jiang and Ding, *Updates on Overseas Investments by China's National Oil Companies: Achievements and Challenges since 2011.*

[149] CNPC website, http://www.cnpc.com.cn/en/Canada/country_index.shtml, accessed 16 February 2015.

would produce a total of 100,000 bpd of synthetic crude oil in 2010 when commercial operations are scheduled to begin. In April 2010, Sinopec spent USD 4,675 in purchasing 50 per cent stake in Canadian oil sands company Syncrude from ConocoPhillips.[150]

In February 2013, CNOOC completed the USD 15.1 billion takeover of Canadian independent Nexen, making Chinese NOCs' largest takeover of an oil and gas company so far. The deal has increased CNOOC's production and reserve base by 20 per cent and 30 per cent respectively, and strengthened CNOOC's positions in Canada's shale and oil sands as well as offshore areas in the Gulf of Mexico, Nigeria and the North Sea. Notably, the political and economic stability was an important consideration for CNOOC, as 90 per cent of the newly acquired reserves are in OECD member countries with more developed and stable operational environment.

4.2.3.8.3 United Kingdom

In 2008, CNOOC purchased 100 per cent stake of Awilco for USD 2.5 billion. Awilco Drilling PLC is a UK based Drilling Contractor owning and operating the two refurbished and enhanced mid-water semi-submersible drilling units WilHunter and WilPhoenix.[151] In June 2009, Sinopec spent USD 8.8 billion in purchasing 100 per cent stake of Addax Petroleum,[152] which is an international oil and gas exploration and production company with a strategic focus on Africa, the Middle East and the North Sea of the UK.[153] In July 2012, Sinopec, through its subsidiary Addax Petroleum UK, acquired a 49 per cent interest in Talisman Energy's North Sea assets for USD 1.5 billion. In 2014, through its subsidiary Nexen, CNOOC acquired

---

[150] Jiang and Sinton, *Overseas Investments by Chinese National Oil Companies: Assessing the Drivers and Impacts.*

[151] Awilco Drilling's website, http://www.awilcodrilling.com/, accessed 20 February 2015.

[152] Jiang and Sinton, *Overseas Investments by Chinese National Oil Companies: Assessing the Drivers and Impacts.*

[153] Addax Petroleum's website, https://www.addaxpetroleum.com, accessed 20 February 2015.

36.5 per cent of the Golden Eagle project, 70 km northeast of Aberdeen, Scotland.[154]

4.3 Conclusion

In the recent years, China's energy diplomacy has been developing fast, as the Chinese government making great efforts in strengthening China's relations and energy cooperation with some energy rich states, while Chinese NOCs quickly expanding their overseas M&A of oil and natural gas assets and transnational operations, becoming emerging significant players in world oil and gas business. Now a key question needs answering: What is the momentum of China's energy diplomacy? Specifically, why does the Chinese government pushes forward its energy diplomacy? And why are Chinese NOCs keen on their overseas venture? Notably, the government and the NOCs are strategic partners in the movement of energy diplomacy, as Beijing's energy diplomacy and the NOCs' ODI and transnational operation are mutually reinforcing. These questions and issues will be addressed in the next chapter.

[154] Jiang and Ding, *Updates on Overseas Investments by China's National Oil Companies: Achievements and Challenges since 2011.*

Chapter

5

National Interests and Corporate Interests behind China's Energy Diplomacy

This chapter identifies the fundamental national and corporate interests behind China's energy diplomacy. As it argues, China's energy diplomacy serves not only China's national interests but also Chinese national oil companies' (NOCs) corporate interests. On the one hand, energy diplomacy improves China's energy security and promotes Beijing's multilateral diplomacy. On the other hand, energy diplomacy is in line with Chinese NOCs' corporate interests, including entering overseas investment markets, generating profits and fulfilling development strategies. The convergence of national and corporate interests is the key momentum of China's energy diplomacy.

This chapter is divided into three sections. The first section briefly illustrates the momentum and process of China's energy diplomacy. The other two sections respectively deal with the major national and corporate interests behind the phenomenon. The discussion of Beijing's energy security and multilateral diplomacy in the second section reflects the realist conception of the vital importance of national interests; while in the third section, the explanation of the NOCs' corporate interests and self-motivation to participate in global business echoes the liberal perspective's emphasis on the role of corporations.

5.1 The momentum and process of China's energy diplomacy

Behind China's energy diplomacy, the Chinese government's national interest consideration motivates it to carry out energy diplomacy throughout many energy rich states; while Chinese NOCs' corporate interest consideration drives them to invest and expand abroad.

As demonstrated in the previous chapter, Beijing's energy diplomatic activities have been facilitating Chinese NOCs' overseas direct investment (ODI) and transnational operation for years. The Chinese government's good relationship with and lobbying efforts on some energy exporting countries, especially those in Central Asia, Africa, Latin America and Southeast Asia, is a positive factor for the NOCs' business operations there. It assists these firms to invest in oil and natural gas recourses and assets in those countries, securing long-term energy supply deals, gain more profits and further transform themselves into world-class oil companies.

Meanwhile, Chinese NOCs' ODI and transnational operation is generally beneficial for Beijing's national interests. Their fast expanding and increasingly diversified investment and production in energy rich countries help improve China's energy security conditions, and support the country's rapid social and economic development. Also, the NOCs' fast expanding business links with many energy rich states give the Chinese government a new and increasingly important economic platform to strengthen its ties with these states. Such relationships help

Beijing safeguard and expand its political and strategic interests and benefits in the international arena, such as securing more diplomatic support in multilateral institutions like the United Nations (UN) and the World Health Organisation (WHO). In particular, the realisation of national interests through this way contributes to China's economic development, social stability and national unity, which are the three pillars of the Chinese Communist Party's (CCP) political legitimacy. Therefore, the Chinese government's energy diplomacy and Chinese NOCs' ODI and transnational operation are mutually reinforcing. The government and the NOCs are strategic partners in the movement. More details about the major national and corporate interests behind China's energy diplomacy will be discussed in the next two sections.

In other words, during the process of energy diplomacy, the two players — the Chinese government and Chinese NOCs — need each other's cooperation and support to realise their respective vital interests. Beijing needs the NOCs' ODI and transnational business to improve China's energy security and provide it with a new tool to engage energy rich states. Simultaneously, just like many international oil companies (IOCs), Chinese NOCs need their government's political and diplomatic support for their overseas operation. Considering their disadvantages in the world energy business competition, the Chinese government's support for them is undoubtedly critical.[1] Therefore, the momentum of China's energy diplomacy and the role of the government and that of the NOCs are briefly illustrated in Figure 5.1.

To begin with, each player's activities realise their own interests. The government's energy diplomacy is motivated by Beijing's

---

[1] Generally speaking, Chinese NOCs are at a disadvantage in the world oil competition because they are latecomers to the global oil business. Compared with their Western counterparts that have been operating worldwide for more than a century, Chinese NOCs have been active abroad for less than two decades. This historical experience gives Western oil companies a competitive edge that other companies are unable to enjoy. Also, in the recent years, some energy rich countries have tightened their state ownership of the domestic energy resources, and increased their take *vis-à-vis* foreign companies. Downs, 'The Fact and Fiction of Sino–African Energy Relations', pp. 51–52.

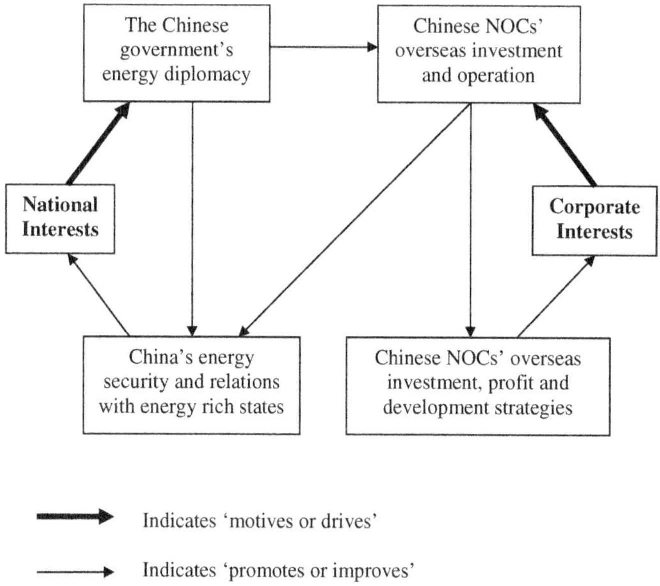

Figure 5.1: Momentum and process of China's energy diplomacy
Source: Author's compilation.

consideration of national interests. It improves China's energy security and relations with many energy rich states, which are beneficial for China's national interests. The NOCs' ODI and transnational operation are pushed by their consideration of corporate interests. The movement enables them to expand abroad, secure more energy supplies, earn greater revenues and implement their ambitious corporate development strategies of trans-nationalisation, which are beneficial for their corporate interests.

Moreover, each player's activities serve and contribute to the other's interests. Beijing's energy diplomacy is supportive for Chinese NOCs' overseas business and finally the realisation of their corporate interests; while these firms' ODI and transnational operation is helpful for China's energy security and external relations and finally the country's national interests. Put it in another way, the government's behaviour in the phenomenon realises both national and corporate interests,

so does the NOCs' behaviour. In this way, the convergence of national and corporate interests is reached in China's energy diplomacy.

5.2 National interests

In general, the national interests behind China's energy diplomacy can be viewed as maintaining the CCP government's political legitimacy[2] which mainly consists of three elements — economic development, social stability and national unification. Energy diplomacy contributes to all the three areas.

Unlike the first and the second generations of the CCP leadership whose political legitimacy largely came from their revolutionary accomplishments, the later generations do not have such credentials. Therefore, the CCP leadership since the 1990s has relied on the three critical resources to sustain the party's political legitimacy. As to economic development, the Chinese government must ensure the continuous distribution of economic benefits that more or less satisfies the growing material demand of the Chinese people. In accordance, socio–economic development remains the CCP's highest priority and the work focus of the government.

Moreover, economic growth is the foundation of the development of China's comprehensive national power (*zonghe guoli*) including economic capability, political clout, military capacity, science and technology strength, etc., because the building of other powers needs significant economic resources and support. Notably, since the beginning of China's economic reforms, Beijing's foreign policy and diplomatic work has consistently been required to serve the country's economic development and modernisation drive. Specifically, the Chinese government has been making efforts in

[2] In history, the CCP's political legitimacy peaked with the establishment of the Peoples' Republic of China (PRC) in 1949, as the CCP was the leader of the contemporary Chinese revolution and just won the civil war. However, after China's reform and opening up, the CCP's political legitimacy faces the risk of waning because of the rampant corruption, relatively slow political reform, as well as the growing social inequality and widening income disparity across the country, although the Chinese economy keeps growing fast.

realising a relatively favourable external environment for China's domestic development. Beijing's energy diplomacy is part of this grand diplomatic strategy.

Apart from economic development, domestic stability and national unification are also critical for the CCP government's political legitimacy. Domestic stability is always closely linked with economic development. For instance, if the country encounters serious economic problems such as stagnation with continuous inflation or widening income disparity, these problems could transform into social and political problems like growing unemployment, grievance towards the government and society, or even social disorder, which could challenge the Chinese government's rule over the country. In this sense, economic development supports social stability.

Also, national unity is an essential pillar to reinforce the CCP's legitimacy. Hong Kong and Macao have been returned to China. These two events have definitely promoted the party's legitimacy among the Chinese people. Now the most intractable problem is the Taiwan issue, which is much more complicated. Beijing always insists on its sovereignty over Taiwan and the goal to reunify the island with mainland China. In addition, the growing separatist activities and extremist violence in some parts of China like Xinjiang Uyghur Autonomous Region and Tibet Autonomous Region have brought more trouble to the Chinese government, directly jeopardising social stability and potentially threatening national unity. How to address these problems is related to the CCP's credibility among the Chinese people. The leadership believe that the maintenance and acceleration of economic development in those unstable and relatively underdeveloped areas are critical for domestic stability and national unity.

The rest of this section consists of two sub-sections. The first part explains how energy diplomacy improves China's energy security; the second part shows the contribution of energy diplomacy to the improvement of China's external relations and Beijing's multilateral diplomacy. The progress made in these two areas is beneficial for economic development, social stability and national unity of China.

5.2.1 *Improving China's energy security*

> 'As an important aspect of economic security, energy security has a direct bearing on national security and social stability'.[3]
>
> *China's Energy Conditions and Policies* (the official White Paper)

In the recent years, energy security has become an urgent challenge in front of the Chinese leadership. Here, energy security can be simply defined as the availability and affordability of energy resources, particularly oil and natural gas. Availability refers to the continuous and sufficient energy supply; and affordability can be understood as the stable and acceptable energy prices. Beijing's energy security concern has been identified in two fields — the domestic aspect as well as the international aspect; in other words, the country faces both domestic and external risks to its energy security. The domestic challenge is expected to be tackled through the reform adjustment and restructuring of China's domestic energy system; while how to deal with the external challenges is related to the country's international relations and foreign policy including energy diplomacy.

Notably, energy security is related to the Chinese government's political legitimacy. The availability and affordability of energy resources is a necessity for China's economic development, social stability and national unity. If the energy supply is insufficient, it could undermine a country's industrial production and economic growth, affect the people's livelihood, increase unemployment, harm national defence, etc. Hence, whether China's energy supply can be guaranteed determines whether the country can maintain its fast social and economic development, satisfy the people's demand for a better life, and maintain social stability and national unity.

However, China's energy security faces several external uncertainties or threats, posing (potential) risks to the availability and affordability of the country's energy import. During the 1970s and the

[3]The State Council Information Office, the People's Republic of China, *China's Energy Conditions and Policies*, 26 December 2007, http://www.china.org.cn/english/environment/236955.htm, accessed 27 December 2007.

1980s when China was self-sufficient in energy supply, its domestic oil production and consumption was planned by the government and had no linkage with the world market; the Chinese economy was largely sheltered from the outside world, and the disruption of the international oil market did not affect China much. But now China has become increasingly dependent on overseas energy supplies; thus, the Chinese economy is increasingly influenced by external factors. There are five major factors affecting China's energy security, namely the instability in some energy exporting regions, the fluctuation of the international oil price, the energy import transportation risk, the international containment against China, and the threat from the United States. The rest of this sub-section discusses each of these factors in turn, before explaining the contribution of energy diplomacy to China's energy security.

5.2.1.1 *The instability in some energy exporting regions*

Firstly, the instability in some major energy exporting regions in the world is always a negative factor for China's energy security. The unstable political and security situation in those areas may affect the regional countries' ability to export energy to China, put Chinese energy production and facilities there at risk, and causes surges in the international oil prices. Some countries exporting a large amount of oil to China are located in areas prone to instability.

Although the Chinese government has relatively good relations with almost all of the countries exporting oil to China, these countries are heavily concentrated in the Middle East and West Africa; and both of these regions are subject to conflicts and wars, raising questions about the stability of oil supply from these regions. As mentioned in Chapter 3, the Middle East is China's largest source of oil import, with about half of China's oil import coming from the region; and West Africa followed the Middle East as China's second largest source of oil import.[4] This situation is unlikely to be changed in the foreseeable future.

[4] British Petroleum, *BP Statistical Review of World Energy June 2014*, http://www.bp.com/content/dam/bp/pdf/Energy-economics/statistical-review-2014/BP-statistical-review-of-world-energy-2014-full-report.pdf, accessed 24 August 2014.

For example, the Middle East is characterised by an unstable geopolitical situation and frequent regional conflicts and domestic disturbances, which have disrupted the oil export from the region for several times in the past, such as the Iran–Iraq War from 1980 to 1988, the continuous Israeli–Palestinian conflicts, the 1991 Persian Gulf War and the 2003 Iraq War. The situation is somewhat similar in Africa.[5] Nigeria is the largest oil producer in Africa, but incidents such as strikes and kidnaps of foreign oil workers often occur in the country. Its social instability has directly affected its oil supplying ability. Its oil production is continuously undermined by factors like armed attacks, posing a threat to the global oil supply. The armed conflict in the country in February 2006 caused a one third reduction of its oil production.[6]

Moreover, some analysts have noted that the Middle East peace process has not achieved significant progress; the domestic situation in Iraq after the Iraq War is far from stable; and some active Islamic forces have carried out continuous terrorist activities there. Also, domestic political problems such as the power succession in some states and the competition among some countries for water resources and territory tend to persist over long periods, posing a threat to the stability of the Middle East situation. In addition, United States anti-terrorist campaign in the Middle East may strain its relationship with the countries in the region, adding to the instability factors there.[7] The political differences among the countries in the region and terrorist activities against energy facilities might pose a threat to the stability of oil export from these regions.

Recently, some events in the Middle East and North Africa have made the regional situation even more challenging. These events

---

[5] Bo Kong, 'An Anatomy of China's Energy Insecurity and Its Strategies', Pacific Northwest National Laboratory, December 2005, http://pnwcgs.pnl.gov/Newsletter/OtherDocs/AnatChinaEnergy.pdf, accessed 16 May 2008, p. 14.

[6] Cui Minxuan, ed., *2007 Zhongguo nengyuan fazhan baogao* (*The Energy Development Report of China 2007*) (Beijing: Shehui kexue wenxian chubanshe, March 2007), p. 93.

[7] Zhao Hongtu, 'Guojia nengyuan anquan' ('National energy security'), in Zhongguo xiandai guoji guanxi yanjiuyuan (China Institute of Contemporary International Relations), *Guojia Jingji Anquan* (*National Economic Security*) (Beijing: Shishi chubanshe, July 2005), pp. 276–277.

include the 'Arab Spring', the Syrian civil war and the rise of the terrorist organisation — the Islamic State of Iraq and the Levant (ISIS), which have been reshaping the geopolitical landscape of the Middle East; the ongoing negotiations on Iran nuclear issue, and the conflict between Sudan and the newly independent South Sudan, as well as violence within South Sudan. Chinese NOCs found themselves caught in the middle of geopolitical crises that have caused production to be shut down and personnel to be evacuated, such as the evacuation of Chinese employees together with other Chinese citizens from Libya in 2011. Also, the conflicts and violence in Sudan and South Sudan have presented a challenge to Beijing's diplomacy as well as China National Petroleum Corporation's (CNPC) long-term investment and operation there.[8]

5.2.1.2 *The fluctuation of the international oil price*

Secondly, the volatility of international oil prices is another factor harming China's energy security. As a Chinese official white paper pointed out, 'Price fluctuations in the international energy market make it more difficult to guarantee domestic energy supply'.[9] Before the 2008 global economic crisis, the international oil price climbed rapidly, increasing from an average of USD 28 per barrel in 2003 to above USD 130 per barrel in mid-2008; and such a high price was often accompanied with substantial fluctuations. Several factors such as the unstable geopolitical situation, the speculation at the international oil market, the depreciation of the U.S. dollar and growing global demand caused the surge in international oil prices at that time.

---

[8] Julie Jiang and Chen Ding, *Updates on Overseas Investments by China's National Oil Companies: Achievements and Challenges since 2011*, International Energy Agency, Partner Country Series, 2011, http://www.iea.org/publications/freepublications/publication/PartnerCountrySeriesUpdateonOverseasInvestmentsbyChinasNationalOilCompanies.pdf, accessed 24 August 2014.

[9] The State Council Information Office, the People's Republic of China, *China's Energy Policy 2012*, October 2012, http://english.gov.cn/archive/white_paper/2014/09/09/content_281474986284499.htm, accessed 14 January 2015.

For example, Strange pointed out that the value of the U.S. dollar is undoubtedly important for the world oil price. There are two volatile prices at work in the world market economy — the price of oil and the dollar's exchange rate. They affect the fortunes of both oil importing and oil exporting countries. Since the 1970s, there has been substantial divergence between the 'real' price of oil and its nominal price. The most important factor to stabilise oil prices is the long-term stability of the U.S. dollar's purchasing power, or pricing oil via a basket of currencies.[10] Similarly, the Organisation of Petroleum Exporting Countries' (OPEC) *World Oil Outlook 2008* noted that 'elements other than supply and demand fundamentals are at play.' The first element driving oil prices is the devaluation of the U.S. dollar in relation to other currencies. It fell from USD 1.3 per euro in August 2007 to about USD 1.6 per euro in June 2008, representing a significant weakening in the value of the U.S. dollar. Another element relates to regulated oil futures and unregulated over-the-counter exchanges. The trade in paper barrels expanded dramatically before 2008. Oil became an attractive financial asset for global investors to diversify their portfolios and increase the returns, with the influx creating upward pressures on oil prices.[11]

Consequently, as a Chinese official White Paper noted in 2008, high oil prices had increased pressure on global inflation, affected the traditional international financial system, and resulted in a heavier economic burden for oil importing countries.[12]

On the one hand, international oil prices surge would bring negative implications for China. For instance, the soar in international oil prices not only reduces the capacity of oil importing states

---

[10] Susan Strange, *States and Markets* (London and New York: Continuum, 1994), p. 179 and p. 207.

[11] OPEC, *World Oil Outlook 2008*, (Vienna: OPEC Secretariat, 2008), http://www.opec.org/library/World%20Oil%20Outlook/pdf/WOO2008.pdf, accessed 6 January 2009.

[12] Xinhua, '08nian zhongguo waijiao baipishu shuban faxing, nengyuan anquan lie shouzhang' (The 2008 edition of China's Foreign Affairs white paper has been published; the issue of energy security is addressed in the first chapter), 18 July 2008, http://politics.people.com.cn/GB/1027/7531802.html, accessed 13 March 2009.

to purchase oil at the international market, but also cause other economic problems. The reason is that energy such as oil and natural gas are among the basic resources widely used in agricultural and industrial production. The increase of oil price will boost the price of most commodities and cause inflation. As some experts noted, the increase of energy prices will not only directly harm those industries heavily relying on energy resources but also affects other areas of the national economy. It will also cause inflation and the increase of unemployment rates.[13]

A Chinese study showed that an increase in oil prices of USD 10 per barrel may slow down China's economic growth by about one per cent.[14] According to some experts, if the international oil price goes up by one per cent and lasts for one year, China's gross domestic products (GDPs) growth will drop by 0.01 per cent on average.[15] Some experts held that every increase of USD 10 per barrel of the international oil price would raise the consumer price index (CPI) in China by 0.4 per cent and reduce China's GDP by 0.8 per cent.[16] If the increase of the CPI is faster than that of salary for the majority of the people, as the case in China from 2007 to 2008, the people would encounter an actual wage decrease and tend to become unsatisfactory about the society and the government. Hence, the skyrocketing international oil prices have been proven negative for China's effort to maintain rapid economic development and social stability.

---

[13] For details, see Ni Jianmin eds., *Guojia nengyuan anquan baogao* (*A Report of National Energy Security*) (Beijing: Renmin Chubanshe, July 2005), pp. 95–96.

[14] People's Daily On-line, 'Tebie cehua: duiyi zhanzheng zhongguo jingji sunshi you duoda?' ('Special report: to what extent has the Iraq War affected the Chinese economy?'), 25 February 2003, http://past.people.com.cn/GB/jinji/31/179/20030225/930064.html, accessed 17 May 2008.

[15] Sohu, 'Zhongguo jianli "shiyou zhanlue shubei" beingfei zhiwei pingyi youjia' ('The establishment of China's "strategic petroleum reserves" does not merely aim at stabilising the oil price'), 11 October 2004, http://business.sohu.com/20041011/n222432572.shtml, accessed 17 May 2008.

[16] Dongfang wang (Eastday), 'Guoji youjia zaichuang lishi xingao, pobai bushi ouran shijian' ('The international oil price reached historical height again, the increase of the price above USD 100 is not an accident'), 28 February 2008, http://finance.eastday.com/m/20080228/u1a3431649.html, accessed 17 May 2008.

Moreover, the three global oil crises in history triggered by surges in the international oil prices (1973–1974, 1979–1980 and 1990) have exerted huge influences on various countries. Such impact is generally favourable for net oil exporting states while causing widespread economic crises for net oil importing states.

Further, for those developed countries which are net oil importers, such impact may be significant at the beginning but gradually reduces later; while for those developing countries which are net oil importers, such impact tends to continuously increase due to their fast economic growth and rising dependence on energy import. In particular, there are six reasons that developing economies are more vulnerable to the international oil price surge: First, some developed countries' advanced energy technologies could reduce their dependence on oil; but many developing countries without these technologies tend to be heavily dependent on oil. Second, the high oil prices particularly increase those poor countries' burden to purchase oil with foreign exchanges. Third, the oil price surge especially promotes inflation in developing countries, because many of them may be eager to develop their economies and adopt expansive economic policies which would result in long-term pressure for inflation, and the high oil prices might add to such pressure. Fourth, high oil prices would increase the debt ratio, increasing the risks to developing countries' financial systems. Fifth, high oil prices often push developing countries to adopt austere monetary policies, which would slow down their economic growth. Sixth, high oil prices tend to raise global interest rates and market risks, and restrain developing countries' economic activities.[17]

On the other hand, however, the dramatic decline of international oil prices may also become a mixed blessing for China, bringing about some negative effects. For example, in the early period of 2008, the international oil prices surpassed USD 100 per barrel and even reached USD 147.27 in mid-July. Then, the global financial and

[17]For details, see Cao Rongxiang, *Jingji anquan: fazhanzhong guojia de kaifang yu fengxian* (*Economic Security: Openness and Risks of Developing Countries*) (Beijing: Shehui kexue wenxian chubanshe, June 2006), pp. 175–183.

economic crisis caused the sudden and substantial drop of the international oil prices by about 70 per cent in less than five months. The speed and scope of the oil price plunge were beyond most people's expectation. Since at that time, about half of China's oil consumption relied on import, while the country's oil and petro–chemical industries purchased crude oil from abroad at the previous rather high international oil prices, the prices of their refined oil products had to be greatly reduced due to the subsequent international oil prices slump. As a result, China's oil enterprises faced much difficulty to transfer the oil import costs to the downstream oil products.[18] But such a situation of low oil prices did not last long.

The most recent case is that in the second half of 2014, the international oil prices plummeted by about 45 per cent. The monthly average Brent crude oil prices fell from USD 112 per barrel in June to USD 62 per barrel in December, the lowest since May 2009. The steep fall of international oil prices in 2014 was largely caused by the supply-demand fundamental, i.e. the combination of weak global oil demand and robust world supply growth. Other important factors include the price competition among major oil producing countries, the weakening outlook for the world economy and oil demand growth, the less concern about geo-political risks, and the appreciation of U.S. dollar as well.

For China, the international oil prices plunge in 2014 may be a double-edged sword. It is generally beneficial to the Chinese economy, as it has reduced the costs of China's oil import and led to the reduction of oil products prices domestically.[19] However, although it

---

[18] Xinhua, '2008 nian zhongguo nengyuan xingshi baogao: Nengyuan fazhan zhanlue touchu xinyi' (The 2008 report on China's energy situation: new development in the energy development strategy), 15 December 2008, http://news.xinhuanet.com/fortune/2008-12/15/content_10506475.htm, accessed 16 February 2009.

[19] From July 2014 to January 2015, the Chinese government lowered the prices for domestic oil products for 13 consecutive times. This would help reduce the costs of oil products consumption, therefore increasing the enterprises' profit margins as well as stimulating domestic consumption. Also, the slump of oil prices could ease inflationary pressure in the Chinese economy, giving more operational leeway to the government's monetary policy.

also allows China to build up its strategic petroleum reserves (SPR) at a lower cost,[20] it could hinder the investment in and development of China's oil, new energy and renewable energy industries, exerting adverse impact on China's energy security in the long run. The sustained low oil prices would cause less profitability in the oil industry, putting restraints on oil companies' investment and affecting the further exploration of China's domestic oil resources. Simultaneously, since oil prices drop, the costs for production of alternative energies become relatively higher. As a result, the impetus for the development of alternative energies may suffer, and China's efforts to adjust its energy consumption mix and improve its ecological environment could be undermined.[21]

5.2.1.3 *The transportation risk to China's energy import*

Thirdly, China's energy import transportation is under various kinds of threats. Currently, its energy import is mainly transported in three ways — shipping, pipeline and railway. Compared with maritime transportation, pipeline and railway transportation is regarded as safer and cheaper ways of energy import, but they are likely to cover a relatively small proportion of China's total energy import. The bulk of China's energy import has to be met by shipping.[22] Currently, about 90 per cent of the country's oil import and all of its liquefied natural gas (LNG) import is transported by sea from both remote regions including the Middle East, Africa and Latin America and adjacent areas of Southeast Asia and Australia.

[20] China started to build its SPR in 2003. The first phase of the project has been finished and put into service, and Beijing is gearing up for the completion of the second and the third phases. The overall target is to store 850 million tons of crude oil by around 2020, by which time the reserves could supply as much as three months of the country's oil imports. At the international oil prices declined, China could take advantage of the low oil prices and import more oil to feed into its SPR.

[21] For more information, see Zhang Chi, 'Changes in the global energy system and their implications for China', EAI Background Brief, No. 1006, East Asian Institute, National University of Singapore, 12 March 2015.

[22] You Ji, 'Dealing with the Malacca dilemma: China's effort to protect its energy supply', *Strategic Analysis* 31: 3, 2007, pp. 467–489.

Therefore, the risks with maritime shipping such as natural disasters, shipping accidents, piracy and potential terrorist attacks could affect China's energy imports by sea, apart from the potential energy import blockade laid by other countries. Three sea areas in the world — the offshore area near Somalia or Horn of Africa, the Straits of Malacca and the Southeast Asian waters — have been identified as particularly dangerous for China's shipping security.[23] In particular, like many other countries, China's commercial shipping including oil import shipping faces the so-called 'deliberate threats to shipping'.[24]

For instance, as some noted, good order at sea is lacked in Southeast Asia. There are some problems in regional waters, including piracy and armed robbery against ships, the threat of maritime terrorism, illicit trafficking in drugs and arms, people-smuggling, pollution, illegal fishing and marine natural hazards like tsunamis and cyclones.[25] In particular, the piracy acts taking place in the form of small-scale hit-and-run robberies have been a major concern in regional maritime security order for a long time. The causes are largely due to factors such as poor domestic governance and dire economic conditions in coastal areas, endemic corruption in local governments, prevailing transactions of small arms, the deficiencies of coastal countries in terms of marine policing, etc.[26]

---

[23] Another saying is that China has identified three locations — Hormuz Strait, Malacca Strait and Horn of Africa — as the energy trade routes which are in danger of disruption by terrorist activities or a blockade by enemy power. See Joshy M. Paul, 'The role of energy security in China's foreign policy: A maritime perspective', *Maritime Affairs* 6: 2, 2010, pp. 49–71.

[24] For details of 'deliberate threats to shipping', see Geoffrey Till, *Seapower: A Guide for the Twenty-First Century* (London and New York: Routledge, second edition, 2009), pp. 290–292.

[25] Sam Bateman and Jane Chan, 'Good order at sea in Southeast Asia', in Konrad-Adenauer Stiftung and European Union, *Maritime Security and Piracy: Common Challenges and Responses from Europe and Asia* (Singapore: Select Books Pte. Ltd., 2014), pp. 69–85.

[26] Hui-Yi Katherine Tseng, 'Maritime security in Southeast Asia: Interfacing regional and extra-regional stakeholder', in Konrad-Adenauer Stiftung and European Union, *Maritime Security and Piracy: Common Challenges and Responses from Europe and Asia* (Singapore: Select Books Pte. Ltd., 2014), pp. 87–105.

Moreover, the spectre of maritime terrorism has refocused attention on the problem of maritime crime since the September 11th. Despite a recent drop in pirate attacks, maritime violence remains a problem in Southeast Asia, where a number of political, geographic and economic factors make the region's seas an opportune place for sea-borne criminals.[27] Terrorists will attack targets which affect the infrastructure of the industrialised world most effectively, such as those choke points with 75 per cent of all maritime transportation activities carried out globally. The strategically important Strait of Malacca is one of the key choke points in the world. It connects the Indian Ocean with the South China Sea and the Pacific Ocean, making itself the most significant trade route between the Far East, the Gulf region and Europe. Each year, 90,000 ships, with one third of the world trade, 80 per cent of oil exports to East Asia and two thirds of LNG exports pass through the strait.[28]

Also, the re-emergence or intensification of the maritime territorial disputes over some islands in South China Sea between China and some Southeast Asian states like the Philippines and Vietnam, with the increasing intervention of outside powers such as the United States, Japan and India, has again reminded the Chinese side about the potential threat to China's key energy shipping lanes in this region.[29] In particular, U.S. pivot to Asia takes the form of picking sides in South China Sea disputes that encourages some claimants to adopt a tougher stance towards China. Thus, the South China Sea has been

---

[27] Ian Storey, 'Securing Southeast Asia's Sea Lanes: A Work in Progress', *Asia Policy* No. 6, July 2008, the National Bureau of Asian Research, http://www.nbr.org/publications/asia_policy/AP6/AP6_E_Storey.pdf, accessed 13 March 2009, pp. 95–127.

[28] Lutz Feldt, Peter Roell and Ralph Thiele, 'Maritime security — perspectives for a comprehensive approach', in Konrad-Adenauer Stiftung and European Union, *Maritime Security and Piracy: Common Challenges and Responses from Europe and Asia* (Singapore: Select Books Pte. Ltd., 2014), pp. 23–45.

[29] In early 2009, the South China Sea disputes received worldwide attention again. After the Malaysian Prime Minister stepped on a disputed island to claim sovereignty, the Philippines passed a law to include a disputed island under its sovereignty, ignoring the strong opposition from Beijing. In response, China has enhanced its inspection in South China Sea. Notably, the United States has been perceived by many Chinese as supporting these Southeast Asian states' confrontation against China for years.

a hotbed for disputes since 2010. The China–Vietnam friction or clash over the No. 981 Oil Rig, which was deployed in 2014 by China National Offshore Oil Corporation (CNOOC) in Xisha (Paracel) islands, is the most recent example.[30]

As some noted, the Indo–Pacific waters, that is the Greater Indian Ocean and the South China Sea, has become the world's energy interstate; and maritime tensions are rising in the South China Sea and in the East China Sea. The territorial disputes in those waters is not only being driven by potential energy reserves and fish stocks in the vicinity, but also by the very fact that these sea lanes of communications (SLOCs) and choke points are of growing geopolitical importance for energy transportation.[31]

There have been cyclic pattern of intermittent flare-ups in the disputed areas in the South China Sea which is described by some people as the new central theatre of conflict. Currently, more than half of the world's annual merchant fleet tonnage and one third of the world's maritime traffic pass through the South China Sea. The oil transported through this sea area to East Asia is more than six times the amount that passes through the Suez Canal, and 17 times the amount that passes through the Panama Canal.[32] Notably, most of China's energy import from the Middle East and Africa are transported by sea via the Indian Ocean through the Strait of Malacca before entering the South China Sea. Even China's oil import from Latin America has to pass through either the Philippines archipelago or the Luzon Strait between the Philippines and Taiwan.[33]

---

[30] You Ji, 'China's civil–military strategies for South China Sea disputes control', EAI Background Brief, No. 1002, East Asian Institute, National University of Singapore, 25 February 2015.

[31] Robert D. Kaplan, 'The geopolitics of energy', *Forbes*, 4 April 2014, http://www.forbes.com/sites/stratfor/2014/04/04/the-geopolitics-of-energy/, accessed 10 February 2015.

[32] Hui-Yi Katherine Tseng, 'The South China Sea disputes: Current state of play and future prospects', in Wang Gungwu and Zheng Yongnian eds., *China: Development and Governance* (Singapore: World Scientific Publishing Co. Pte. Ltd., 2013), pp. 515–526.

[33] Zhao Hong, 'China's efforts to enhance its energy security', in Wang Gungwu and Zheng Yongnian eds., *China: Development and Governance* (Singapore: World Scientific Publishing Co. Pte. Ltd., 2013), pp. 507–514.

Another factor is the significant increase of pirate attacks against commercial ships, including oil tankers, near Somalia and the Gulf of Aden in the recent years, which has seriously affected the international commercial shipping activities in that part of the Indian Ocean. Several Chinese ships were hijacked and some of the crew killed. After the United Nations authorised individual states to send warships to patrol along the SLOCs and protect maritime transportation in the region, the Chinese People's Liberation Army (PLA) Navy have carried out a dozen times of such mission with some of its best warships; and such missions will continue in the future.

Moreover, China's simplistic oil import shipping routes result in potential risks to its energy imports. For instance, about 80 per cent of China's oil import is shipped through the Straits of Malacca. That is to say, China's oil import SLOCs are not only threatened by piracy and terrorist attacks but also controlled by the U.S. Together with the South China Sea dispute, shipping risk is a growing threat to China's energy security.[34] An author described the importance of the Straits of Malacca for China as the following:

> 'The Straits of Malacca, linking the Indian and Pacific Oceans, is the shortest sea route between the Persian Gulf and China. It is the key chokepoint in Asia...The narrowest point is the Phillips Channel in the Singapore Strait, only 1.5 miles wide at its narrowest. Each day, more than 12 million barrels in oil super tankers pass through this narrow passage, most en route to the world's fastest-growing energy market, China, or to Japan. If the strait were closed, nearly half of the world's tanker fleet would be required to sail further. Closure would immediately raise freight rates worldwide. More than 50,000 vessels per year transit the Straits of Malacca. The region from Myanmar to Banda Aceh in Indonesia is fast becoming one of the world's most strategic

[34] Chen Fengying, 'Zhongguo nengyuan anquan de zhanlue sikao' ('The strategic thinking about China's energy security'), in Zhongguo xiandai guoji guanxi yanjiuyuan (China Institute of Contemporary International Relations), *Quanqiu nengyuan daqiju* (*Global Energy Structure*) (Beijing: Shishi chubanshe, January 2005), pp. 329–333.

> chokepoints. Who controls those waters controls China's energy supplies'.[35]

Beijing is concerned about the increasing military presence and activities around South China Sea and the Straits of Malacca, particularly by the United States. In late 2003, the then Chinese President Hu Jintao expressed concern on the strategic vulnerability of the Strait of Malacca, which 'certain major powers' were trying to control. Notably, the United States will remain the most formidable power with control over the energy shipping routes on the Gulf, the Indian Ocean and the South China Sea, which are all vital to China's energy security.[36] In the recent years, the United Sates has been implementing its strategy of 'pivot to Asia' or 'rebalancing the Asia–Pacific', while other powers like India, Japan, etc., have been increasing their intervene into South China Sea disputes, making the Chinese government more concerned that China's over-dependence on the Strait to import energy is a significant challenge to its energy security. Once China's relations with those countries turn sour, its energy and trade lifeline may be disrupted or cut by those states.

Also, with the significant growth of the world energy trade by sea, there are more and more oil shipping accidents. These accidents are mainly caused by natural factors like varied topography, thick fog, stormy waves and artificial factors such as the congestion in some SLOCs. Due to the limit of topographical and natural factors, there are only a limited number of sea lanes in the world suitable for the passage of super oil tankers. The fast development of international trade has led to the rapid growth of the number of oil tankers and other ships, which has caused chaotic traffic conditions in some key SLOCs, causing more uncertainties to maritime shipping security.[37]

---

[35] F William Engdahl, 'The geopolitical stakes of 'Saffron Revolution'', Asia Times On-line, 17 October 2007, http://www.atimes.com/atimes/Southeast_Asia/IJ17Ae01.html, accessed 6 April 2008.

[36] Zhao Hong, 'China's efforts to enhance its energy security'.

[37] Zhao Hongtu, '"Maliujia kunju" yu zhongguo nengyuan anquan zai sikao' ('"Malacca dilemma" and the rethinking of China's energy security'), *Xiandai guoji guanxi* (*Contemporary International Relations*) No. 6, 2007, pp. 38–39.

In particular, the narrowest point of the Straits of Malacca — the Phillips Channel in the Singapore Strait — is only 1.5 miles wide at its narrowest point, which is a natural bottleneck with more possibilities of a potential collision, grounding or oil spill. Any of these accidents may force the Strait to be closed and seriously affect China's energy import.[38]

Besides maritime shipping, pipelines and railways play a supplementary role in China's energy imports. In particular, pipeline transportation is becoming increasingly important with more oil and gas pipelines linking China and some of its neighbouring countries such as Myanmar, Russia, Kazakhstan, Turkmenistan, etc. have been built or are under construction or consideration, as shown in the previous chapter. However, there are also obvious risks with pipeline and railway transportation. Although they could avoid some risks associated with maritime shipping such as piracy and maritime accidents, they are not immune to onshore terrorist attacks and local riots, not to mention military strikes. For example, the growing terrorist and separatist activities in Central Asia pose a direct threat to the operation of Kazakhstan–China Oil Pipeline and Central Asia–China Gas Pipelines; and the China–Myanmar Oil and Gas Pipelines pass through some Myanmar areas where political situations are complex and unstable from time to time. Part of China's oil import from Russia is transported by railway. But compared with maritime shipping or pipeline transportation, the capacity of railway transportation is rather limited.

5.2.1.4 *Beijing's perception of the international energy containment against China*

Fourthly, Beijing has its own particular energy security concern. China's search for overseas energy supplies has met with great difficulties from outside forces, especially in the first a few years of this

[38] Bo Kong, 'An Anatomy of China's Energy Insecurity and Its Strategies', Pacific Northwest National Laboratory, December 2005, http://pnwcgs.pnl.gov/Newsletter/OtherDocs/AnatChinaEnergy.pdf, accessed 16 May 2008.

century. Such containment consists of three elements — strategic, business and ideological. The strategic containment worsens China's international environment and restricts its access to some overseas energy resources; the business containment directly interrupts Chinese NOCs' overseas mergers and acquisitions (M&A) activities; while the ideological containment not only damages China's international image but also provides moral justification for strategic and business containment.

The first is the strategic containment. Some countries such as the United States and Japan view China's overseas search for energy as a threat to their national interests, and have counteracted this movement strategically. For example, as a Chinese scholar pointed out, China's overseas energy investment encountered 'malicious competition' from Japan.[39] Also, some have noted that the risks in China's surrounding areas are growing. U.S. military presence in several Central Asian states has been considered as a risk to the energy cooperation between China and Kazakhstan. The arms race in the Asia–Pacific has accelerated and intensified regional geopolitical instability. Regional states like Japan, South Korea, Vietnam, the Philippines and India have quickly increased their military spending and attached great importance to international oil resources competition, which may cause threat to China's oil import shipping and overseas oil assets. Moreover, Islamic extremism, separatism and terrorism have gathered in Central Asia, forming a potential threat to China's energy import from that region.[40]

The second is the business containment. As some authors have noted, Chinese NOCs face competition for scarce oil and natural gas

---

[39] Chen Fengying, 'The strategic thinking about China's energy security', pp. 336–339.

[40] Zhao Hongtu, 'Guojia nengyuan anquan' ('National energy security'), in Zhongguo xiandai guoji guanxi yanjiuyuan (China Institute of Contemporary International Relations), Guojia Jingji Anquan (National Economic Security) (Beijing: Shishi chubanshe, July 2005), pp. 257–281; and Pablo Bustelo, 'China and the geopolitics of oil in the Asian Pacific region', Working Paper, 38/2005, Real Instituto Elcano (Elcano Royal Institute), http://129.3.20.41/eps/othr/papers/0511/0511005.pdf, accessed 24 June 2008, p. 19.

resources both at home and abroad.[41] Compared with their Western counterparts, Chinese NOCs are at a disadvantage in the global business competition. Since over 80 per cent of the world's high quality oil and natural gas resources have been controlled by Western oil corporations like Exxon Mobil, British Petroleum and Royal Dutch/ Shell, the oil fields Chinese NOCs can enter are often those that have already been developed and discarded by Western oil companies, those located in regions with high risk and continuous conflicts, or new oil fields. Even in these areas, Chinese NOCs' operation is disturbed by some Western oil companies.[42]

Also, Chinese NOCs' financial, technological and managerial ability lags behind their Western counterparts. As some have noted, Chinese NOCs are latecomers in the world oil arena, and are competing with major IOCs from the United States and Europe which over several decades have developed long-term relationships, exploration and production expertise, and strong investment positions in all of the most accessible and promising oil rich states offering investment opportunities.[43] Chinese NOCs are trying to catch up, but are facing fewer opportunities and more obstacles. Their transnational operations have encountered business and political obstacles.[44]

For instance, in May 2003, Royal Dutch Shell and five other companies used their shareholders preferential purchasing rights to block the joint offer of CNOOC and Sinopec to purchase shares of an oil field in the Caspian Sea owned by Kazakhstan.[45] Also, the failed

---

[41] Steven W. Lewis, 'Chinese NOCs and World Energy Markets: CNPC, Sinopec and CNOOC', James A. Baker III Institute for Public Policy of Rice University, 2007, http://www.rice.edu/energy/publications/docs/NOCs/Papers/NOC_CNOOC_Lewis.pdf, accessed 13 March 2009, p. 2.

[42] Chen Fengying, 'The strategic thinking about China's energy security', pp. 339–340.

[43] Lieberthal, Kenneth and Mikkal Herberg, 'China's search for energy security: implications for U.S. policy', NBR Analysis, 17: 1, the National Bureau of Asian Research, April 2006, http://www.nbr.org/publications/analysis/pdf/vol17no1.pdf, accessed 20 June 2008.

[44] Zha Daojiong, 'China's energy security: Domestic and international issues', *Survival* 48: 1, 2006, p. 182.

[45] Joseph Y. S. Cheng, 'A Chinese view of China's Energy Security', *Journal of Contemporary China* 17: 55, 2008, pp. 297–317.

CNOOC–Unocal deal in 2005 is a typical example. On the excuse of safeguarding national security, the United States House of Representatives effectively prevented CNOOC that made an unsolicited cash bid of USD 18.5 billion to purchase the American company Unocal, paving the way for Chevron to acquire Unocal for USD 17.7 billion. This case has convinced many Chinese that 'the U.S. — despite rhetoric to the contrary — does not always live up to the free-market rhetoric it broadcasts to the rest of the world'. Furthermore, some Chinese even suspect that 'the U.S. is committed to slowing down the pace of China's development by keeping energy prices high and limiting the role of Chinese companies in the global energy market'.[46] Indeed, these business failures were considered by many Chinese as part of Western effort to contain China. Especially, the unsuccessful bid by CNOOC to purchase Unocal has reinforced Beijing's perception that China cannot depend on the international energy market for its energy supply because Washington will not allow China reliable access to that market.[47]

Thirdly, China's energy diplomacy faces ideological containment. Some international media spreads the so-called 'China energy threat theory' or 'China oil and gas threat theory', which refers to China's worldwide search for energy as a very dangerous and negative phenomenon that could use up the world's oil resources, damage the humanitarian and democratic efforts in some areas, disrupt the current international order and harm some countries' national security. Some Western observers viewed China's energy demand as a threat to the global environment and the balance of world oil and gas geopolitics;[48] and some Western reports even refer to Chinese investment and operation in some resource rich states in Africa as 'neo-colonialism'.

---

[46] Jiang Wenran, 'Beijing's "New Thinking" on Energy Security', China Brief, Volume 6, Issue 8, 12 April 2006, the Jamestown Foundation, http://www.jamestown.org/china_brief/article.php?articleid=2373181, accessed 7 April 2008.

[47] Jeffrey A. Bader and Flynt L. Leverett, 'Oil, the Middle East and the Middle Kingdom', 16 August 2005, http://www.brookings.edu/opinions/2005/0816globalenvironment_bader.aspx, accessed 24 June 2008.

[48] Zhao Hongtu, 'National energy security', p. 276.

For example, China is criticised for carrying out energy cooperation with some energy exporting states in tense relationships with the West, such as the former Sudan, Myanmar and Iran; and China's role in the Darfur issue was even linked to the 2008 Beijing Olympic Games. After the failed Unocal bid, Chinese NOCs have made a series of high-risk energy investments elsewhere, such as in Africa, the Middle East and Latin America. 'Thus when the Chinese read Western media accounts of Beijing dealing with dictators or 'rogue states' as defined by the US, they feel especially bitter'.[49] The ideological containment against China's energy diplomacy has not only put more pressure on China, but also provided a moral justification for some people and groups to condemn and contain China's worldwide search for energy.

5.2.1.5 *The United States factor*

Among those factors causing China's energy security concern, the United States factor is called 'Beijing's biggest worry'.[50] Some Chinese scholars and ordinary people feel that the United States poses a big threat to China's energy security.[51] The United States controls all the key SLOCs in the world, and is the single power capable of launching an oil embargo against China or seriously damaging China's oil import system. Moreover, economic sanction is always an important tool of U.S. foreign policy in the post-Cold War era. Since China is somewhat vulnerable to U.S. economic pressure and is relatively lack of allies, oil sanctions could appeal to the U.S. which may

[49] Jiang Wenran, 'Beijing's "New Thinking" on Energy Security'.

[50] You Ji, 'Dealing with the Malacca Strait dilemma: China's efforts to enhance energy transportation security', EAI Background Brief, No. 329, East Asian Institute, National University of Singapore, 12 April 2007, pp. 3–4.

[51] For example, according to a research, university students in China perceived that the country has an energy crisis. They fear China will be controlled due to energy dependence and consider the United States as China's primary energy competitor. See David Zweig and Shulan Ye, 'A Crisis is Looming: China's energy challenge in the eyes of university students', *Journal of Contemporary China* 17: 55, 2008, pp. 273–296.

apply this measure to punish China 'on a variety of issues ranging from human rights abuses to arms sales.'[52] In a broader sense, as Avery Goldstein wrote, the United States is China's principal strategic problem. 'Washington has already tapped the world's premier economy to deploy military force of a size and type that give it unparalleled power projection capabilities throughout East Asia.' U.S. military power not only threatens China's core security interests, but offsets its ability to exert influence over continental and maritime developments in the region beyond its borders.[53]

Indeed, Beijing has good reasons to be alert against the United States which may threaten China's energy security. China has always been in a passive position in the bilateral relationship; the United States has enhanced its hedging policy towards China and accelerated implementing its strategy of 'rebalancing the Asia–Pacific'; and the two countries are divergent over some strategic interests. Although interdependence between China and the United States is deepening, the concern over a potential China–U.S. conflict has been expressed by some people in both countries. The two powers may have common interests in bilateral energy cooperation; but obstacles for such cooperation are also obvious. The following paragraphs explain this in detail.

To begin with, China has always been in a passive position in China–U.S. relations. Although the interaction between the two nations against the background of economic globalisation and regionalisation is extremely complex and beyond the reach of this research, China's passivity in the bilateral relations can be easily reflected by a few facts: The United States is the world's only superpower, owning the world's hard currency — the U.S. dollar, unmatchable military strength, more benefits from international regulations and institutions, the leading scientific and technological capacity, and the moral high ground as a mature democracy. But China possesses none of the above advantages, although its economy and national strength have been growing fast. In accordance, it is often the United States that

---

[52] Erica S. Downs, *China's Quest for Energy Security* (Santa Monica, CA: Rand Corporation, 2000), p. 45.

[53] Avery Goldstein, *Rising to the Challenge: China's Grand Strategy and International Security* (Stanford, California: Stanford University Press, 2005), p. 206.

takes the initiative in China—U.S. relations. Since the end of the Cold War, China's policy towards the United States has been reactive. To a great extent, the changes of China's U.S. policy depend on the changes of the United States' China policy.[54] This situation is likely to continue in the foreseeable future.

Also, the United States has reinforced its hedging policy towards China since the end of the Cold War,[55] and has accelerated its containment strategy against China under the guise of the global anti-terrorism campaign since the beginning of this century.[56] The United States has shifted its strategic focus from Europe to the Asia–Pacific, and enhanced its military deployment in East Asia. It has strengthened its security alliance with Japan, Taiwan and Australia, establishing a 'mini NATO in Asia,' which is viewed by the Chinese as a military coalition for the purpose of containing China.[57] The Chinese are especially

---

[54] Yongnian Zheng, *Discovering Chinese Nationalism in China: Modernization, Identity, and International Relations* (Cambridge: Cambridge University Press, 1999), pp. 126–131.

[55] During the Cold War, cooperation between Beijing and Washington was based on the single common strategic interest of counterbalancing their common enemy — the Soviet Union. Moreover, China's reform and opening up policy were mainly in accordance with U.S. economic interests at that time. However, with the end of the Cold War and the collapse of the Soviet Union, such common strategic interest suddenly disappeared. In addition, the balance in China–U.S. trade began to change dramatically: From a U.S. surplus to a U.S. deficit. Many Americans began to view China as another mercantilist Asian country that was only out to hurt American economic interests. Yongnian Zheng, *Discovering Chinese Nationalism in China*, p. 127.

[56] Some argued that the Bush administration's grand strategy was not aimed primarily at anti-global terrorism, the incapacitation of rogue states, or the spread of democracy in the Middle East. These issues just dominated the rhetorical arena and were the focus of immediate concern; but they were not the reasons for key decisions regarding the long-term allocation of military resources. 'The truly commanding objective — the underlying basis for budgets and troop deployments — is the containment of China.' Michael T Klare, 'Containing China: The U.S.'s real objective,' Asia Times On-line, 20 April 2006, http://www.atimes.com/atimes/China/HD20Ad01.html, accessed 30 December 2007.

[57] Zheng Yongnian, 'Zhongguo dang lixing huiying yazhouban "beiyue",' ('China should appropriately react to the Asian "NATO"'), Lianhe zaobao wang (United Morning Post On-line), 27 March 2007, http://www.zaobao.com/special/forum/pages5/forum_zp070327.html, accessed 30 December 2007.

concerned on U.S. strategic intention towards China when Washington recently increased its military deployment and action around the Straits of Malacca. Another example is that only a few years ago, the Chinese still considered Central Asia as a region where energy supply to China was less vulnerable to U.S. influence. Some Chinese analysts held that although the United States was a competitor for energy resources there, its capacity to threaten China's oil import from the region was limited.[58] Later, however, the United States significantly promoted its military deployment in Central Asia in the name of anti-terrorism after the September 11 attacks, China's energy interests in the region is also directly within U.S. strategic reach. Besides, Washington has been carrying out the so-called transformational diplomacy. It means toppling some current regimes and establishing pro-American regimes through non-military means, where domestic opposition forces and non-governmental organisations play vital roles. The series of 'colour revolutions' that happened in the recent years across many regions in the world are good examples. This strategy is also regarded as targeting some non-democratic countries including China.[59] The Chinese leadership is nervous about this phenomenon.

With specific regard to the global competition for oil resources between China and the United States, some authors argued that Chinese NOCs' aggressive pursuit of oil supply around the world has created tension between Beijing and Washington. China will probably continue to assertively access oil resources worldwide and 'it will seek to make new inroads into areas that the United States has historically viewed as falling within its sphere of interest.' Although a conflict between the two powers is far from inevitable, there is 'considerable risk that inept handling could transform competition for oil into much more serious geo-strategic conflict'.[60] Especially, both China

---

[58] Downs, *China's Quest for Energy Security*, p. 46.

[59] Zheng Yongnian, 'Meiguo zhuanxing waijiao dui zhongguo de hanyi' ('The impact of the U.S. transformational diplomacy on China'), Lianhe zaobao wang (United Morning Post On-line), 18 April 2006, http://www.zaobao.com/special/china/sino_us/pages6/sino_us060418a.html, accessed 30 December 2007.

[60] Peter Hatemi and Andrew Wedeman, 'Oil and conflict in Sino–American relations', *China Security* 3: 3, 2007, pp. 95–118.

and the United States have critical energy and other strategic interests in Africa, one of the two hot spots in the world[61] where future energy competitions or even strategic conflicts among big powers may occur, and in the recent years Chinese NOCs have been expanding their operation on the continent, posing a challenge for American and other Western oil companies. Therefore, the worry about a potential energy conflict between Beijing and Washington is not without any sense.

Moreover, the two countries have some conflicting strategic interests. Among them, the rise of China and the Taiwan issue are two critical issues. The Chinese strategic interest of pursuing economic rise and strengthening the nation is not in line with U.S. strategic interest of maintaining the U.S. hegemony. On the one hand, the rise of China is an unavoidable and ongoing phenomenon. China's long term goal is to realise the so-called revival of the Chinese nation. This is regarded by the Chinese people as a legitimate and essential interest. Although the Chinese government has been referring to the concept of a 'peaceful rise' or 'peaceful development' in order to ease international concerns about the rising China, how to accommodate or cope with China's rise is an urgent challenge to the United States. On the other hand, one of the key U.S. strategic interests is the prevention of the rise of a big power that is able to challenge its hegemony, not to mention a big power with a different ideology. Many Americans believe that the rise of China will change the *status quo* of the international system and the distribution of power and interests, posing a threat to Washington's strategic interests. Such divergence of fundamental interests between the two sides is likely to exist for a long time.

As for the Taiwan issue, the fundamental interests of China also conflict with those of the United States. From Beijing's perspective, Taiwan is an inalienable part of China; thus, the best final result of the issue is 'peaceful reunification' of the Chinese mainland and Taiwan. If this is unrealistic, the last resort is 'reunification by force.' But from

---

[61] The other hot spot for future energy competitions among big powers is the area near the North Pole, which has been proven to have abundant reserves of oil and natural gas.

Washington's perspective, Taiwan is an important U.S. ally and one of U.S. strategic strongholds in the West Pacific. This island is viewed by many Chinese as used by the United States to check the rise of China. In this sense, the best result for Washington is Taiwan's peaceful independence. Although China and the United States have reached an essential temporary compromise in maintaining the *status quo* across the Taiwan Straits, that is a *de facto* but not *de jure* independent Taiwan, this is merely makeshift and the two powers' ultimate goals conflict with each other. Hence, Beijing might tend to perceive that whenever Taiwan takes provocative actions to gradually move towards *de jure* independence, Washington merely pays lip service and condemns the behaviour, of course under pressure from Beijing, but actually connives with the Taiwan authorities to forge ahead towards the island's *de jure* independence. Also, since the United States still has a flexible defence commitment with Taiwan according to the *Taiwan Relations Act*, while China's flexible *Anti-Secession Law* determines that Beijing would resort to force to settle the issue if Taiwan pursues *de jure* independence, the potential for a China–U.S. conflict across the Taiwan Straits still exists.[62] Thomas J. Christensen even argued that the strategic history of the PRC provides 'cautionary warnings about the potential for future conflict across the Taiwan Strait and, by association, across the Pacific' sometime in the first two decades of the 21st century.[63]

The most important counteractive factor against a conflict between China and the United States is the interdependence between

[62] Furthermore, Beijing may tend to believe that the United States and Taiwan are on the same side while the Chinese mainland is on the other. No matter how Taipei provokes Beijing, it is still under the protection of Washington; and the differences between Washington and Taipei are merely disagreements within a coalition. But the differences between Washington and Beijing are often conflicts over strategic interests. Finally, since the United States is the only power in the world that might interrupt mainland China's efforts to reunify Taiwan, it is natural for Beijing to be concerned about U.S. role in the Taiwan issue.

[63] Thomas J. Christensen, 'Windows and War: Trend Analysis and Beijing's Use of Force', in Alastair Iain Johnston and Robert S. Ross ed., *New Directions in the Study of China's Foreign Policy* (Stanford, California: Stanford University Press, 2006), pp. 50–85.

the two countries.[64] However, interdependence does not eliminate the possibility of conflict.[65] Specifically, in the Asia–Pacific today, although China and the United States are highly interdependent economically, in front of the geopolitical competition and struggle for core national interests, such high-degree of economic interdependence does not necessarily guarantee the maintenance of peace. Although the current China–U.S. relationship is relatively stable and optimistic, the possibility of conflict between the two powers still exists. Therefore, for those Chinese analysts who take into consideration of the worst situation, if China–U.S. relations deteriorate to a certain degree, it is not impossible that disrupting China's oil imports could be on Washington's agenda; or the United States may just use

---

[64]The interdependence between China and the United States has been deepening in the recent years. Economically, the two nations are so highly interdependent that the collapse of either one would definitely affect the other. Politically, the two countries need the support from each other to address many international issues. It seems that neither of the two countries would benefit from a China–U.S. conflict. Specifically, if the United States interrupts China's oil import and harm the Chinese economy, the U.S. economy would also be hurt. Hence, some people hold that a U.S. oil embargo against China is rather unlikely.

[65]Interdependence does not definitely prevent conflicts. History shows that interdependence does not always work. Prior to the First World War, some scholars and politicians in Europe believed that the increasing interdependence among European countries, such as trade and economic links, would prevent a war from breaking out among these states. But their optimism has proven unrealistic. Now, there is a similar discourse advocating that China–U.S. interdependence can prevent conflict between the two nations. Whether it is realistic still remains to be seen. Admittedly, many things have changed over the last century. The economic globalisation and regionalisation have been developing fast. Economic interdependence among various countries has been deepening. Especially, the European Union has moved ahead of the rest of the world regarding regional integration, through the establishment of a series of institutions and regulations, which has largely made a future military conflict among its member states unlikely. However, in the Asia–Pacific region, there is no mature and stable institution coordinating different interests among regional states. On the contrary, due to various historical and current reasons, insufficient mutual trust as well as conflicting strategic interests among some regional countries make the building of an effective coordination institution very difficult. Therefore, at least in the Asia–Pacific, the argument that deepening interdependence can prevent conflicts seems a bit too optimistic.

this as a bargaining chip in addressing the bilateral relations. Actually, as long as China lacks an effective energy security emergency institution, it has to continue living under the Sword of Damocles of energy import interruption.

Apart from China–U.S. interdependence, another factor that might ease the tension between the countries regarding energy issues is the bilateral energy cooperation. But there are apparent obstacles to such cooperation. Although the two countries have common interests in reducing China's energy consumption, their policies regarding China's role in the international energy market are conflicting. What the United States wanted was China not only reduces energy consumption but also shrinks its overseas energy investment; while what China needed was its worldwide energy search continues and the United States transfers advanced energy conservation technologies to China.[66] Also, Gaye Christoffersen pointed out that although there has been more than three decades of energy cooperation between China and the United States, it is only recently that there has been a concerted U.S. effort to create a framework for bilateral energy relations that would build capacity in China and nest the bilateral relationship in multilateral regimes.[67]

Also, the two sides have different interests in some specific issues of energy cooperation, such as the transformation of up-to-date

---

[66] It seems that Washington and Beijing have conflicting attitudes towards China's energy diplomacy. Washington's policy is that, on the supply side, it tries to resist China's worldwide search for energy which is viewed as adding to the competition for energy resources as well as political influence with the United States; as U.S. Congress and the U.S. government have been using their legislative and administrative resources to work with U.S. oil companies to interrupt Chinese NOCs' overseas investment and acquisition, as indicated by U.S. Congress' blockade of the CNOOC–Unocal deal and the U.S. government's intervention in Sinopec's energy deal with Iran; and on the demand side, the United States urges China to improve its energy efficiency and reduce its energy consumption. However, Beijing views energy diplomacy as legitimate behaviour, and has been calling for the United States to transfer advanced energy conservation technologies to China to promote China's energy efficiency.

[67] Gaye Christoffersen, 'U.S.–China energy relations and energy institution building in the Asia–Pacific', *Journal of Contemporary China* 19: 67, 2010, pp. 871–889.

energy conservation technologies.[68] As one scholar noted, the United States and China face serious common challenges, the energy security of both can be enhanced through cooperation, and extensive bilateral energy dialogues exist at both policy and working levels. But 'sovereignty issues remain, and both sides continue their desire for energy independence.'[69] Similarly, according to another author, although there is far more commonality in the energy requirements of both China and the United States than is often acknowledged, 'this overlap is often obscured by domestic political agendas, corporate and bureaucratic interests, and the increasing tendency to view energy as a defence planning issue, in particular with reference to future Chinese

---

[68] Energy cooperation between China and the United States is a complex issue. Some people, especially some Chinese scholars and officials, have argued for a long time that the United States can transfer advanced energy conservation technologies to China. But apart from those official talks to push bilateral energy cooperation, few concrete deals or projects have been reached and implemented. It seems that the Chinese proposal of energy technological cooperation has not been matched with effective responses from the U.S. Perhaps there are several reasons for the reluctance of the United States. For example, the U.S. side especially those U.S. corporations possessing advanced energy conservation technologies may not be very active in transferring the technologies to their Chinese counterparts. After all, the energy technological cooperation between the two countries has to be implemented by those companies. However, although the Chinese government and some Chinese companies have similar interests in getting those technologies from the United States, the U.S. government and some U.S. firms do not necessarily have similar interests in that. Even if the U.S. government reaches an agreement with the Chinese government to transfer the technologies, whether the agreement can be implemented depends on the attitude of those U.S. companies possessing the technologies. But these companies may have their reasons not to cooperate. They could be concerned about China's practice of intellectual property rights, and might be afraid that once they export such technologies to China, some Chinese companies would copy their technologies; or they just unwilling to sell the technologies to China unless they can get a windfall profit from the deals. In addition, taking into consideration of the high economic cost, even if those U.S. firms are willing to sell the technologies to China, whether the Chinese side can afford to purchase them and widely apply them in China's industrial sectors is another question. As long as these obstacles exist, the proposed China–U.S. energy cooperation is unlikely to gain significant achievements soon.

[69] June Teufel Dreyer, 'Sino–American Energy Cooperation', *Journal of Contemporary China* 16: 52, 2008, pp. 461–476.

and U.S. maritime strategies.' The conditions to promote China–U.S. energy cooperation exist, and some important initial steps have been undertaken toward this end. However, sustained and more active governmental and institutional interactions are needed to prevent the potential antagonism in the longer-term energy futures of both countries.[70] As a result, although the two governments have expressed their willingness for bilateral energy cooperation, such expression is still on paper to some extent and needs matching with concrete actions.

In the recent years, China–U.S. relations have entered a more challenging phase with Washington strengthening its presence in the Asia–Pacific on several fronts. On the political front, the United States joined the East Asian Summit in 2011 as a full member and raised the sensitive issue of the South China Sea disputes despite Beijing's objections. Washington also approached Myanmar and further eased sanctions on the country. On the economic front, the United States is pushing for a Trans-Pacific Partnership Agreement that goes beyond the traditional trade and investment items to include broader issues like labour standards, intellectual property rights and environmental protection. Beijing regards it as an attempt to change the rules of the game by raising the development bar for China. On the security front, the United States has enhanced its military alliance with Australia and the Philippines, and approached Vietnam. It is clear that the U.S. pivot to the Asia–Pacific is viewed by some in China as a move to contain the rise of China. So far, China has not over-reacted and instead stressed the importance of continued cooperation with the United States.[71]

---

[70] Jonathan D. Pollack, 'Energy Insecurity with Chinese and American Characteristics: Implications for Sino–American relations', *Journal of Contemporary China* 17: 55, 2008, pp. 229–245.

[71] In fact, a fundamental issue affecting the dynamics of China–U.S. relations is a China that is on the rise and a United States that is in relative decline. During this process of adjustment, the United States is likely to be more sensitive to perceived moves on the part of the rising China to challenge it. Fortunately, Beijing has been cautious on this front. It still recognises the United States' leading role on the world stage and its military prowess. This awareness on China's part is important as it provides some assurance and time for the United States to find its niche in a reinvigorated role. Lye Liang Fook, 'China–U.S. relations: Coping with a U.S. pivot to the Asia–Pacific region', in Wang Gungwu and Zheng Yongnian eds., *China: Development and Governance* (Singapore: World Scientific Publishing Co. Pte. Ltd., 2013), pp. 399–404.

5.2.1.6 *The contribution of energy diplomacy to China's energy security*

Energy diplomacy improves China's energy security conditions. Admittedly, energy diplomacy itself is unable to fundamentally eliminate the external threats listed above. But it contributes to China's energy security in the following three ways:

First, energy diplomacy contributes to the diversification of China's overseas energy suppliers. It helps China secure energy imports from not only the Middle East and Southeast Asia which are traditionally important regions exporting oil to China, but also Africa, Australia, Central Asia, Russia and Latin America which are relatively new and increasingly important energy suppliers for China. The energy import diversification drive is expected to reduce the proportion of oil from the Middle East in China's overall oil import by raising that from other regions including Africa, Central Asia, Russia and Latin America; and alleviates the over-reliance of China's energy import transportation on the Straits of Malacca, through which oil from the Middle East and Africa comes, by importing energy from Central Asia, Russia, Australia, Southeast Asia and Latin America. Hence, if one or two energy exporting regions suspend their oil exports to China for any reason, the other regions may still export oil to China; if the Straits of Malacca is blocked, energy imports through other shipping lanes and pipelines and railways may still reach China. Also, the diversification of China's energy imports makes the potential placement of a comprehensive energy embargo against China more and more difficult and unlikely, as it is perhaps impossible to blockade all of China's oil import routes.

Second, Chinese NOCs' production of equity oil abroad increases the world's total oil output as well as the oil supply to China. Chinese NOCs can develop some oil fields located in countries to which Western oil companies usually do not have access due to political reasons such as sanctions, unfavourable natural conditions, or have been abandoned by Western oil companies. Thus, the transnational operations of Chinese NOCs increase the overall amount of oil supply in the world. As Downs wrote, Chinese NOCs are actually increasing the amount of oil available to other oil importing countries through their operations overseas, especially through their development of oil

fields that other oil companies are unable or unwilling to invest in.[72] The growth of the world's total oil output is beneficial for all oil importers including China, relieving their worries about the shortage of overseas oil supply.

Although Chinese NOCs do not send all of their overseas production of equity oil to China, their operation is helpful for global oil supply as well as China's oil import security.[73] Notably, not all of the equity oil produced by Chinese NOCs abroad is shipped back to China. They actually sell most of their equity oil produced abroad in the international market instead of sending it home, due to various economic reasons such as selling the oil for better prices on the world market, the oil may be more economically costly to be shipped back to China due to the long distance, that type of oil is not suitable to be refined in China, etc.

As a report published by the International Energy Agency (IEA) noted, Chinese NOCs' overseas energy production does not directly and exclusively translate into energy supplies flowing to China. Decisions about the marketing of Chinese equity oil produced abroad are mainly based on commercial considerations such as market prices, and in some cases carried out by marketing subsidiaries located outside the headquarters of Chinese NOCs. For instance, almost all the Chinese equity production of oil in the Americas was sold locally instead of being shipped to China. Considering the geographical distances, it is more costly to ship that oil to China by sea. Also, not all of CNPC's equity oil produced in Kazakhstan is transported to China via Kazakhstan–China Oil Pipeline. In some cases, selling the oil to other players can be more profitable for Chinese NOCs. Of course, if there were a shortage of oil supply in China's domestic market due to Chinese equity oil being sold outside China, Chinese NOCs would

---

[72] Erica S. Downs, 'The Fact and Fiction of Sino–African Energy Relations', *China Security* 3: 3, 2007, p. 47.

[73] Some people tend to argue that Chinese NOCs' overseas investment and transnational operation does not contribute to China's oil import security, because they did not shipped all of their overseas production of equity oil back to China. But this opinion is not comprehensive enough.

purchase more oil from the international market, such as from Middle Eastern exporters, to fill the gap.[74]

But this does not indicate that the NOCs' production of equity oil is not helpful for China's oil imports. The reason is that the globalised world oil market means that the more oil one country purchases from a certain source, the less it needs to purchase from other sources in the world. In accordance, Chinese NOCs' selling of their equity oil at the open market to a third country may reduce this country's need to purchase oil from some sources that are also supplying oil to China. For example, if a Chinese NOC sells part of its equity oil produced in Latin America to a third country, say Japan, such deal would in theory relieve Japan's need to import oil from other regions in the world like the Middle East and Africa; and then China can import more oil from these regions. In that case, it just seems that China and Japan swap oil with each other in the world market. Just as the Chinese official white paper mentioned, 'Ninety per cent of Chinese enterprise-invested energy resources abroad are sold locally, thus increasing and diversifying supplies in the global energy market'.[75] A report by the IEA also maintained that Chinese NOCs' investment in global upstream assets has and will continue to contribute significantly to expanding global oil and gas supplies.[76]

Third, energy diplomacy helps stabilise prices for China's oil products domestically, which is important to maintain rapid economic development and social stability. China's production of equity oil overseas and the conclusion of long-term oil supply contracts with some energy exporting countries are expected to provide part of its oil import with relatively predictable and stable prices, alleviating China's over-reliance on the international oil market, where the prices

---

[74] Jiang and Sinton, *Overseas Investments by Chinese National Oil Companies: Assessing the Drivers and Impacts.*

[75] The State Council Information Office, the People's Republic of China, *China's Energy Policy 2012*, October 2012, http://english.gov.cn/archive/white_paper/2014/09/09/content_281474986284499.htm, accessed 14 January 2015.

[76] Jiang and Ding, *Updates on Overseas Investments by China's National Oil Companies: Achievements and Challenges since 2011.*

often fluctuate and are largely determined by the United States or OPEC.

5.2.2 *Promoting Beijing's multi-lateral diplomacy*

Apart from improving China's energy security, energy diplomacy enhances Beijing's ability to carry out multilateral diplomacy and secure more international support. Since China is hungry for overseas energy supply while many developing countries are abundant in energy resources, China's energy diplomacy and increasing energy trade and investment in these countries provide Beijing with a new approach through which it can more effectively engage them and win their political and diplomatic support. As Aaron Friedberg noted, China's need to seek natural resources worldwide and increasing intention and ability not only to buy them but also to invest in their extraction, 'is giving Beijing new instruments with which to exert influence.' These tools are especially useful for Beijing when it deals with countries which are heavily dependent on natural resource exports, such as some developing countries in Asia, Africa and Latin America.[77]

In the recent years, many developing countries have frequently played a prominent role in frustrating efforts by some Western countries to bring about a formal condemnation on China's human rights record in the United Nations Commission on Human Rights; and the votes from many developing countries helped two Chinese cities, Beijing and Shanghai, win their bids to host the 2008 Olympic Games and the 2010 World Expo respectively. Moreover, the political support from some developing countries is critical for Beijing to tackle the Taiwan issue that is directly related to the CCP's political legitimacy.

The Taiwan issue is of vital importance to the CCP. The diplomatic backing from other countries is critical for the Chinese

[77] Aaron L. Friedberg, '"Going out": China's pursuit of natural resources and implications for the PRC's grand strategy', NBR Analysis, 17: 3, the National Bureau of Asian Research, September 2006, http://www.nbr.org/publications/analysis/pdf/vol17no3.pdf, accessed 20 June 2008.

government's strategy to ease the domestic pressure and address the issue peacefully. Beijing has never given up the possibility of settling this problem by force, as expressed in the *Anti-Secession Law*. The situation across the Taiwan Strait was unstable from the mid-1990s to 2000s, marked by five events — the Taiwan Strait crisis in 1996, Taiwan's pro-independence Democratic Progressive Party's (DPP) rise to power in 2000, the re-election of the DPP in 2004, the DPP authorities' abandonment of the *Guidelines for National Unification* in 2006, and the DPP authorities' unsuccessful campaign to join the United Nations (UN) under the name of Taiwan in 2007. Especially, during 2000 to 2008 when Taiwan was ruled by the DPP, Taipei often provoked Beijing and embarrassed Washington, and was viewed as a 'trouble maker'. Therefore, hawkish voices and radical nationalism rose in mainland China and many people believed that a military solution may be the only way to resolve the Taiwan issue and that the hope of 'peaceful reunification' is unrealistic. Then, the relations across the Taiwan Strait have been stable since the current Kuomintang (KMT) authorities came into power in Taiwan in 2008. But at present, considering the KMT's landslide defeat to the DPP in the 'Nine in One' election in late 2014, the possibility of the pro-independence DPP's returning to power in 2016 looms large, adding to the uncertainties of the cross-strait relations in the near future.

The Chinese government usually faces mounting domestic pressure when the cross-strait relations are unstable. The Taiwan issue is critical for their credibility among the Chinese people and the political legitimacy of the CCP. But Beijing does not want to be involved in a military conflict across the Taiwan Strait, which would result in U.S. and Japanese intervention and affect China's economic development. In the past, what it tried to do was to show the people both inside and outside China that they were able to settle this issue gradually through peaceful measures, while at the same time building its military deterrence and preparing for a potential showdown across the Strait. One of their strategies was to reduce Taiwan's international scope of activity and ensure that Taiwan was not able to establish *de jure* independence because of insufficient international support, as reflected by the small number of countries maintaining diplomatic ties

with Taipei instead of Beijing. Therefore, decreasing the number of states recognising Taiwan and impeding Taiwan's efforts to join international organisations such as the UN and the WHO, both intergovernmental bodies composed of sovereign states, were important indicators of the progress of the CCP's management of the Taiwan issue. Since almost all of the countries having diplomatic links with Taiwan are developing countries, the Third World was a critical diplomatic battlefield between Beijing and Taipei.

Energy diplomacy helps strengthen Beijing's relationship with many developing countries, which supports its competition with Taipei for diplomatic recognition.[78] In the recent years, Beijing has obviously got the upper hand. For example, mainland China progressed boldly in Africa. In the past, a number of African countries recognised Taiwan rather than the mainland. The competition for diplomatic recognition between the two sides across the continent is well-known. Beijing's energy diplomacy and rising economic and political influence there has assisted its competition with Taipei. In the 2000s, Liberia switched its diplomatic recognition to Beijing in 2003, Senegal in 2005, Chad in 2006 and Malawi in 2007. Now, among the 54 African countries, only four — Burkina Faso, Gambia, São Tomé & Príncipe and Swaziland — still maintain official links with Taiwan. The Beijing–Taipei diplomatic war was also fierce in Latin America and the Caribbean. Currently, 12 out of the 33 countries in the region have diplomatic ties with Taipei (see Table 5.1), making up more than half of all the countries that recognise Taiwan.[79] Besides, many developing countries have supported mainland China to block Taiwan's efforts to join the UN and the WHO. For instance, the 60th World Health Assembly rejected Taiwan's application to

[78] Some people in Taiwan also viewed energy diplomacy as a useful approach to fight the diplomatic war with Beijing, and argued that energy diplomacy can offer diplomatic interests, and Taiwan's increasing contact with oil-producing countries may be its 'best bet' in the diplomatic fight with Beijing. See Taipei Times, "Energy diplomacy' could offer diplomatic relief', 26 March 2006, http://www.taipeitimes.com/News/taiwan/archives/2006/03/26/2003299279, accessed 7 April 2008.

[79] There are altogether 23 countries that recognise Taipei, four in Africa, one in Europe, 12 in Latin America and the Caribbean, and six in the Asia–Pacific.

Table 5.1: Diplomatic recognition between Beijing and Taipei in Latin America and the Caribbean

- ★ Antigua and Barbuda
- ★ Argentina
- ★ Bahamas
- ★ Barbados
- ◆ Belize
- ★ Bolivia
- ★ Brazil
- ★ Chile
- ★ Colombia
- ★ Costa Rica (Switched to Beijing in 2007)
- ★ Cuba
- ★ Dominica (Switched to Beijing in 2004)
- ◆ Dominican Republic
- ★ Ecuador
- ◆ El Salvador
- ★ Grenada (Switched to Beijing in 2005)
- ◆ Guatemala
- ★ Guyana
- ◆ Haiti
- ◆ Honduras
- ★ Jamaica
- ★ Mexico
- ◆ Nicaragua
- ◆ Panama
- ◆ Paraguay
- ★ Peru
- ◆ Saint Kitts and Nevis
- ◆ Saint Lucia (Switched to Taipei in 2007)
- ◆ Saint Vincent and the Grenadines
- ★ Suriname
- ★ Trinidad and Tobago
- ★ Uruguay
- ★ Venezuela

★ Countries having diplomatic relations with the People's Republic of China (PRC)
◆ Countries having diplomatic relations with the Taiwan authorities

become a member of the WHO, which is the 11th time in a row that the Assembly has turned down such a proposal.

In particular, China's energy diplomacy and energy cooperation with Africa is an important element of China's engagement with the continent. It provides Beijing with a new platform to reinforce the PRC relationship with African states and secure their diplomatic support in multilateral institutions. Africa is always critical for Beijing's multilateral diplomacy. The continent is the largest single regional grouping of states with the tendency towards 'bloc voting' in multilateral institutions such as the UN and its agencies. African governments have proven to be a reliable source of political support whenever China's behaviour is criticised.[80] He Wenping wrote that international relations are a kind of game, and Africa is an important player instead

[80] Chris Alden, *China in Africa* (London and New York: Zed Books, 2007), p. 22.

of a pawn. It is a key force for developing countries. African countries account for about half of the non-aligned countries and one third of UN member states, demonstrating themselves as 'reliable supporters of China's position in opposing hegemony and power politics.'[81]

For example, the year 2006 was also the 35th anniversary of the UN resolution 2758 which resumed the rights of the PRC in the UN. The support of African states was vital in this. Of the 76 votes for the resolution, 26 were from Africa.[82] The continent is also a critical battlefield for diplomatic recognition between Beijing and Taipei. Competition between mainland China and Taiwan for diplomatic recognition has been intense for years. In 1956 when the PRC first established diplomatic relationships in Africa, many countries in the continent still recognised the KMT regime in Taiwan instead of the communist authorities in Beijing as the government of China. Hence, the CCP government had to promote its legitimacy in Africa.[83] Beijing's decades of engagement with Africa has gradually brought it diplomatic dominance over Taipei across the continent. In particular, Beijing has made significant progress in the recent years. Now among the 53 African countries, 49 have diplomatic ties with the Chinese mainland, compared with only four for Taiwan. Moreover, during the past decade, African states that hold 15 of the 53 seats at the UN Commission on Human Rights have played a prominent role in frustrating efforts by some Western countries to bring about a formal condemnation of China's human rights record in the commission 11 times. Beijing could not have done this without the diplomatic support from African nations.[84] Further, 16 African states have recognised China's full status

[81] He Wenping, 'The Balancing Act of China's Africa Policy', *China Security* 3: 3, 2007, pp. 23–40.

[82] People's Daily Online, 'Roundup: Africa, China usher in 50th anniversary of sincere cooperation', 31 October 2006, http://english.peopledaily.com.cn/200610/31/eng20061031_316924.html, accessed 18 January 2007.

[83] Jianjun Tu, 'China's New National Energy Commission and Energy Policy', China Brief, 8: 7, March 2008, the Jamestown Foundation, http://www.jamestown.org/programs/chinabrief/single/?tx_ttnews%5Btt_news%5D=4820&tx_ttnews%5BbackPid%5D=168&no_cache=1, accessed 13 March 2009.

[84] Li Anshan, 'China and Africa: Policy and Challenges', *China Security* 3: 3, 2007, pp. 69–93.

as a market economy, a crucial marker after China's entry into the World Trade Organisation (WTO), helping it to avoid accusations of dumping. Seven of these countries announced their recognition during the Beijing summit of the Forum on China–African Cooperation (FOCAC), bringing the total number of countries supporting China's market economy status to 64.[85] Besides, China and African nations have united in fighting for 'fair and equitable international economic trade rules' such as in the WTO's negotiation over agricultural issues.[86]

Two recent examples of the essential diplomatic backup from Africa for Beijing's multilateral diplomacy are the dispute over the expansion of the UN Security Council in 2005 and the election of a new Director-General of the WHO in 2006. To a large extent, both cases indicated China's securing of African support to impede Japan's diplomatic moves. The chilly political relationship between Beijing and Tokyo is accompanied by fierce diplomatic competition between them. The following paragraphs discuss these two examples in details.

The proposed UN Security Council expansion was part of a broader UN reform plan spearheaded by the former UN Secretary General Kofi Annan.[87] Japan is one of the countries that are likely to become a permanent member of the Security Council if such an enlargement plan were implemented. However, historical and territorial disputes intertwined with current rivalries make Beijing unwilling to see Tokyo get a permanent seat in the Security Council. Although as a permanent member of the Council, China could block Japan's proposal at the second stage, it has successfully done so much earlier,

---

[85] Xinhua, 'Shangwubu: yiyou 64 ge guojia chengren zhognguo wanquan shichang jingji diwei' ('Ministry of Commerce: There have been 64 countries recognising China's full status as a market economy'), 7 November 2006, http://news.xinhuanet.com/fortune/2006-11/07/content_5302214.htm, accessed on 18 January 2007.

[86] He Wenping, 'The Balancing Act of China's Africa Policy', p. 27.

[87] According to Article 108 of the UN charter, the Security Council expansion procedure is as follows: At the first stage, the General Assembly should vote in favour of a proposal by a two-thirds majority; then at the second stage, the vote has to be ratified by all the permanent members of the Security Council. See BBC, 'Battle joined on Security Council reform', 12 July 2005, http://news.bbc.co.uk/2/hi/americas/4675823.stm, accessed 18 January 2007.

even before the procedure began, thanks to the cooperation of African countries.

There were three competing proposals in 2005. The first and best prepared plan was from the Group of Four (G4) established in September 2004 and was promoted by Japan, Germany, India and Brazil. They proposed adding ten new members in the Security Council with six permanent ones and four non-permanent ones. They hoped to occupy the Asian, West European and Latin American slots and leave two seats for Africa.[88] The second proposal was from a group called Uniting for Consensus, which proposed adding ten non-permanent seats. Although the G4 countries worked very hard trying to win support from Africa, the African Union (AU) voted at a summit in Ethiopia to reject the G4 plan and ratify their own alternative proposal calling for adding 11 members and proposed two permanent and two non-permanent seats for African states.[89] Without support from the AU, it is unlikely that the G4 plan could be approved with a two-thirds majority in the first round of voting by the General Assembly, not to mention proceeding to a second round. As a result, the G4 countries have to shelve their bid for now.[90]

Another case is Chinese candidate Dr Margaret Chan who was elected Director-General of the WHO in 2006, following the sudden death of her predecessor Dr Lee Jong-wook. Eleven candidates were recommended by their respective governments of whom five were selected as finalists. Among them was Margaret Chan, the former

[88] Britain and France supported the G4 plan while China and the United States opposed it.

[89] A big difference between the G4 plan and the AU plan is that the latter insisted that those new permanent members should have veto power while the former did not mention it.

[90] BBC, 'Africa to seek UN Council seats', 3 July 2005, http://news.bbc.co.uk/2/low/africa/4645545.stm, accessed 18 January 2007; BBC, 'UN debates new Security Council', 12 July 2005, http://news.bbc.co.uk/1/hi/world/americas/4673977.stm, accessed 18 January 2007; Xinhua, 'AU rejects G4 proposal on UN reform', 5 August 2005, http://news3.xinhuanet.com/english/2005-08/05/content_3311971.htm, accessed 18 January 2007; Xinhua, 'Japan to give up UNSC bid for now', 21 August 2005, http://news3.xinhuanet.com/english/2005-08/21/content_3385064.htm, accessed 18 January 2007.

health chief of China's Hong Kong Special Administrative Region. She joined the WHO in 2003 and became the agency's top official for pandemic influenza as well as the Assistant Director-General for communicable diseases. Her strongest challenger was Japanese candidate Dr Shigero Omi, head of the WHO's operations in the Western Pacific region.[91]

As an experienced senior official of the WHO, Omi was widely expected to succeed Lee as the next Director-General. But in view of Taiwan's frequent attempts to join the WHO as a sovereign state in the recent years and Japan's vague attitude towards the Taiwan issue, Beijing was concerned that the election of Omi would boost Taiwan's case among WHO members and harm China's national interests. Therefore, the election of Chan was a setback for Japan as well as Taiwan's pro-independence Democratic Progressive Party government.

The Chinese government spared no effort in securing ballots for Chan. For instance, then Chinese Vice-premier Wu Yi, Health Minister Gao Qiang and ambassador to the UN in Geneva Sha Zukang publicly supported or lobbied for Chan. The election took place soon after the Beijing summit of the FOCAC and Africa's support certainly contributed to Chan's success. Among the 34 countries on the WHO Executive Board, nine were African — Djibouti, Kenya, Lesotho, Liberia, Libya, Madagascar, Mali, Namibia and Rwanda[92] — making up more than one quarter of the board. After four rounds of secret balloting by the WHO Executive Board, Chan was nominated as the next Director-General and the nomination was formally approved by the World Health Assembly, the WHO's top decision-

---

[91] The other three candidates were Dr Julio Frenk from Mexico, Dr Kazen Behbehani from Kuwait and Ms Elena Salgado Mendez from Spain. See World Health Organisation, 'Dr Margaret Chan to be WHO's next Director-General', 9 November 2006, http://www.who.int/mediacentre/news/releases/2006/pr66/en/index.html, accessed 18 January 2007; and Xinhua, 'China's Margaret Chan elected WHO chief', 10 November 2006, http://news.xinhuanet.com/english/2006-11/10/content_5311568.htm, accessed 18 January 2007.

[92] World Health Organisation, 'Dr Margaret Chan nominated to be WHO Director-General', 8 November 2006, http://www.who.int/mediacentre/news/releases/2006/pr65/en/index.html, accessed 18 January 2007.

making body, with a two-thirds majority. Chan took office in January 2007 for a five-year term, and became the first Chinese national to head a UN specialised body.[93]

In brief, the major contribution of energy diplomacy to the CCP's political legitimacy is that it improves China's energy security and securing more international backup for Beijing's multilateral diplomacy, which supports China's economic development, social stability and national unification. In addition, some of China's international energy cooperation projects are also beneficial in this regard. China's energy cooperation with Central Asia is a typical example.[94] It is supposed to promote economic development in China's western areas such as Xinjiang, contributing to the country's social stability and national unity.[95]

---

[93] Chan had been leading in all the four rounds of balloting while Omi was eliminated in the third round. Chan finally got 24 ballots in the last round and won. Xinhua, 'China's Margaret Chan elected WHO chief'; and World Health Organisation, 'Dr Margaret Chan takes office as Director-General of the World Health Organization', 4 January 2007, http://www.who.int/mediacentre/news/releases/2007/pr01/en/index.html, accessed 17 January 2007.

[94] As some authors have noted, China's initial involvement in Central Asia was not primarily driven by energy consideration, but as a means to enhance the security of its western border that faces the threat from Islamic separatists. See Xuanli Liao, 'Central Asia and China's Energy Security', *China and Eurasia Forum Quarterly* 4: 4, 2006, pp. 61–69.

[95] To some extent, the relationship between the Uigur population and the Han population in Xinjiang has been tense in the recent years largely due to the relatively slow economic development there. Moreover, the Islamic separatists in Xinjiang have been increasingly active since the end of the Cold War, trying to establish an independent Islamic state there. They have launched several terrorist attacks against the Han people in the recent years. For the Chinese government, social stability and economic prosperity in Xinijang are essential to China's national unity. The oil and gas pipeline programmes coincide with China's long-term strategy of Western Region Development, with the purpose of accelerating socio–economic development in the western part of China and narrowing down the development gap between the relatively rich eastern provinces and the less rich western provinces. Thus, the building of the pipelines and auxiliary oil infrastructures, such as refineries in Xinjiang, is expected to enhance the industrial infrastructure in the region, provide more jobs for local people, and improve the people's livelihood there. This will help boost the local development and enhance the relationship between the Han people and ethnic minorities, contributing to China's domestic stability and national unity.

5.3 Corporate interests

Chinese NOCs have their own corporate interests behind energy diplomacy. They are not commanded by the government to invest abroad. Rather, they are the protagonist of their overseas expansion. Their corporate interests are not subordinate to the state's national interests. As Lieberthal and Herberg noted, China's worldwide quest for energy is much less state-directed, coherent and strategic than generally assumed by U.S. policymakers. Chinese NOCs 'are not driven solely by state exhortation and diplomatic and financial support.' They gain experience in their global business and are 'torn between their traditional fealty to state interests and their expanding commercial instincts'. Just like their foreign counterparts, Chinese NOCs are engaging in transnational business in response to commercial competitive pressure and resource constraints.[96] The overseas activities of the NOCs are not always part of China's national strategy for energy security. The NOCs' overseas investment and purchase of equity oil are often linked to their commercial interests rather than grand strategic interests.[97] Also, as Xu Xiaojie who is a researcher with CNPC pointed out, the company's overseas operations are driven by its corporate strategy instead of the government's policy.[98] Chinese NOCs' outward direct investment (ODI) in and M&A of overseas oil and gas assets are largely driven by commercial incentives to take advantage of opportunities in the global market.[99]

Indeed, the NOCs' corporate interest consideration is the fundamental and key factor motivating them to participate in global business. They expand overseas to explore new investment markets, generate

---

[96] Lieberthal and Herberg, 'China's search for energy security', pp. 10–18.

[97] James Tang, 'With the Grain or Against the Grain?: Energy Security and Chinese Foreign Policy in the Hu Jintao Era', the Brookings Institution, October 2006, http://www.brookings.edu/~/media/Files/rc/papers/2006/10china_tang/tang2006.pdf, accessed 20 June 2008.

[98] Xu Xiaojie, 'Chinese NOCs' Overseas Strategies: Background, Comparison and Remarks', James A. Baker III Institute for Public Policy of Rice University, 2007, http://www.rice.edu/energy/publications/docs/NOCs/Papers/NOC_ChineseNOCs_Xu.pdf, accessed 13 March 2009, p. 21.

[99] Jiang and Sinton, *Overseas Investments by Chinese National Oil Companies: Assessing the Drivers and Impacts.*

further profits and fulfil their development strategies. This section is divided into two sub-sections. The first sub-section tackles the capital surplus in China's domestic market and the NOCs' capital accumulation which fundamentally push them to look for new investment opportunities abroad. The second sub-section shows the importance of overseas operations for the NOCs' profits making and development strategies.

5.3.1 *Capital accumulation and overseas investment*

Capital accumulation is an essential promoter of Chinese NOCs' ODI activities and overseas expansion. It consists of two aspects. One is the capital accumulation and surplus at China's home market, which is a general motivator for many Chinese enterprises to invest abroad. The other is the NOCs' capital accumulation, which provides them with growing financial capabilities and incentives to expand overseas. The two parts below deal with them respectively.

5.3.1.1 *The capital surplus in the home market*

A vital but long neglected factor pushing more and more Chinese enterprises to invest abroad is the capital surplus in China's home market. Thanks to its three decades of economic reform, China has changed from a capital shortage economy to a capital surplus one. The major reasons for this phenomenon are the following:

To begin with, China became the world's largest exporting country and top trading nation in 2009 and 2013 respectively. The country's overall trade in goods volume reached USD 4.16 trillion in 2013, within which export volume was USD 2.21 trillion and import volume was USD 1.95 trillion.[100] Currently, China holds the largest foreign currency reserves in the world. According to National Bureau of Statistics of the PRC, at the end of 2014, its foreign exchange reserves amounted to USD 3.843 trillion, an increase of USD 21.7 billion compared with the previous year. At the end of the year, the exchange

[100] Xinhua, 'Zhongguo yueju shijie diyi huowu maoyi daguo' (China has become the world's largest trading nation in goods), 2 March 2014, http://news.xinhuanet.com/fortune/2014-03/02/c_119565636.htm, accessed 10 March 2015.

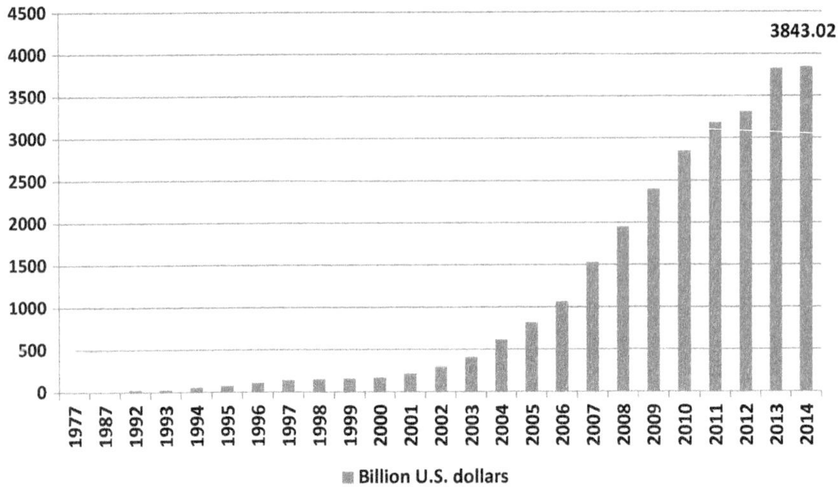

Figure 5.2: The accumulation of China's foreign exchange reserves, 1977–2014 (USD Billions)

Sources: Based on statistics from the website of the State Administration of Foreign Exchange, the People's Republic of China, http://www.safe.gov.cn, accessed 10 March 2015.

rate was RMB 6.1428 to USD 1 trillion, an appreciation by 0.8 per cent over that of 2013.[101] Figure 5.2 shows China's accumulation of foreign exchanges reserves in the past three to four decades.[102]

[101] Guojia tongjiju (National Bureau of Statistics, the People's Republic of China), '2014 guomin jingji he shehui fazhan tongji gongbao' (Statistical Communiqué of the People's Republic of China on the 2014 National Economic and Social Development), 26 February 2015, http://www.stats.gov.cn/english/PressRelease/201502/t20150228_687439.html, accessed 8 March 2015.

[102] There was modest growth of China's foreign exchange reserves from 1977 to 1992. After Deng Xiaoping's 'southern tour' in 1992, which pushed China's economic reform and development, the country's accumulation of foreign exchange reserves has been steadily and fast except for the three years between 1998 and 2000, due to the 1997 Asian financial crisis. China's entry into the WTO in 2001 has led to a fast growth of its trade surplus. Thus, the country's accumulation of foreign currency reserves increased more quickly in the past decade or so. Niu Tiehang and Lye Liang Fook, 'The challenges of managing China's huge foreign reserves: From Huijin to China Investment Corporation', EAI Background Brief, No. 352, East Asian Institute, National University of Singapore, 1 October 2007.

Notably, the reserves have been increasing fast since the mid-1990s, particularly since the beginning of this century, echoing the rapid development of Chinese enterprises including NOCs' ODI and M&A activities. The United Nations Conference on Trade and Development (UNCTAD) pointed out that such reserves have been accumulated mainly through China's sustained surplus in both of its current and capital accounts since the mid-1990s. In the recent years, capital inflows driven by the expectation of the RMB appreciation have also contributed to the increase of China's foreign currency reserves. Using the example of Japan whose rapid accumulation of foreign currency reserves in the 1980s had led to a surge of overseas investment, UNCTAD estimated that a similar situation could occur in China. The large and ever-increasing quantity of 'China dollars' has pressed the Chinese government to promote overseas investment and take concrete measures to speed up internationalising Chinese enterprises. Consequently, the strong growth of Chinese ODI activities is expected to continue in the coming years and China is likely to become an even more important outward investor in the near future.[103]

It is reported that in February 2009, officials in Beijing agreed to consider using China's foreign currency reserves to establish a special fund for the three leading Chinese NOCs to purchase oil and gas equities abroad. The NOCs are expected to benefit from low-interest loans or direct capital injections. The timing was good for their overseas expansion because the dramatic drop in international oil prices in 2008 and the concerns about recession have dragged down the share prices of many energy firms in the world.[104] In consequence, as demonstrated in the previous chapter, Chinese NOCs' ODI and M&A activities and overseas production have realised substantial and fast development since 2009.

---

[103] United Nations Conference on Trade and Development, *World Investment Report 2006* (New York and Geneva: United Nations, 2006), http://www.unctad.org/en/docs/wir2006_en.pdf, accessed 21 March 2008.

[104] Telegraph On-line, 'China Prepares to Buy up Foreign Oil Companies', 22 February 2009, http://www.telegraph.co.uk/finance/newsbysector/energy/4781037/China-prepares-to-buy-up-foreign-oil-companies.html, accessed 25 February 2009.

Also, the huge flow of foreign direct investment (FDI) into China's domestic market has pushed China's ODI. When the country lacked FDI in the past, the Chinese government offered foreign investors very favourable conditions and treatment. A typical example is that the government previously imposed skewed corporate income tax rates on enterprises operating in China — one rate for foreign invested companies and the other for domestic invested companies. For foreign companies, the nominal rate was 15 per cent and the actual rate was 11 per cent, while for domestic companies, the nominal rate was 33 per cent and the actual rate was 23 per cent — more than double the rate for their foreign counterparts (see Figure 5.3).[105] The actual rate was lower because of preferential tax policies in China. The purpose of such a tax policy was to encourage the inflow of FDI, relatively advanced technologies and qualified personnel. Nevertheless, this discriminatory policy had a profoundly negative impact on the survival and development of domestically invested enterprises in the Chinese market. Since there are a fast growing number of foreign invested firms operating in China and they enjoyed the advantage of tax policy, Chinese domestically invested firms were under tremendous pressure.[106]

---

[105] People's Daily On-line, 'Liangshui hebing: neiwaizi qiye chongfan gongping jingzheng' ('The unification of the two tax rates: foreign invested enterprises and domestic invested enterprises are expected to compete fairly again'), 30 October 2006, http://finance.people.com.cn/GB/4972421.html, accessed on 17 January 2007.

[106] The unfair tax policy has also led to many Chinese companies registering overseas and then investing back in the Chinese market as foreign enterprises, in order to enjoy the lower tax rates, resulting in much purported FDI that is in fact domestic in origin. After years of criticism from Chinese scholars, businessmen and some government officials that the corporate income tax policy was unfair for domestic entities, the Chinese government finally ended its dual corporate income tax rates. In March 2007, the 10th National People's Congress approved a bill that would eventually unify corporate income tax rates for domestic and foreign enterprises at 25 per cent. See Xinhua, 'Jinrenqing: tongyi hou de neiwaizi qiye suodeshui shuilu dingwei 25%' ('Jinrenqing: the unified corporate income tax rate in 25%'), 8 March 2007, http://www.lianghui.org.cn/2007lianghui/2007-03/08/content_7924484.htm, accessed 23 March 2008. This change took effect on 1 January 2008.

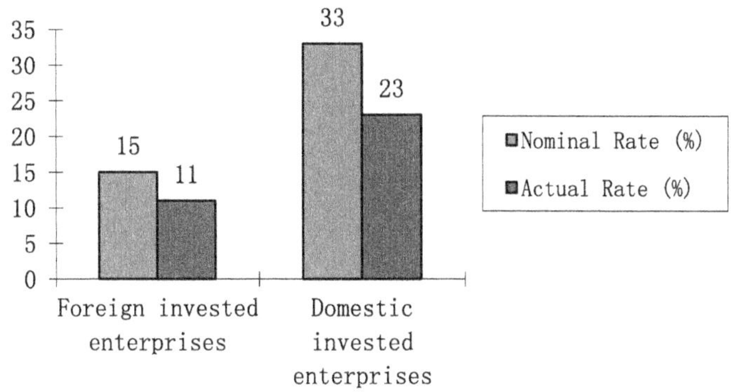

Figure 5.3: The previous dual corporate income tax rates in China (Percent)

Moreover, John Wong noted that there has been massive influx of foreign capital into China in the recent years, despite the fact that the country has already become a capital-surplus economy because of its persistent twin surpluses on both capital and current accounts.[107] Indeed, as Yu Yongding pointed out, China's long-term twin surpluses were caused by the underdevelopment of the domestic financial market and the presence of market distortions created by preferential FDI treatment and trade promotion policies implemented in the past 25 years. The persistent twin surpluses have led to the steady growth of China's foreign exchange reserves; and the capital account surplus has been a more important contributor to the country's accumulation of foreign exchanges reserves than trade surplus in the past decade. Such twin surpluses represent a case of resource misallocation that results in the drainage of national wealth. Especially, China's huge domestic savings makes the country a net capital exporter.[108]

Today, China is one of the major recipients of FDI in the world. For example, according to *World Investment Report 2014* published

[107] John Wong, 'China's Economic Growth in East Asian Context', in John Wong and Wei Liu ed., *China's Surging Economy: Adjusting for More Balanced Development* (Singapore: World Scientific Publishing Co. Pte. Ltd., 2007), pp. 31–53.

[108] Yu Yongding, 'China's "twin surpluses": Causes and remedies', EAI Background Brief, No. 283, East Asian Institute, National University of Singapore, 27 April 2006.

by UNCTAD, in 2013, the Chinese mainland received USD 124 billion of FDI inflows, recording its largest ever FDI inflows, and maintaining its position as the second largest recipient in the world only after the United States (received USD 188 billion of FDI inflow in 2013), and the largest FDI recipient among all developing and transition economies.[109]

Actually, since China's accession to the WTO in 2001, domestic companies have been facing tougher competition because China has largely reduced the barriers for foreign companies to enter the Chinese market. Consequently, as profit margins fall and competition increases within China, the economic pressures of trade liberalisation combined with the entry of foreign firms into the home market have pushed Chinese firms to pursue internationalisation.[110] The increasing presence of foreign enterprises in China has greatly eroded Chinese enterprises' scope for doing business at home. Therefore, if they want to survive and develop, they have to find a place where they can compete with others. Therefore, an increasing number of Chinese enterprises have to invest abroad.

In the past two decades, Chinese transnational corporations (TNCs) have been expanding worldwide, covering trade, transportation, resources exploration, tourism, manufacturing and other activities in 160 countries and regions. Access to natural resources, markets, and the acquisition of strategic assets are the main driving force.

Recently, China has become a major outward direct investor in the world. For instance, China's ODI reached USD 101 billion in 2013, a 15 per cent increase from the previous year, driven by a number of megadeals in developed countries, such as CNOOC's takeover of the Canadian upstream oil and gas company, Nexen for

---

[109] United Nations Conference on Trade and Development (UNCTAD), *World Investment Report 2014: Investing in the SDGs: An Action Plan* (New York and Geneva: United Nations, 2014), http://unctad.org/en/PublicationsLibrary/wir2014_en.pdf, accessed 11 March 2015.

[110] Friedrich Wu, 'The Globalization of Corporate China', NBR Analysis 16: 3, December 2005, the National Bureau of Asian Research, http://www.nbr.org/publications/analysis/pdf/vol16no3.pdf, accessed 13 March 2009.

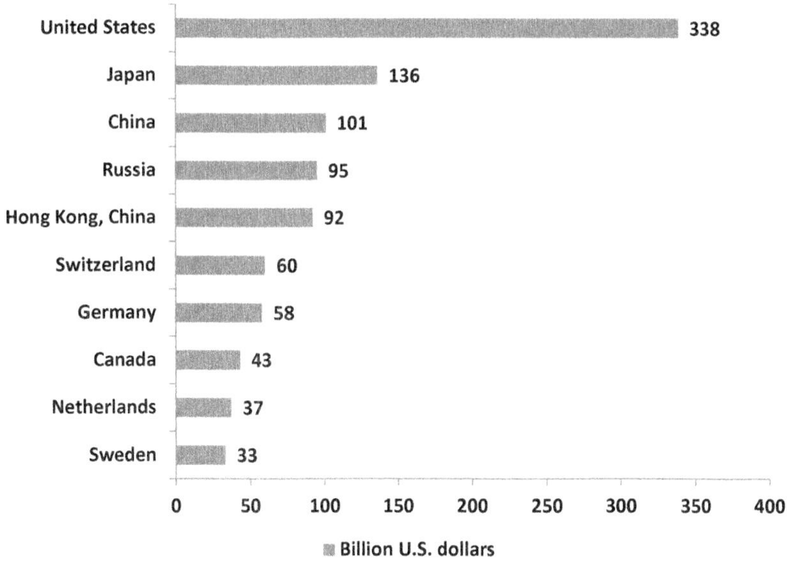

Figure 5.4: Top ten home economies of FDI outflow, 2013 (USD Billions)

Source: United Nations Conference on Trade and Development (UNCTAD), *World Investment Report 2014: Investing in the SDGs: An Action Plan* (New York and Geneva: United Nations, 2014), http://unctad.org/en/PublicationsLibrary/wir2014_en.pdf, accessed 11 March 2015.

USD 19 billion. This has made the country the third largest home economy of FDI outflow in the world (see Figure 5.4). If taking into consideration of FDI outflow from Hong Kong, China, the fifth largest home economy of FDI outflow in 2013, the combined FDI outflow from the Chinese mainland and Hong Kong is the second largest source of FDI in the world. Also, according to the UNCTAD, China's ODI is expected to surpass its FDI inflows within two to three years.[111]

According to surveys conducted by the UNCTAD, China is consistently ranked the most promising source of FDI and top prospective host economies for TNCs, together with the United States (see Figures 5.5 and 5.6).

[111] United Nations Conference on Trade and Development (UNCTAD), *World Investment Report 2014: Investing in the SDGs: An Action Plan.*

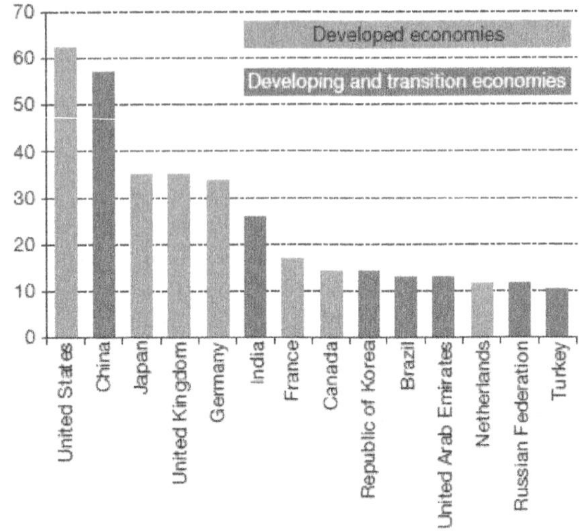

Figure 5.5: Most promising investor home economies for FDI, 2014–2016 (Per centage of respondents)

Source: United Nations Conference on Trade and Development (UNCTAD), *World Investment Report 2014: Investing in the SDGs: An Action Plan.*

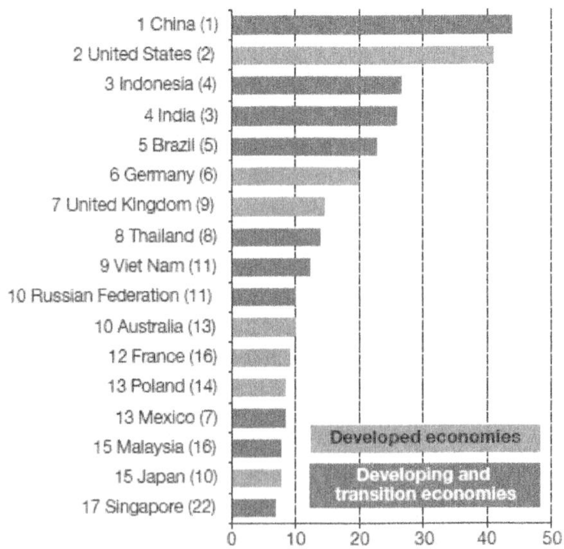

Figure 5.6: TNCs' top prospective host economies, 2014–2016 (Per centage of respondents)

Source: *Ibid.*

Table 5.2: The Chinese companies among the top 100 non-financial TNCs from developing and transition economies, ranked by foreign assets, 2012 (USD Millions)

Rank	Company	Foreign assets	Total assets
2	CITIC Group	85,721	103,715
6	China Ocean Shipping (Group) Company	43,452	56,126
7	*CNOOC*	*34,276*	*129,834*
24	*CNPC*	*19,284*	*541,083*
35	Sinochem Group	14,704	45,488
44	Lenovo Group Ltd	11,962	16,882
61	China Mobile Ltd	8,349	166,972
68	China Electronic Corporation	7,784	29,047
77	Cofco Ltd	5,952	41,264
85	*Sinopec*	*5,030*	*201,027*
89	China Minmentals Corp	4,885	39,225
100	China Railway Construction Corporation Ltd	3,761	76,282

Sources: Website of the United Nations Conference on Trade and Development (UNCTAD), unctad.org/Sections/dite_dir/docs/WIR2014/WIR14_tab29.xls, accessed 11 March 2015.

Notably, 12 companies from Mainland China were listed by the UNCTAD among the top 100 TNCs from developing and transition economies in 2012, ranked by their foreign assets (see Table 5.2).[112] Most of them are SOEs, including the three leading Chinese NOCs–CNPC, Sinopec and CNOOC.

5.3.1.2 *Chinese NOCs' capital accumulation*

In the recent years, Chinese NOCs have accelerated their capital accumulation and become stronger financially, which is the fundamental factor enabling and motivating them to carry out more and more ODI and M&A activities. The reasons for their rapid accumulation of capital and financial strength include their monopoly in China's

[112] United Nations Conference on Trade and Development (UNCTAD), *World Investment Report 2014: Investing in the SDGs: An Action Plan.*

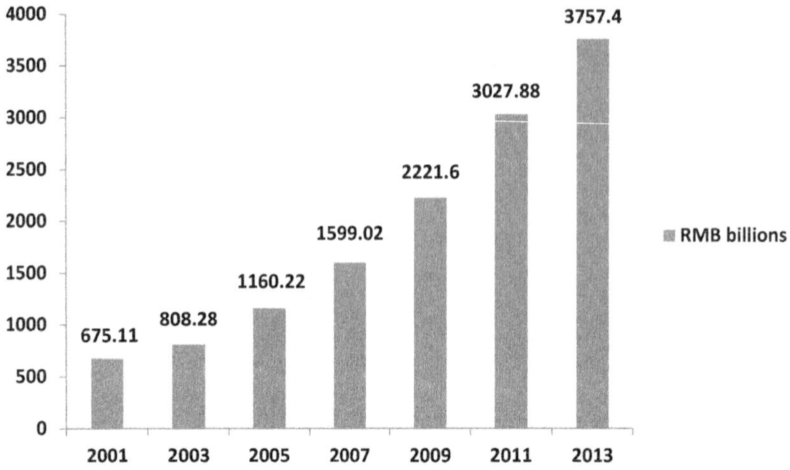

Figure 5.7: Total assets of CNPC Group, 2001–2013 (RMB Billions)

Sources: Various issues of CNPC's annual report.

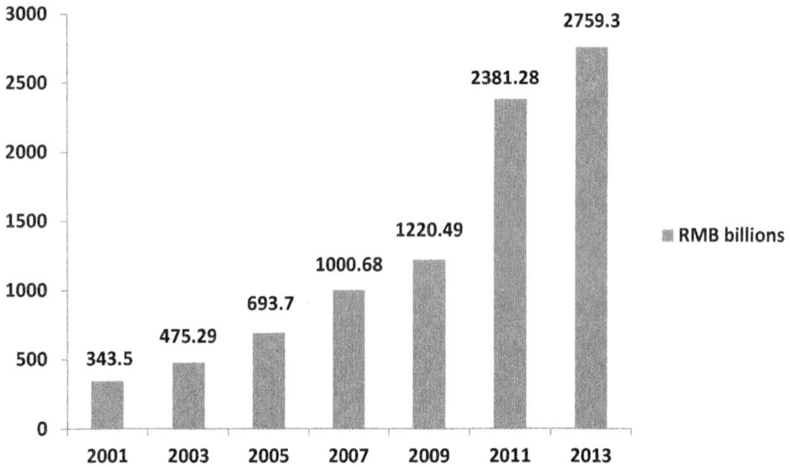

Figure 5.8: Operating income of CNPC Group, 2001–2013 (RMB Billions)

Sources: Various issues of CNPC's annual report.

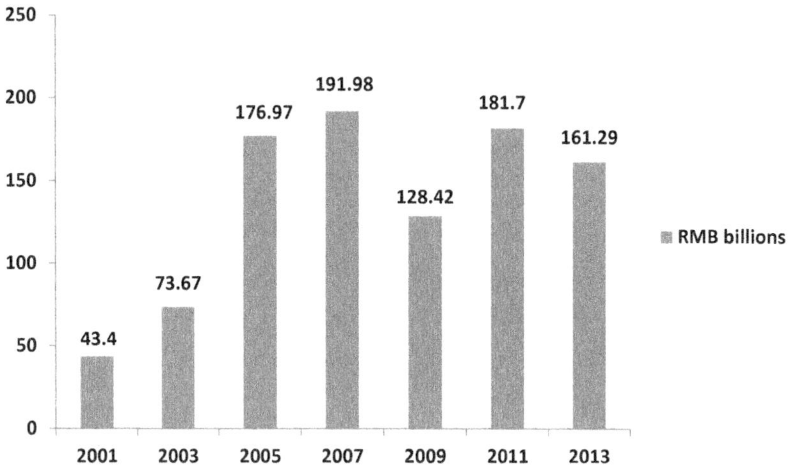

Figure 5.9: Operating profit of CNPC Group, 2001–2013 (RMB Billions)

Sources: Various issues of CNPC's annual report.

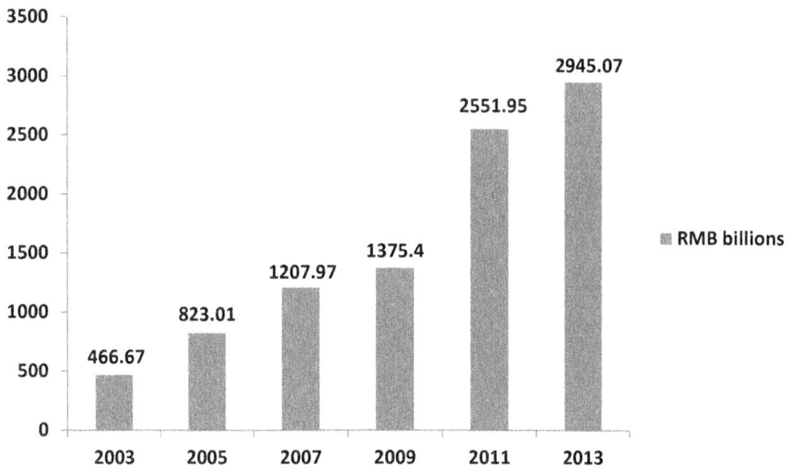

Figure 5.10: Operating income of Sinopec Group, 2003–2013 (RMB Billions)

Sources: Various issues of Sinopec's annual report.

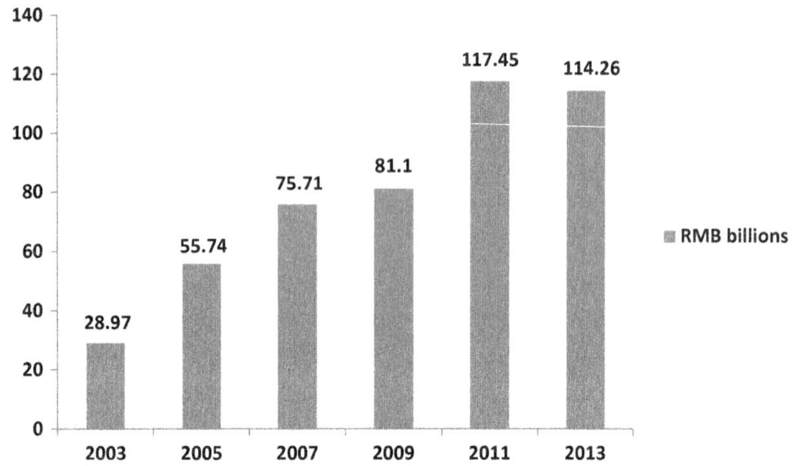

Figure 5.11: Operating profit of Sinopec Group, 2003–2013 (RMB Billions)

Sources: Various issues of Sinopec's annual report.

market, their receipt of almost all after-tax profits, the state's subsidy for them, the rising oil prices, etc. The details will be tackled in the next chapter where the NOCs' expanding capability and influence are explained. But some evidence of the NOCs' capital accumulation is provided here. For example, the figures below demonstrate the fast growth of CNPC and Sinopec's total assets, operating income and operating profit in the past few years.

5.3.2 The importance of transnational business for Chinese NOCs' profits making, capital accumulation and development strategies

Apart from the capital accumulation factor which is the fundamental motivator for Chinese NOCs' overseas expansion, their transnational operations are beneficial for their revenue generation, capital accumulation and corporate development strategies.

Notably, although the NOCs are already among the most profitable companies in China, they have a very strong desire to further maximise their profits;[113] while the upstream industry (e.g. exploration and production) is traditionally the most important and profitable part of the oil industry. Take CNOOC for example, upstream business accounted for 71.88 per cent of its operating profit in 2013 (see Figure 5.12).[114]

Nowadays, Chinese NOCs have limited domestic energy resources for further upstream production, and enjoy relatively less profit from downstream production because of the Chinese government's control on prices for domestic oil products. Thus, they are seeking reserve holdings, production, revenues and clout by expanding overseas.[115]

---

[113] Just like other big Chinese SOEs, the NOCs have very strong desires to maximise their profits. The reasons include the following: First, the SOE bosses are different from entrepreneurs in Western companies. They are actually politicians rather than entrepreneurs. They are transformed from governmental officials and do not have many professional skills in managing enterprises. SOEs are mainly managed by those professional managers under the bosses' leadership. Both the bosses and the professional managers have their own desire to maximise the enterprises' profits. On the one hand, those bosses and politicians are under economic as well as political pressure to maximise SOEs' profits. They must promote the enterprises' commercial achievements, which are also their personal political achievements related to their official career. On the other hand, those professional managers in SOEs also have their own key interests in maximising the enterprises' profits. Usually, these managers and their whole families work in SOEs for their career and are highly loyal to their enterprises. As employees, they do not have many opportunities to step into the political circle or gain political interests. Since these SOEs are very profitable, their employees enjoy high salaries and good welfare. In order to maintain such favourable treatment, they have to maintain the SOEs' high profits. In other words, high profits propel these firms to pursue more profits. Therefore, although some SOEs such as the NOCs are already among the most profitable companies in China, they need to make more profits. My interview with Dr Jiang Yong, Director of Centre for Economic Security, China Institute of Contemporary International Relations, Beijing, China, July 2008.

[114] 2013 Annual Report of CNOOC.

[115] Daniel H. Rosen and Trevor Houser, 'China Energy: A Guide for the Perplexed', China Balance Sheet, A Joint Project by the Center for Strategic and International Studies and the Peterson Institute for International Economics, May 2007, http://www.petersoninstitute.org/publications/papers/rosen0507.pdf, accessed 3 May 2007.

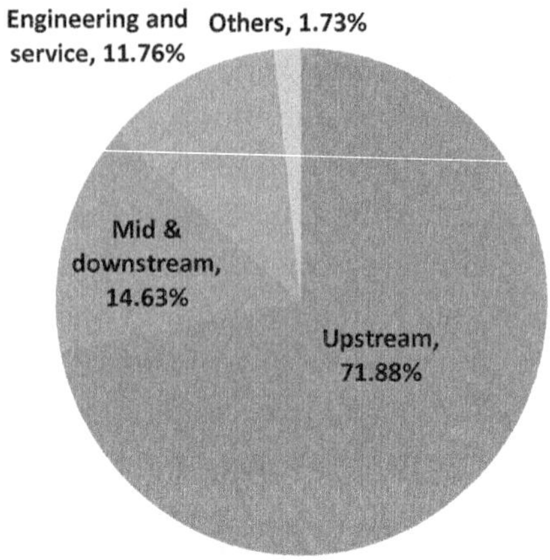

Figure 5.12: The composition of CNOOC's operating profit, 2013 (Percentage)
Source: 2013 Annual Report of CNOOC.

These firms especially their listed subsidiaries are following the strategies of foreign oil companies to seek profits through ODI and M&A of upstream assets abroad. They mainly gain the profits existing between the cost of producing a barrel of oil and the price of a barrel of oil in the international market. Indeed, considering the soaring international oil price in the past few years, the upstream sector is much more profitable than downstream business (e.g. refinery and the sale of oil products). For example, CNPC Group viewed overseas investment as a way to quickly bolster its outputs and wanted to sell its foreign production at the international market, because the international oil price is higher than the government-set oil price in China's domestic market.[116]

Moreover, although established much later than their Western counterparts, Chinese NOCs have the ambition and strategies to be among the world's top oil companies. In order to realise this aim, they must be internationalised and transformed into TNCs with

[116] Erica S.Downs, 'China', *Energy Security Series*, the Brookings Foreign Policy Studies, the Brookings Institution, December 2006, http://www.brookings.edu/~/media/Files/rc/reports/2006/12china/12china.pdf, accessed 8 April 2008.

strong international competitiveness. Hence, the development of their ODI and transnational operation is essential.

For instance, CNPC has the ambition to become 'a world-leading integrated international energy company with the general business portfolio of an international energy corporation and the characteristics of China's national energy corporation.'[117] The company has also declared its goal 'to achieve industry-leading operational performance with enhanced competitiveness and profitability by 2020 through strategic development initiatives, technological innovation, and an accelerated shift to a new development mode highlighting quality and efficiency.'[118] Its corporate strategy, as shown on its website, includes three aspects: The first is increasing resources through maximising and diversifying oil and gas resources, realising an orderly replacement of reserves, and maintaining the dominance in the upstream sector. The second is expanding market through integrating upstream and downstream operations to seek an advantage in the market and maximise profits. The third is pursuing a greater international role in the oil and gas business through actively and prudently enhancing international cooperation, strengthening capital operations, and expanding international oil and gas trading.[119]

Similarly, Sinopec is dedicated to building itself into 'a world-leading energy and chemical company'.[120] CNOOC also promised to strengthen its capabilities in operational management, resource integration and risk control in order to promote its international development.[121] In addition, overseas investment is part of Chinese NOCs' competition with each other to obtain economic and political

[117] Website of CNPC, http://www.cnpc.com.cn/en/companyprofile/companyprofile.shtml, accessed 11 March 2015.

[118] Website of CNPC, http://www.cnpc.com.cn/en/ourgoal/about_common.shtml, accessed 11 March 2015.

[119] Website of CNPC, http://www.cnpc.com.cn/en/ourstrategy/about_common.shtml, accessed 11 March 2015.

[120] Website of Sinopec, http://www.sinopecgroup.com/group/en/companyprofile/, accessed 11 March 2015.

[121] Website of CNOOC, http://www.cnooc.com.cn/col/col8041/index.html, accessed 11 March 2015.

interests from the Chinese state. Overseas oil assets acquirement helps these firms influence key energy officials and win financial support from state-owned banks. 'The more assets a company acquires, the more likely it is to obtain support for subsequent acquisitions'.[122]

5.4 Conclusion

From the discussion above, it can be seen that both the Chinese government and Chinese NOCs have essential interests behind China's energy diplomacy. On the one hand, this movement improves China's energy security and provides Beijing with a new platform to improve its international relations as well as multilateral diplomacy, contributing to the country's economic development, social stability and national unification, which are beneficial for the CCP's political legitimacy. On the other hand, China's energy diplomacy assists Chinese NOCs' to access new investment markets abroad, generating more profits and fulfilling their long-term corporate development strategies. Therefore, the convergence of national and corporate interests is realised behind China's energy diplomacy, which promotes the cooperation between the government and the NOCs in searching energy worldwide. This is the key and sustainable momentum of China's energy diplomacy. As long as there is overlapping national and corporate interests, China's energy diplomacy is likely to keep progressing. Then, another important aspect of the domestic dynamics of China's energy diplomacy is the relationship and interaction between the government and the NOCs, which is a fundamental factor of this movement. This issue will be dealt with in the next chapter.

[122] Downs, 'China', p. 36.

Chapter

6

Interaction Between the Government and the NOCs

This chapter attempts to explore the interaction of the Chinese government and Chinese national oil companies (NOCs), in particular the government's institutionalised approach to manage and supervise the NOCs and the NOCs' ability and measures to influence the government's policy making. During China's economic reform, the decentralisation process has transferred much power from the central government to the NOCs. This has enabled them to become increasingly independent and powerful. The Chinese government oversees the oil industry and the NOCs mainly through oil supervising agencies in the State Council, the nomenclatural system and the investment approval process; while Chinese NOCs possess substantial material, institutional and ideological resources to exert considerable

impact upon the government's policy making. To certain extent, the recent years have witnessed the growth in the NOCs' power *vis-à-vis* the state's declining authority over the oil sector.

This chapter consists of four sections. The first section deals with the background of the government–NOC relationship — China's decentralisation that has led to a shift of power from the central government to these firms. The second section demonstrates key measures with which the government administers the NOCs, especially the state's oil administration organisation that has been evolving with China's continuous bureaucratic restructurings. The third section discusses the resources for the NOCs' leverage over the government. The fourth section explains how decisions for the NOCs' outward direct investment (ODI) projects are made, before reaching a conclusion.

In this chapter, the interaction between the government and the NOCs is in line with the new diplomacy idea's general observation on the state-firm interaction. Also, China's decentralisation process echoes Susan Strange's argument about the trend of power shift from the state to firms in the post-war period. Moreover, the description of China's pluralised oil decision making organisation coincides with liberalism's description of various interests groups in a society. Finally, the explanation of the three critical resources of the NOCs' influence follows a Marxist idea — Robert Cox's discourse of the three elements of the hegemony.

6.1 Decentralisation

An essential background for the interaction between the government and the NOCs is the decentralisation of authority from the Chinese government to Chinese NOCs during China's economic and enterprise reform. Here, decentralisation refers to the process whereby power flows from the state to society and from the government to enterprises during China's transition from a planned to a market economy. With specific regard to the government–NOC relationship, the Chinese central government has decentralised a significant part of its power over the oil sector to those newly established NOCs. In

general, China's economic and enterprise reform since the 1980s has not only created the three leading Chinese NOCs but also promoted their power and autonomy, while greatly reducing the government's intervention in these firms' operations, reflecting the country's transition from a planned economy to a market economy.

Therefore, Chinese NOCs are no longer the previous type of state enterprises that served as agencies of the government, but rather independent firms with their own corporate consideration and interests. In the past, those state enterprises had important socio–economic responsibilities apart from their production function. They were just like mini-welfare states that provided their employees with welfare services such as housing, medical care, education, pension, etc. They enjoyed soft budgets since the government would always subsidise their losses. Also, the relationship between the government and state enterprises in the pre-reform period is quite different from the relations between the government and state-owned enterprises (SOEs) today. Under the planned economy, enterprises were part of an enormous bureaucracy. They 'did not possess any of the strategic planning, marketing, logistics, or personnel capabilities that we associate with a market business'. Instead, they served as multifunctional social unit called '*danwei*'. The government, which exercised the real power of ownership through its hierarchically organised bureaucracy, took all the enterprises' profits and made up their losses. Enterprise managers were kept busy fulfilling different commands and tasks assigned by the planners in the government, and were under overlapping official controls and restrictions. 'But none of the disciplines under which managers laboured motivated them adequately to increase firm productivity or profitability'.[1]

China started its enterprise reform in the 1990s. When Zhu Rongji was still China's Vice-premier in the early 1990s, he began to make initiatives towards economic reform, which was supported and encouraged by Deng Xiaoping's 'southern tour' in 1992. State power was decentralised to those state enterprises, as reflected in the change

[1] Barry Naughton, *The Chinese Economy: Transition and Growth* (Cambridge, MA: The MIT Press, 2007), pp. 308–309.

of the names of those enterprises. In 1993, China's amended Constitution which was approved by the Eighth National People's Congress led to the formal change of the name of *guoying qiye* or 'state enterprise' (literally 'state-run enterprise') to *guoyou qiye* (literally 'state-owned enterprise/SOEs'). 'This simple change in name was significant as it distinguishes between enterprise ownership and enterprise management, with the clear implication that the government is no longer obliged to be directly involved in enterprise management.' SOEs are officially and theoretically owned by the state. But they are technically to be managed by themselves.[2] In 1994, China adopted the Company Law, which provided a legal framework for converting traditional SOEs into corporations that are more appropriate to the market economy.[3] More importantly, Zhu carried out a nationwide SOE restructuring, as Zheng Yongnian described:

> 'In early 1995, Zhu's SOE reform efforts crystallised further into a more explicit strategy of *Zhuada fangxiao* or 'nurturing the big into giant conglomerates while letting the small SOEs face the forces of the market'. Reformist leaders believed that while they could "let go" of the 240,000 or so small, mainly local-level SOEs via various forms of restructuring including reorganisation, mergers and takeover, leasing and management contract, conversion into shareholding companies, or even outright closure, they had to retain the 1,000 large SOEs belonging to the central government, for obvious economic and social reasons. These key SOEs are still of strategic importance as they constitute the backbone of China's industrial economy in terms of total capitalisation and employment'.[4]

Meanwhile, China's SOE reform since the 1990s has been aimed at building a modern enterprise system (*xiandai qiye zhidu*). The core task of the reform is *zhengqi fenkai*, or literally the disarticulation of the government's management from the enterprises' operation,

---

[2] Zheng Yongnian, *Globalization and State Transformation in China* (Cambridge: Cambridge University Press, 2004), p. 131.

[3] Naughton, *The Chinese Economy*, p. 301.

[4] Zheng Yongnian, *Globalization and State Transformation in China*, pp. 131–132.

which means that governmental departments at various levels should significantly reduce their intervention in SOEs' business operation. After the mid-1990s, under pressure from the changing competitive and financial conditions, the Chinese government launched a second wave of industrial reforms, centring on corporatising SOEs and improving their governance mechanisms.[5] Moreover, since the 1980s, many governmental units that directly run important industries previously have been transformed into large SOEs. The three major Chinese NOCs — China National Petroleum Corporation (CNPC), China Petroleum and Chemical Corporation (Sinopec) and China National Offshore Oil Corporation (CNOOC) — are among them. Especially, the power of the NOCs increased dramatically between 1993 and 2003 when Zhu was in charge of China's economy. In the second half of the 1990s, he deliberately enhanced the NOCs' financial and administrative autonomy, in order to make them more efficient in preparation for the listing of their subsidiaries on international stock exchanges.[6]

After a series of economic and enterprise reform, much of the government's power over the industrial operation had been handed over to the SOEs. With the establishment of more and more subsidiaries under these SOEs, the parent companies usually become the largest shareholder of their subsidiaries, while the subsidiaries have their own autonomy and interests to a great extent. Hence, as Figure 6.1 shows, decentralisation has led to the flow of power from the central government to the SOEs including the NOCs, and also from the SOE parent companies to their subsidiaries.

Further, during the economic reform period, the internationalisation of the Chinese state and Chinese enterprises provided a foundation for the transformation of some big SOEs, including the three leading NOCs, into transnational corporations (TNCs). As Zhang Yongjin pointed out, it has taken China a long time to come to terms

---

[5] Naughton, *The Chinese Economy*, p. 298.

[6] Erica S. Downs, 'China', Energy Security Series, the Brookings Foreign Policy Studies, the Brookings Institution, December 2006, http://www.brookings.edu/~/media/Files/rc/reports/2006/12china/12china.pdf, accessed 8 April 2008.

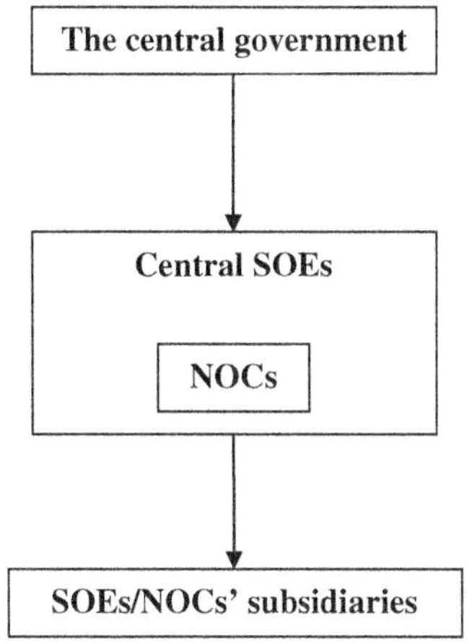

Figure 6.1: Decentralisation of power from the Chinese government to Chinese NOCs

Source: Author's compilation.

with TNCs, accepting intellectually and ideologically this capitalist-dominated institution and embracing it into the Chinese economic development strategy. The country experienced a winding process from 1979 to 1992 to accept the market orientation of its economic reforms. In the late 1980s, the emerging regulatory framework attempted to encourage, regulate and control China's ODI activities. The 14th Chinese Communist Party (CCP) National Congress in 1992 finally and officially endorsed the transnationalisation of Chinese enterprises through ODI; and the government focused on institutionalising or fine-tuning the existing regulatory framework. In the 1990s, a policy regime conducive to China's ODI and Chinese companies' transnational operation was more dependent on the transformation of the Chinese economy into a market economy.[7]

In a word, the decentralisation process, which was initiated by the Chinese state, has led to the decline of state authority over many SOEs. Specifically, the government has withdrawn much of its

[7] Zhang Yongjin, *China's Emerging Global Business: Political Economy and Institutional Investigations* (New York: Palgrave Macmillan, 2003), pp. 47–80.

institutional power from the NOCs, and these firms have become highly independent players that manage their own operations. Compared with the previous era of planned economy, it is now much more difficult for the government to control or manage the NOCs. They have gained huge benefit from their transnational operations and ODI activities, as mentioned in the previous chapter, and desire to continue expanding their global business. As David Zweig wrote, domestic and global interests such as new corporate organisations benefiting from their transnational linkages can undermine strong states' control over their global exchange as they open to the outside world. It was easy for states to keep control when domestic actors did not know their interests. But once states allow transnational flows, the companies would demand more global exchange and transnational linkages.[8]

To some extent, decentralisation has enabled the NOCs to become increasingly powerful at the cost of the government's leverage over the oil sector. Erica S. Downs also noted that the liberalisation and decentralisation of China's energy sector and the country's bureaucratic restructuring over the past two decades have resulted in a shift of power and resources away from the central government to the NOCs, a substantial reduction in the government's ability to monitor these firms, and a fragmented institutional structure of authority over the energy sector. When it comes to deciding which overseas assets to acquire, the NOCs are in the driver's seat and the government is just along for the ride with little idea of the final destination. Thus, contrary to popular opinion, the NOCs' overseas investment is not driven from the 'top–down' but rather from the 'bottom–up'.[9] With the development of China's oil industry and the NOCs, the government has also been adjusting its management on these firms. The next section shows the major approaches and institutions through which the government oversees the oil industry and the NOCs.

---

[8] David Zweig, *Internationalizing China: Domestic Interests and Global Linkages* (Ithaca and London: Cornell University Press, 2002), p. 277.

[9] Erica Downs, 'China's Quest for Overseas Oil', Far Eastern Economic Review, September 2007, pp. 52–56; and Downs, 'China'.

6.2 The government's administration over the NOCs

This section discusses the Chinese government's approaches to administrate Chinese NOCs. The first and foremost aspect is that the government manages the NOCs through some oil supervising departments and institutions in the State Council, as will be demonstrated in the first part of this section. In addition, the second part explains that the government also manages and supervises the NOCs in other important ways such as the nomenclatural system and the investment approval process.

6.2.1 *Oil administrative agencies in the State Council*

The Chinese government mainly oversees China's oil sector through various oil supervising bodies in the State Council. An examination of the oil policy making organisation is provided in this sub-section. Since the beginning of China's economic reforms in 1978, the government has carried out six rounds of bureaucratic restructurings in the years of 1982, 1988, 1993, 1998, 2003 and 2008, respectively. These changes are in line with the country's transition from planned to market economy.[10] In accordance, China's oil administrative structure has experienced several waves of restructuring.

Notably, there are three major reasons that such bureaucratic restructuring or adjustments are periodical and frequent. Firstly, each National Congress of the CCP since the 1980s determines new goals and guiding principles for the work of the Chinese government, which need translating into real changes and developments of the government. Thus, the formation of a new administration in the next year is often accompanied by a round of bureaucratic restructuring, in order to fulfil the tasks set by the CCP National Congress and accommodate to the country's economic reform. Secondly, a wave of

[10] Xinhua, 'Jingbing jianzheng, zhuanbian zhineng — xinzhongguo chengli yilai de lici zhengfu jigou gaige' ('Streamlining the administration and transforming the function: the bureaucratic restructurings since the establishment of the People's Republic of China'), 6 March 2003, http://news.xinhuanet.com/ziliao/2003-03/06/content_761776.htm, accessed 3 May 2008.

bureaucratic restructuring means a round of power redistribution within the State Council and a personnel reshuffle, during which the new Premier can create or reinforce his power base which often wields significant leverage over the economically and strategically important oil affairs.[11] Thirdly, it has been noted that in order to enhance its management on energy affairs, the government has made efforts to recentralise some power over the energy sector. However, such efforts have been crisis-driven, incremental and limited in effect. The enormous difficulty in redistributing power among various departments and balancing the interests of various stakeholders has made the progress of the restructuring of China's energy administration system slow and piecemeal.[12]

The rest of this sub-section is divided into nine parts. The first to the eighth parts demonstrate the Chinese government's oil administration organisations and major oil supervising agencies in different periods. The ninth part discusses the oil supervising agencies' insufficient authority over the NOCs.

6.2.1.1 *The period of 1949–1978*

During the pre-economic reform era from 1949 to 1978, there were three major institutions in charge of the oil affairs in the central government, namely the Ministry of Fuel Industry (MFI), the State Planning Commission (SPC) and the Ministry of Petroleum Industry (MPI).

6.2.1.1.1 The Ministry of Fuel Industry (1949–1955)

When the Peoples' Republic of China (PRC) was established in 1949, the central government set up the MFI, with Chen Yu as its Minister. At that time, the development level of the country's energy sector was very low. Within the MFI, there were General Bureau of Coal

---

[11] For example, Premier Zhu Rongji's power base was the SETC and his successor Premier Wen Jiabao's power base was the NDRC. Both commissions are the essential oil supervising agencies in their respective cabinets.

[12] Downs, 'China'.

Management, General Bureau of Electric Power Management, General Bureau of Petroleum Management and General Bureau of Hydropower Construction.[13]

6.2.1.1.2 The State Planning Commission (1952–1998)

The critical governmental institution that oversees China's energy sector was the SPC which existed for a long time throughout the history of the PRC. Basically, the SPC decided whether a major energy construction project should be approved or rejected. The SPC was set up in 1952, and Gao Gang, a Vice Chairman of the Central People's Government of the PRC, became the SPC Chairman. Other politically prominent figures who have headed the SPC include Li Fuchun, Yu Qiuli, Yao Yilin, Song Ping, etc. Generally, the SPC played an essential role in the management of China's previous centrally planned economy. Primarily focused on macro-economic management, the SPC guided and organised the production and distribution of major commodities and important construction projects. It was also the key body making the country's Five-year Plan for socio–economic development.

Especially, as Kenneth Lieberthal and Michel Oksenberg described, the SPC had three energy-related functions: First, it formulated China's mid-term and long-term energy plans including construction, production and conservation. Second, it approved or decided which projects to construct, conducted feasibility studies and determined the project's size and speed of development. Third, it coordinated the construction of energy projects. Usually, the line ministries and the SPC's provincial bodies submitted plans to the SPC. The plans included requests for new projects and the studies to support these requests. Since the requests far outstripped the country's capacity for new construction, the SPC determined which proposals would be

[13] Zhongguo shiyou xinwen zhongxin (CNPC News On-line), "'Nengyuanju" shengji "guojiaju"' ('"The Energy Bureau" will be promoted to be a "state bureau"'), http://news.cnpc.com.cn/system/2008/03/19/001163500.shtml, 19 March 2008, accessed 12 April 2008.

approved, taking into consideration their technical and economic feasibility. Due to its lack of staff and expertise, the SPC often drew on outside expertise to help it consider the proposals more scientifically. Its decision to permit a project produced enforceable obligations for various relevant ministries and units to do their respective work. Also, after the project was approved and included into the official plan, the SPC coordinated the work of relevant ministries during the construction of the project.

The SPC had experienced ups and downs throughout its lifetime. It played a critical role during the First Five-Year Plan. But the Great Leap Forward (1958–1961), which emphasised decentralisation, eroded the SPC's role. It was revived between 1962 and 1965. Then, however, the Cultural Revolution (1966–1976) undermined its ability to guide the national economy again during this period. The SPC was fully rehabilitated and reinforced after 1976.[14] Since then, it had been the essential administrator over the country's energy sector until the bureaucratic restructuring in 1998.

6.2.1.1.3 The Ministry of Petroleum Industry (1955–1988)

Thanks to the PRC's first five-year plan (1953–1957), China's petroleum, coal and electric power sectors grew and expanded rapidly. Thus, in 1955, the central government abolished the MFI and set up three new ministries — the MPI, the Ministry of Coal Industry and the Ministry of Electric Power Industry.[15] Notably, while the MPI grew from the General Bureau of Petroleum Management in the previous MFI, it also drew heavily from the Chinese People's Liberation Army (PLA) and its ties with the PLA General Logistics Department were intimate. In 1958, against the background of the deterioration of Beijing's relationship with Moscow, Mao Zedong decided to give greater priority to self-reliance and assigned Deng Xiaoping to be the

[14] Kenneth Lieberthal and Michel Oksenberg, *Policy Making in China: Leaders, Structures, and Process* (Princeton: Princeton University Press, 1988), pp. 64–70.
[15] CNPC News On-line, '"The Energy Bureau" will be promoted to be "state bureau"'.

Vice Premier in charge of this sphere.[16] In the same year, Yu Qiuli who was Political Commissar of the PLA General Logistics Department was appointed the MPI minister. Since then, he had been a key leader in the establishment and development of China's oil industry. In the 1970s, the MPI experienced three rounds of restructuring. It was restructured into the Ministry of Fuel and Chemical Industry in 1970, later into the Ministry of Petroleum and Chemical Industry in 1975, then into the MPI again in 1978.[17]

The MPI, which was in charge of the exploration, development and production of crude oil and natural gas in China, was a very successful, influential and publicised ministry. Without the MPI, China's industrialisation and modernisation drive would have been severely impaired. The MPI not only turned the country from an oil importer to an oil exporter, but also significantly enriched the state's coffers. For instance, Daqing oil field led the nation as the factory or enterprise with the largest financial payment to the state. In the mid-1980s, Daqing was the largest single source of the state revenue, accounting for no less than three per cent of the total revenue the state earned from all enterprise profits and taxes. Also, oil export at that time was a major source of the country's foreign currency earning. For example, the country's oil export in 1978 contributed to USD 1.3 billion or 11 per cent of China's total foreign exchange earnings in that year, and these figures grew to USD 6.4 billion and 23 per cent in 1985. The MPI's prominent records from the 1960s to the 1980s promoted its self-confidence. With its production growing nearly twenty times, its position and reputation rose dramatically. It was considered a good place to work throughout the 1980s and reportedly attracted the best and the brightest. Since the MPI was earning a lot for the state and retained part of its revenue, it was able to provide better welfare, treatment and service for its employees than many other ministries.[18]

---

[16] Kenneth Lieberthal and Michel Oksenberg, *Policy Making in China: Leaders, Structures, and Process* (Princeton: Princeton University Press, 1988), p. 84.

[17] The website of CNPC Group, http://www.cnpc.com.cn/CNPC/gsjs/fzlc/, accessed 9 April 2008.

[18] Lieberthal and Oksenberg, *Policy Making in China*, pp. 82–83.

6.2.1.2 *The period of 1978–1982*

During the transitional period from 1978 to the first round of the bureaucratic restructuring in 1982, there were four institutions in the State Council with major authority over the oil affairs — the MPI, the SPC, the State Economic Commission (EC) and the State Energy Commission (SEC).

6.2.1.2.1 The State Economic Commission (1978–1988)

The EC was originally established in 1956 and merged into the SPC in 1970, then resumed in 1978.[19] Compared with the SPC, the EC played a relatively minor role in the energy policy making and the construction of energy projects. But it was critical in enabling an energy project to function effectively. The core task of the EC was to organise the implementation of production plans made by the SPC through coordinating the work of relevant ministries. Also, one of the EC's functions was assuring that adequate raw materials and energy were available to meet plan demands.

Specifically, the EC had significant influence on China's energy industry in four areas: First, it had the responsibility to ensure the up-to-standard operation of energy projects. Second, it played an essential role in creating industrial enterprises to replace the former governmental units and increasing these enterprises' autonomy from higher administrative control, such as increasing the autonomy of oil fields *vis-à-vis* the MPI. Third, it played a key role in the technological update of energy plants and in energy conservation. Fourth, it was responsible for managing energy production in the face of energy shortages. In addition, within the EC, the Energy Research Institute participated in long-term energy planning and forecasting. Under the EC, there were two bureaus — the Production Management Bureau

---

[19] Quanjing Wang (Quanjing On-line), 'Jigou gaige sanshi nian jingjian weibiao, zhuanxing weili, minsheng weizhong' ('The bureaucratic restructuring in the past three decades with simplification as its surface, transformation as its substance, and the improvement of the people's livelihood as its emphasis'), 14 March 2008, http://www.p5w.net/news/gncj/200803/t1546372.htm, accessed 13 April 2008.

and the Energy Bureau — that primarily dealt with energy issues. The Production Management Bureau directed the total energy production process on the basis of daily reports; and the Energy Bureau focused on policy issues and provided advices for the formulation of annul plans.[20]

6.2.1.2.2 The State Energy Commission (1980–1982)

After the Cultural Revolution, China faced an acute shortage of energy production such as coal, electricity and petroleum. Thus, the Chinese government needed to recentralise authority over the energy sector and improve its overall energy administration, in order to ensure sufficient energy supply for economic development. Against this background, the SEC was established in 1980 and Vice Premier Yu Qiuli became the SEC Chairman. Meanwhile, the ministers of the MPI, the Ministry of Coal Industry and the Ministry of Electric Power Industry acted as SEC Vice Chairmen, so as to integrate and coordinate various energy management agencies in the government. The SEC was abolished during the 1982 restructuring. It is said that the main reason was that the responsibility and function of the SEC were in serious conflict with those of the SPC.[21]

6.2.1.3 *The period of 1982–1988*

In 1982, the first wave of China's bureaucratic restructuring during the economic reform era took place. In order to improve both the efficiency and the responsiveness of state institutions, the restructuring ran against vested local, unit and factional interests. It reduced the number of vice-premier from 18 to three and that of organisations in the central government from 100 (52 ministries and commissions) to 61 (43 ministries and commissions). The ideological background of this round of restructuring was that the 12th CCP National Congress convened in September 1982 decided that China should build a

[20] Lieberthal and Oksenberg, *Policy Making in China*, pp. 72–78.

[21] CNPC News On-line, "“The Energy Bureau” will be promoted to be “state bureau”".

mixed economy with the planned economy as the principal element and a market economy as the supplementary. Thus, during the bureaucratic restructuring, the EC was restructured and reinforced, becoming a powerful unit promoting the market economy. Together with ten other organisations, the SEC was merged into the EC.[22] In accordance, an Energy Bureau was created within the EC. Hence, from 1982 to 1988, there were three major oil supervising agencies in the central government — the MPI, the SPC and the EC. In addition, two state oil companies were created during this period. In 1982, a newly established firm — CNOOC — was separated from the MPI. In 1983, another newly created company — China Petroleum and Chemical Industry General Corporation — was separated from the MPI. It was the predecessor of Sinopec Group.[23]

6.2.1.4 *The period of 1988–1993*

The second wave of bureaucratic restructuring was implemented in 1988. The ideological background of this restructuring was that in 1987, the 13th CCP National Congress made progress to build a market economy by deciding that the country should combine planned and market economies, and declared that China was still at an early stage of socialism. The market economy was granted equal status with the planned economy. This round of restructuring reduced the number of ministries and commissions from 45 to 41.[24] In particular, the EC was merged into the SPC, which could be viewed as a setback for the reformist effort, due to the reaction of conservative forces. In

[22] The other ten organisations are the old State Economic Commission, the State Agriculture Commission, the State Infrastructure Construction Commission, the State Machinery Industry Commission, the Finance and Trade Group of the State Council, the State Standard Bureau, the State Measurement Bureau, the State Bureau of Medicine Management, the State Patent Office and the State Bureau of Constructing Material Industry.

[23] The website of CNPC Group, http://www.cnpc.com.cn/CNPC/gsjs/fzlc/, accessed 15 March 2008.

[24] There were 61 organisations including 43 ministries and commissions after the 1982 restructuring. But the number increased to 72 organisations with 45 ministries and commissions in 1987. The restructuring experienced a cycle of contraction and expansion.

general, the EC was considered as promoting a market economy and the SPC was seen as supporting the planned economy. Notably, the MPI was abolished and the Ministry of Energy (ME) was created. But it had not possessed much actual power and authority over the oil sector since its establishment. In brief, from 1988 to 1993, there were two major oil supervising agencies — the SPC and the ME.

Also in 1988, major estates and energy assets previously under the control of the abolished MPI were transformed into a newly established company — China Petroleum and Natural Gas General Corporation, which was the predecessor of CNPC Group. This was a market-oriented move. As a huge SOE and the largest state oil company, China Petroleum and Natural Gas General Corporation was mainly involved in the upper stream production of oil and natural gas, simultaneously playing part of the role of governmental management and control over the oil sector.[25]

6.2.1.4.1 The Ministry of Energy (1988–1993)

During the 1988 bureaucratic restructuring, the MPI was one of the three ministries (the other two ministries were the Ministry of Coal Industry and the Ministry of Nuclear Industry) [26] that were merged into the newly created ministry — the ME. Actually, the previous energy ministries' management and production functions went to the newly established state energy enterprises,[27] and their administrative functions were incorporated into the ME.[28] The SPC Vice-chairman Huang Yicheng became the ME minister.

---

[25] CNPC News On-line, '"The Energy Bureau" will be promoted to be "state bureau"'; Downs, 'China'; and the website of CNPC Group, http://www.cnpc.com.cn/CNPC/gsjs/fzlc/, accessed 15 March 2008.

[26] During this restructuring, the Ministry of Water Resources and Power Industry handed over its power sector to the newly established ME, and was renamed the Ministry of Water Resources.

[27] These state energy enterprises created during the 1988 bureaucratic restructuring were China National Petroleum and Natural Gas General Corporation, China National Nuclear Industry General Corporation and China National Unified Distribution Coal Mining General Corporation.

[28] Downs, 'China'.

The ME was described as an 'experimental field' of the bureaucratic restructuring of the State Council. But due to various reasons, this experiment failed and the life of the ME came to an end five years later during the next round of bureaucratic restructuring. The reason for the abolishment of the ME includes: First, those state energy enterprises were created from former government ministries and enjoyed almost the same bureaucratic level — the ministerial level — with the ME, which means the ME hardly had any absolute authority over the energy companies. Second, the power to examine and approve proposed energy projects, which was the essential power for the administration over the oil industry, was with the SPC instead of the ME. Third, two big state energy companies, China Petroleum and Natural Gas General Corporation and China Nuclear Industry General Corporation, were not willing to be subordinated to the ME. These factors had made the authority of the ME over China's energy sector rather weak and ineffective.[29] Fourth, the ME met strong resistance from interest groups of the previous energy supervising ministries especially the powerful former Ministry of Coal Industry which petitioned to have their ministries reconstituted, and it was beyond the ME's reach to coordinate the relationship among different interest groups and manage the work of various energy sectors like coal, electric power, petroleum and nuclear industry.[30] Fifth, the authority of the ME overlapped with that of both the SPC and the state energy enterprises. As a result, the ME was only active in the electric power sector because the other energy sub-sectors 'refused to coordinate planning and investment activities'.[31]

6.2.1.5 *The period of 1993–1998*

The third wave of bureaucratic restructuring took place in 1993. The ideological background was that Deng Xiaoping's 'southern tour' in

[29] CNPC News On-line, '"The Energy Bureau" will be promoted to be a "state bureau"'.

[30] Xinlang (Sina), 'Huang Yicheng', 13 July 2005, http://finance.sina.com.cn/manage/cfrw/20050713/17011792952.shtml, accessed 26 April 2008.

[31] Downs, 'China'.

1992 pushed the 14th CCP National Congress held in that year announced that the goal of China's economic reform is to build a socialist market economy. Accordingly, a new round of bureaucratic restructuring was carried out to meet the demand of the market economy. The number of ministries and commissions decreased from 41 to 40. The ME was expanded into two ministries — the Ministry of Coal Industry and the Ministry of Electric Power Industry. Therefore, from 1993 to 1998, the SPC was the only significant oil supervising agency in the State Council, while the state oil companies possessing part of management function in the oil sector.

6.2.1.6 *The period of 1998–2003*

In 1998, the fourth round of bureaucratic restructuring occurred. The background was that the 15th CCP National Congress in 1997 decided that the SOE reform would be at the top of the government's reform agenda. In this wave of restructuring, the number of ministries and commissions was cut down from 40 to 29. The SPC which was viewed as a symbol of the old planned economy was restructured and became the State Development Planning Commission (SDPC). Its importance was greatly decreased and it became merely a research institute focusing on the country's long-term development plans. Moreover, the State Economic and Trade Commission (SETC), which was created during the 1993 restructuring, was significantly empowered and replaced the abolished SPC as the essential oil administrator. It was viewed as then Chinese Premier Zhu Rongji's power base and the so-called 'mini-State Council'. In addition, the Ministry of Land and Resources (MLR) was established and exerted important leverage over oil affairs. Thus, from 1998 to 2003, there were two major oil supervising agencies in the State Council — the SETC and the MLR.

6.2.1.6.1 The SETC (1993–2003)

The SETC was created in the 1993 restructuring and headed by Zhu who was the Vice-premier at that time. It became an essential oil supervising agency only after the 1998 restructuring. During this round of

restructuring, China Petroleum and Natural Gas General Corporation was transformed into CNPC Group and officially put under the domain of the SETC; China Petroleum and Chemical Industry General Corporation was transferred into Sinopec Group and also placed under the supervision of the SETC. Thanks to the restructuring, the new SETC took over seven existing commissions and ministries,[32] two councils[33] and two large SOEs (CNPC Group and Sionpec Group), making itself the most powerful institution within the State Council.[34] However, the SETC's lifetime ended in 2003 with the finish of the Zhu era.

6.2.1.6.2 The MLR (1998–present)

Moreover, during the 1998 restructuring, a new ministry — the MLR — was established through merging the Ministry of Geology and Mineral Resources, State Land Administration, State Oceanic Administration and the State Bureau of Surveying and Mapping.[35] The MLR is the administrator of the country's mineral and energy resources. It is responsible for the planning, administration, protection and rational utilisation of the China's natural resources such as land, mineral and marine resources. For example, the MLR has the function to 'supervise the examination, approval, registration and licensing of the rights to explore and to mine the mineral resources and the transfer of the rights to examine and approve blocks open to foreign investment'.[36] Zhou Yongkang who was President of China

[32] These seven ministries and commissions are the old SETC, the Ministry of Power, the Ministry of Coal Industry, the Ministry of Metallurgical Industry, the Ministry of Machine-Building, and the Ministry of Internal Trade.

[33] The two councils are Textile Industry Council and Light Industry Council.

[34] The website of CNPC Group, http://www.cnpc.com.cn/CNPC/gsjs/fzlc/, assessed 9 April 2008; and the website of Sinopec Group, http://www.sinopec-group.com/summarize/, accessed 9 April 2008.

[35] The State Oceanic Administration and the State Bureau of Surveying and Mapping are state bureaus under the administration of the MLR. See the MLR website, http://www.mlr.gov.cn/mlrenglish/about/history/, accessed 11 April 2008.

[36] Website of the Ministry of Land and Resources, http://www.mlr.gov.cn/mlrenglish/about/mission/, accessed 11 April 2008.

Petroleum and Natural Gas General Corporation was appointed the MLR minister.

6.2.1.7 *The period of 2003–2008*

After taking the premiership from Zhu Rongji in 2003, Wen Jiabao carried out the fifth wave of bureaucratic restructuring or adjustment, reducing the number of ministries and commissions from 29 to 28. The ideological background was that the 16th CCP National Congress in 2002 decided to further the market-oriented reforms. The most important development was the birth of three commissions and ministries, namely the National Development and Reform Commission (NDRC), the State-owned Assets Supervision and Administration Commission (SASAC) and the Ministry of Commerce (MOC). The SETC was dissolved and its function was taken by these three newly created institutions.[37] The NDRC became the most important energy responsible agency. Also, a high level coordinating institution — the State Energy Leading Small Group (SELSG) — was established in 2005 to improve the government's administration over energy affairs. Hence, from 2003 to 2008, the NDRC, the SASAC, the MLR and the SELSG were the major oil decision making bodies.[38]

---

[37] Xinhua, 'Guojia jingmao wei pingjing miandui "danchu"'('State Economic and Trade Commission calmly faces its "fading out"'), 11 March 2003, http://news.xinhuanet.com/zhengfu/2003-03/11/content_770480.htm, accessed 9 April 2008.

[38] Apart from these major oil supervising agencies, the State Council's authority over the energy affairs is also shared by the MOC, which issues licences for oil imports and exports, oversees Chinese firms' overseas investment, and organise and coordinate China's 'going global' strategy; the Ministry of Finance, which provides financial support such as tax and fiscal policies to promote the government's energy objectives; and the MFA, which provides support for the NOCs' bids to gain trade and investment opportunities abroad, 'and works to ensure that the deals pursued by the NOCs do not run counter to other foreign policy objectives'. See Downs, 'China'; and website of the Ministry of Commerce, http://wms.mofcom.gov.cn/aarticle/gywm/200203/20020300003218.html and http://hzs.mofcom.gov.cn/aarticle/gywm/200307/20030700105079.html, accessed 11 April, 2008.

6.2.1.7.1 The NDRC (2003–present)

During the 2003 bureaucratic restructuring, the SDPC was merged with the State Council Office for Restructuring the Economic System and part of the SETC and restructured into the NDRC.[39] It was the most powerful institution within the State Council or the new 'mini-State Council'. The NDRC was designated with the function to formulate 'annual plans, industrial policies and price policies', examine and approve major construction projects, guide the development of industry, and 'formulate plans for the development of the energy sector and manage national oil reserve'.[40] It was in charge of 'planning long-term energy development, setting energy prices, and approving investment in domestic and international energy projects'.[41] Moreover, there were several departments or bureaus under the NDRC with critical authority over the energy sector. For example, the Bureau of Energy (BE) had the following function:

> 'The Bureau of Energy is responsible for studying energy development and utilisation both at home and abroad, and putting forward energy development strategies and major policies; formulating development plans of the energy sector, and making recommendations on system reform in the energy sector; administering oil, natural gas, coal, power and other parts of the energy sector and national oil reserve; and formulating policy measures for energy conservation and renewable energy development.'[42]

Apart from the BE, the Bureau of Economic Operations had the responsibility to coordinate the supply of key raw materials and energy such as coal, oil and electricity; and there was a Division of

---

[39] The National Development and Reform Commission (NDRC) website, http://www.ndrc.gov.cn/jj/default.htm, accessed 9 April 2008.

[40] The NDRC website, http://en.ndrc.gov.cn/brief/default.htm, accessed 10 April 2008.

[41] Downs, 'China'.

[42] The NDRC website, http://en.ndrc.gov.cn/mfod/t20050519_0901.htm, accessed 10 April 2008.

Petroleum and Chemistry within this bureau.[43] The Department of Trade had the function of formulating plans for the total volume of import and export of important raw materials, supervising the implementation of these plans and adjusting them in accordance with the performance of the national economy.[44] The prices for the domestic oil products were under the supervision of the Division of Petroleum and Industrial Products in the Department of Price.[45]

6.2.1.7.2 The SASAC (2003–present)

Like many large SOEs in China, the NOCs are theoretically controlled by the central government. They are under the direct administration and supervision of the SASAC. Established in June 2003, the SASAC exercises the central government's ownership over the state-owned assets in those so-called central SOEs (zhongyang qiye), which are big non-financial SOEs including China's strategic industries such as oil and natural gas, nuclear power, aviation and spaceflight, electricity, telecommunications and arms industry.[46]

As Naughton wrote, since the Chinese government's ownership of SOEs has been redefined as shareholding, it is necessary to create an effective organisation to actually exercise the government's ownership rights. After some unsuccessful experiments, China established the SASAC whose 'core mission is to carry out the government's functions as investor and owner of state assets, and thus separate these tasks from the government's role as public manager of society as a whole.' The SASAC's function lies in three important dimensions: First, the SASAC 'is tasked to monitor enterprise operations in order

---

[43] The NDRC website, http://yxj.ndrc.gov.cn/jgsz/default.html, accessed 10 April 2008.

[44] The NDRC website, http://en.ndrc.gov.cn/mfod/t20050520_0894.htm, accessed 10 April 2008.

[45] The NDRC website, http://jgs.ndrc.gov.cn/jgsz/default.html, accessed 10 April 2008.

[46] The State-owned Assets Supervision and Administration Commission (SASAC) website, http://www.sasac.gov.cn/n1180/n1196/n3145/n5738/index.html, accessed 11 April 2008.

to protect the rights of the government owner; to appoint members of boards of directors and establish procedures for appointing managers; and to approve major decisions in enterprise operation.' Second, the central SASAC[47] has authority over a specific list of enterprises and exercises direct ownership only over those firms. Third, the central SASAC's demarcation of authority freed local governments to set up their own local SASACs exercising ownership of SOEs in every province.[48]

6.2.1.7.3 The SELSG (2005–2008)

Basically, the background of the establishment of the SELSG included three aspects: The first was the inability of the BE under the NDRC to effectively manage China's energy sector. The second was the energy crises during 2003 and 2004. The third was the strong advocacy within China to establish a high-level energy management agency. First of all, although the BE under the NDRC was the most important energy responsible agency in the State Council, its lack of manpower, financial resources and political clout had limited its capacity to manage the energy sector. Particularly, as a bureau under the NDRC, the BE did not have the necessary authority to coordinate those actors that enjoy higher bureaucratic ranks and more political power than the bureau, such as the NOCs and other ministries. Second, China experienced relatively small-scaled energy crises from 2003 to 2004. The widespread energy shortage presented an urgent need for the Chinese leadership to improve its energy administration through institutional change. Also, the energy crisis prompted the NDRC officials to express to the top leadership their frustration with the NDRC's lack of authority to manage the energy sector without

[47] There is a central SASAC in the State Council and local SASACs in the provinces.

[48] Naughton, *The Chinese Economy*, p, 316. Apart from the central SASAC which administrates and supervises central SOEs, there are local SASACs under each provincial-level government administrating and supervising local SOEs. For more information about local SASACs, see the SASAC website, http://www.sasac.gov.cn/n1180/n20240/n2454922/2459714.html, accessed 21 February 2008.

support from the top leaders. Although the NDRC is the most powerful department in the State Council, it does not have the authority to coordinate some other stakeholders such as the Ministry of Foreign Affairs (MFA).[49] Third, many Chinese scholars, energy experts and government officials had called for a long time for the establishment of an independent energy responsible agency with a higher ranking (ministerial ranking or higher) to solve the country's long existing and serious energy management problems, which has exerted continuous pressure on the Chinese leadership. Indeed, the conflicts or divergent interests regarding China's energy security and energy diplomacy among various departments within the State Council, such as the differences between the MFA and some other oil administrative bodies, have existed for years and need coordinating. Therefore, the establishment of a coordinating institution above the ministerial level was necessary.

In 2005, the Chinese government created the SELSG which was headed by then Premier Wen Jiabao and included heads of 13 governmental and military units. The members of the SELSG as in 2005 are shown in Table 6.1. Premier Wen convened the first SELSG meeting in June 2005. The SELSG was designated to be a high-level platform for dialogue on energy issues and coordination among various stakeholders.

In China, Leading Small Groups (*lingdao xiaozu*) are supraministerial coordinating and consulting bodies formed to build consensus on some crucial and strategic issues when the existing bureaucracy is unable to do so. They provide a mechanism for top decision makers to exchange views and formulate guiding principles for policies. Since Leading Small Groups have no permanent staff, they have to rely on their administrative offices to manage their daily operations and for research and policy recommendations. Hence, the effectiveness of a Leading Small Group often depends on that of its office. In

---

[49] People's Daily On-line, 'Guojia nengyuan lingdao xiaozu chengli, Wen Jiabao ren zuzhang' ('The State Energy Leading Small Group has been established and is headed by Wen Jiabao'), 30 May 2005, http://politics.people.com.cn/GB/1026/3426700.html, accessed 27 April 2008; and Downs, 'China'.

Table 6.1: Members of the SELSG in 2005

Name	Title
Wen Jiabao (Head of the SELSG)	Premier
Huang Ju (Deputy Head of the SELSG)	Executive Vice-Premier
Zeng Peiyan (Deputy Head of the SELSG)	Vice-Premier
Ma Kai (Head of the SELSG Office)	Chairman of the NDRC
Li Zhaoxing	Minister of Foreign Affairs
Jin Renqing	Minister of Finance
Zhang Yunchuan	Chairman of the Commission of Science, Technology and Industry for National Defence
Li Rongrong	Chairman of the SASAC
Bo Xilai	Minister of Commerce
Sun Wensheng	Minister of Land and Resources
Chai Songyue	Chairman of the State Electricity Regulatory Commission
Xu Guanhua	Minister of Science and Technology
Du Qinglin	Minister of Agriculture
Xie Zhenhua	Director of the State Environmental Protection Administration
Li Yizhong	Director of the State Administration of Work Safety
Ge Zhenfeng	Deputy Chief of the General Staff of the Chinese PLA

Source: Based on Downs, 'China', p. 19.

accordance, the administrative body of the SELSG is the SELSG Office which was granted the vice-ministerial rank and had three sub-groups and 24 employees. The SELSG had to rely on the SELSG Office to execute its decisions and manage its daily administrative affairs. Then NDRC Chairman Ma Kai became Director of the Office; the two Deputy Directors were Ma Fucai who was previously President of CNPC Group and Xu Dingming who was head of the BE. Such arrangement reflected their expertise on energy issues and also gave the NOCs another critical channel to reach and influence

the top energy leadership and administrative institution. Moreover, it is said that the SELSG Office was designed to serve the energy companies as much as possible while ensuring their behaviour does not harm other areas. Indeed, the Office served as a vehicle both for the NOCs to influence the government's energy policy making and for the government to attempt to manage the NOCs.[50]

The major function of the SELSG Office included: First, carrying out the routine duties of the SELSG and supervising the implementation of the decisions made by the SELSG; second, monitoring the country's energy security condition, predicting and providing early warnings of big energy problems and providing suggestions on solutions to the problems for the SELSG; third, organising relevant institutions to carry out research on energy strategy and plans; fourth, preparing important policies on energy development and conservation, energy security and emergency measures, international energy cooperation, etc.[51]

In addition, after the creation of the SELSG, there was still a strong and prevalent voice in China calling for the establishment of an independent ME or the re-establishment of the ME, or advocating that the SELSG Office should be upgraded to the future ME. For instance, a researcher with the Energy Research Institute under the NDRC said that the BE under the NDRC was originally created as a transitional agency; it was not able to solve the issues in the management of the energy sector. The establishment of the vice-ministerial SELSG Office was also a transitional arrangement. If it is proved by the practice that an ME is necessary, the re-establishment of the ME would be realised in the future.[52] Such advocacy to set up an inde-

---

[50] Xinhua, 'Guojia nengyuan lingdao xiaozu bangongshi zuo chengli, Wen Jiabao qinren zuzhang' ('The State Energy Leading Small Group Office was established yesterday and Wen Jiabao became the Head of the Group'), 3 June 2006, http://news.xinhuanet.com/fortune/2005-06/03/content_3040477.htm, accessed 27 April 2008; Downs, 'China'; and Zhang Libin and Jason Lee, 'Untangling China's Energy Policy', *China Security* 4: 3, 2008, pp. 58–61.

[51] The website of the SELSG Office, http://www.chinaenergy.gov.cn/category_news_4_0.html, accessed 11 April 11, 2008.

[52] People's Daily On-line, 'The State Energy Leading Small Group has been established and is headed by Wen Jiabao'.

pendent and powerful ME might continue in the future. Considering the proved inability of the NDRC to effectively deal with China's energy problems, the Chinese leadership had to do something to improve the government's administration on oil affairs.

6.2.1.8 *The period of 2008–present*

The next round of bureaucratic restructuring was carried out in 2008 under the leadership of Wen Jiabao who served as the Premier for the second term. The number of ministries and commissions was reduced from 28 to 27. The background was that the CCP 17th National Congress in 2007 determined that the 'scientific development' concept was the guiding principle for the work of the government, reflecting the Hu Jintao-Wen Jiabao leadership's attempt to direct China away from its blind and ruthless pursuit of GDP growth to a more balanced path with a more sustainable form of development.[53] During this wave of restructuring, the BE under the NDRC was upgraded to the vice-ministerial level National Energy Administration (NEA), but still under the leadership of the NDRC. The SELSG and its Office were abolished. The SEC, which existed between 1980 and 1982, was re-established and headed by then Vice-premier Zhang Dejiang. Therefore, since 2008, there have been four institutions in the State Council with major authority over oil affairs, namely the NEA under the NDRC, the SEC, the SASAC and the MLR.

6.2.1.8.1 The National Energy Administration (2008–present)

During the 2008 restructuring, the BE under the NDRC was upgraded to the NEA enjoying a vice-ministerial level ranking but still under the administration of the NDRC. The NEA incorporated the NDRC's functions of the administration over energy sectors, the NDRC's

---

[53] Zheng Yongnian and Zhengxu Wang, 'China's National People's Congress 2008: New Administration, Personnel Reshuffling and Policy Impacts', Briefing Series — Issue 38, China Policy Institute, The University of Nottingham, March 2008, http://www.nottingham.ac.uk/shared/shared_cpi/documents/policy_papers/Briefing_38_NPC_2008.pdf, accessed 3 May 2008.

sub-institutions with such functions, the functions and responsibilities of the SELSG Office, and those of the previous Commission of Science, Technology and Industry for National Defence[54] over the nuclear power sector.[55] The NDRC Vice-Chairman Zhang Guobao, who was a full ministerial level official, held the concurrent post as Commissioner of the NEA. Actually, it is a convention that NEA Commissioner is simultaneously one of the Vice-Chairmen of the NDRC. This reflects the NDRC's administration over the NEA.

The purpose of establishing the NEA was to strengthen the unified administration of energy sectors, cope with the increasingly challenging energy issues in and outside China, and ensure the sustainable, stable and healthy development of the national economy. The NEA's detailed functions included: First, formulating energy development strategies, programmes and policies, and providing suggestions for relevant institutional reform; second, carrying out administration over energies like oil, natural gas, coal and electric power (including nuclear power); third, managing the national oil reserves; fourth, formulating policy measures for the development of new energies and energy conservation; fifth, carrying out international energy cooperation. The NEA is required to strengthen research on energy issues and promote the state's capability to safeguard national energy security.

In 2008, there were nine departments (*si*) and 112 personnel in the NEA. The nine departments were: Department for Comprehensive Affairs, Department for Policies and Regulations, Department for Development Planning, Department for Energy Conservation and Science and Technological Equipments, Department for Electric Power, Department for Coal, Department for Oil and Natural gas,

[54] The Commission of Science, Technology and Industry for National Defence was abolished during the 2008 restructuring. Its functions and responsibilities besides nuclear power management were incorporated into the newly established Ministry of Industry and Information, under which a ministerial-level State Bureau of Science, Technology and Industry for National Defence was created.

[55] Xinhua, 'Guowuyuan jigou gaige fangan' ('The bureaucratic restructuring programme of the State Council'), 15 March 2008, http://news.xinhuanet.com/misc/2008-03/15/content_7794932.htm, accessed 3 May 2008.

Department for New Energies and Renewable Energies and Department for International Cooperation.[56]

The period between 2008 and 2015 witnessed a frequent change of NEA Commissioners, as four officials held the post one by one. When Zhang Guobao became NEA Commissioner in 2008, he held several concurrent posts as NDRC Vice-chairman, member of the CCP NDRC party group, Director of Office of the State Council Leading Small Group for Revitalising the Old Industrial Base in North–Eastern China, member of the National Defence Mobilisation Committee, First Political Commissar of the Gold Forces of the Chinese People's Armed Police, member of the National Stereotypes Committee, member of the Standing Committee of the 11th Chinese People's Political Consultative Conference (CPPCC), and member of the 11th CPPCC Economic Committee. In 2011, he retired from the posts of NDRC Vice-chairman and NEA Commissioner at the age of 67. Liu Tienan, who had been NDRC Vice-chairman and member of the CCP NDRC party group since 2008, succeeded Zhang as NEA Commissioner. In 2013, Liu was arrested and put under investigation for accusation of corruption and violation of the CCP discipline. Later in the year, he was expelled from the CCP and dismissed from his post. In 2014, he was sentenced to life imprisonment for taking bribe. After Liu, Wu Xinxiong, who was previously Chairman of State Electricity Regulatory Commission, took over the posts in March 2013 as NDRC Vice-chairman and NEA Commissioner, enjoying full ministerial level ranking. Actually, Wu's assumption of duty in NEA happened with the latest round of bureaucratic restructuring of the State Council, in which the State Electricity Regulatory Commission was merged into the NEA. In December 2014, Wu retired from the posts of NDRC Vice-chairman and NEA Commissioner. In the same month, Nur Bekri, who was previously Chairman of Xinjiang Uyghur Autonomous Region, succeeded Wu as NDRC Vice-chairman and NEA Commissioner, also enjoying full ministerial level ranking and

[56] Renmin Ribao (People's Daily), 'Guojia nengyuanju "sanding" fang'an zhengshi huopi' ('The restructuring programme of the State Energy Bureau has formally been approved'), 30 July 2008, p. 9.

becoming one of the highest ranked ethnic-minority officials in the Chinese government.

After the bureaucratic restructuring in 2013, the NEA currently has 12 departments. Apart from the nine departments that have existed since 2008, three departments were created under the NEA in 2013, namely Department for Nuclear Electricity, Department for Market Supervision and Department for Electricity Safety Regulatory. Notably, since 2008, the number of personnel in the NEA has more than doubled. There are 240 personnel in the NEA including one Commissioner and four Deputy Commissioners.[57]

6.2.1.8.2 The new State Energy Commission (2008–present)

The SEC, which existed from 1980 to 1982, was re-established during the 2008 restructuring, and was reported to be headed again by a Vice-premier. The former SEC was headed by Yu Qiuli and the new SEC was said to be headed by Zhang Dejiang. Similar to the abolished SELSG, the SEC is a high level institution for dialogue on energy issues and coordination of various stakeholders. The work of the SEC is undertaken by the NEA.[58] Such work presumably includes executing decisions made by the SEC and managing the SEC's daily affairs. The SEC is responsible for formulating national energy development strategies, discussing important issues regarding energy security and energy development.[59]

Nevertheless, it was not until 2010 that the new SEC was formally established. Then Chinese Premier Wen Jiabao, instead of any of the Vice-premiers, became Chairman of the SEC. Then Executive Vice-premier Li Keqiang was the Vice-chairman of the SEC. In April 2010, Premier Wen chaired the first meeting of the SEC.[60]

---

[57] Website of National Energy Administration, http://www.nea.gov.cn/n_home/n_nyjjj/index.htm, accessed 14 March 2015.

[58] Xinhua, 'The bureaucratic restructuring programme of the State Council'.

[59] People's Daily, 'The restructuring programme of the State Energy Bureau has formally been approved'.

[60] Xinhua, 'Wen Jiabao zhuchi zhaokai guojia nengyuan weiyuanhui diyici huiyi' (Premier Wen Jiabao chaired the first meeting of the State Energy Commission), 22 April 2010, http://news.xinhuanet.com/photo/2010-04/22/c_1250287.htm, accessed 15 March 2015.

Therefore, the new SEC finally enjoyed a higher ranking and political status than its predecessor in the 1980s. Then, however, the SEC was scarcely reported and mentioned by the public for four years, until April 2014 when the second meeting of the SEC was chaired by Premier Li Keqiang who succeeded Wen Jiabao as Premier one month ago. This happened after the CCP 18th National Congress in 2012 and the assumption of duty by the new administration and the latest round of State Council bureaucratic restructuring in 2013.[61]

In this sense, it seems that in fact the SEC finally became a loose and powerless state institution designed to coordinate various interests and players beyond the State Council, which largely resembled the abolished SELSG from 2005 to 2008, rather than the previous SEC within the State Council in the 1980s. To a certain extent, this might be viewed as a setback for those advocating a significant reform in China's energy management system through the establishment of a powerful SEC with effective administration over the NOCs as well as the energy sector.

Notably, this setback is similar to the fate of the previous short life SEC from 1980 to 1982. At that time, the termination of the short life SEC was because its responsibility and function were in serious conflict with the SPC which is the predecessor of today's NDRC. Indeed, in the 1980s, there was a zero-sum game between the SEC and the SPC. If the SEC were empowered and became the major energy regulator and administrator, the SPC's power and interests would be seriously eroded. In the 2000s, there could be a similar situation between the proposed new SEC and the existing NDRC. Therefore, it is likely that the failure to establish a SEC with real power and authority over the energy sector was due to the strong opposition from the NDRC.

In 2013, the SEC had more than 22 members including heads or senior officials from 19 governmental and military units. The members of the SEC as in 2013 are shown in Table 6.2.

The Office of the SEC was also established, with Chairman of the NDRC as its Head and Commissioner of the NEA as its Deputy

[61] Xinhua, 'Li Keqiang zhuchi zhaokai xinyijie guojia nengyuan weiyuanhui shouci huiyi' (Premier Li Keqiang chaired the first meeting of the State Energy Commission in the new administration), 20 April 2014, http://news.xinhuanet.com/photo/2014-04/20/c_126411441.htm, accessed 15 March 2015.

Table 6.2: Members of the State Energy Commission in 2013

Name	Title
Li Keqiang (Chairman)	Premier
Zhang Gaoli (Vice-chairman)	Executive Vice-Premier
Xiao Jie	Secretary General of the State Council
Liu He	Director of Office for the CCP Central Leading Small Group on Financial and Economic Affairs
Wang Yi	Minister of Foreign Affairs
Xu Shaoshi	Chairman of the NDRC
Wan Gang	Minister of Science and Technology
Miao Wei	Minister of Industry and Information Technology
Geng Huichang	Minister of State Security
Lou Jiwei	Minister of Finance
Jiang Damin	Minister of Land and Resources
Zhou Shengxian	Minister of Environmental Protection
Yang Chuantang	Minister of Transport
Chen Lei	Minister of Water Resources
Gao Hucheng	Minister of Commerce
Liu Shiyu	Vice-president of the People's Bank of China
Jiang Jiemin	Chairman of the SASAC
Wang Jun	Commissioner of State Administration of Taxation
Yang Dongliang	Director of the State Administration of Work Safety
Shang Fulin	Chairman of China Banking Regulatory Commission
Wang Guanzhong	Deputy Chief of the General Staff of the Chinese PLA
Wu Xinxiong	Vice-chairman of the NDRC and Commissioner of National Energy Administration

Source: Based on Xinhua, 'Guowuyuan tiaozheng guojia nengyuan weiyuanhui zucheng renyuan' (The State Council has adjusted the members in the State Energy Commission), 11 July 2013, http://news.xinhuanet.com/2013-07/11/c_116499600.htm, accessed 15 March 2015.

Head. The administrative function and daily affairs of the SEC Office are expected to be fulfilled and carried out by the NEA.[62]

Figure 6.2 and Table 6.3 respectively illustrate the current structure of the State Council's administration over the oil sector and the

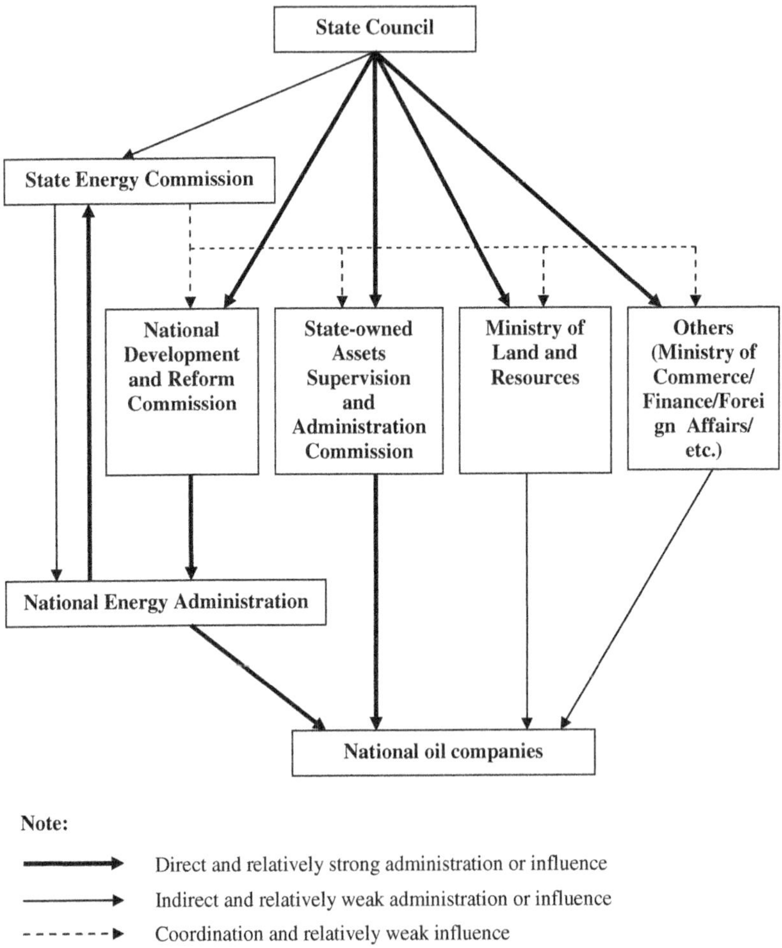

Figure 6.2: The State Council's oil administration structure (2008–present)

Source: Author's compilation.

[62] Xinhua, 'Guowuyuan tiaozheng guojia nengyuan weiyuanhui zucheng renyuan' (The State Council has adjusted the members in the State Energy Commission), 11 July 2013, http://news.xinhuanet.com/2013-07/11/c_116499600.htm, accessed 15 March 2015.

Table 6.3: Major oil supervising agencies in the State Council during various periods

Year	Major oil supervising agencies
1949–1952	• Ministry of Fuel Industry (MFI)
1952–1955	• MFI • State Planning Commission (SPC)
1955–1970	• Ministry of Petroleum Industry (MPI) • SPC
1970–1975	• Ministry of Fuel and Chemical Industry • SPC
1975–1978	• Ministry of Petroleum and Chemical Industry • SPC
1978–1980	• MPI • SPC • State Economic Commission (EC)
1980–1982	• MPI • SPC • EC • State Energy Commission (SEC)
1982–1988	• MPI • SPC • EC
1988–1993	• Ministry of Energy • SPC
1993–1998	• SPC
1998–2003	• SETC • MLR
2003–2005	• NDRC • SASAC • MLR
2005–2008	• SELSG • NDRC • SASAC • MLR
2008–present	• SEC • National Energy Administration under NDRC • SASAC • MLR

Source: Author's compilation.

NOCs and the evolution of major oil supervising agencies in the State Council from 1949 to present.

6.2.1.9 *Oil supervising agencies' limited authority over the NOCs*

Although the various governmental departments and institutions are the administrative bodies over the NOCs, they do not necessarily have enough authority or capacity to manage or control the firms. This is a big problem for the government's supervision of the NOCs' activities. For example, as the most important oil supervising agency from 2003 to 2008, the BE under the NDRC lacked political clout because it was merely a bureau-level agency, which is lower than the bureaucratic ranking of the NOCs. CNPC and Sinopec enjoy ministerial standing and CNOOC has a lower ranking of general bureau or vice-ministerial level, all of which are higher than the standing of the BE. Also, the BE suffered from insufficient financial resources and staff to oversee the energy sector. Besides, it is said that energy companies often undermined the authority of the BE by circumventing it to hold face-to-face discussions with China's top leadership.[63]

After the 2008 bureaucratic restructuring, the BE was promoted and enlarged to become the vice-ministerial level NEA and allocated more resources. Even if such progress may give the NEA enough financial and manpower for the management of the oil sector, the NEA still faces the old problem of political authority. It is still under the administration of the NDRC, rather than a fully independent administration. In fact, to a certain degree, it is a creature of the NDRC. The appointment of a NDRC Vice-chairman Zhang Guobao as head of the NEA and its party group indicated that this new agency's room to manoeuvre could be constrained by the NDRC. Moreover, the NEA's autonomy was limited by the fact that key tools

[63] Erica S. Downs, 'China's Energy Policies and Their Environmental Impacts', http://www.brookings.edu/testimony/2008/0813_china_downs.aspx, accessed 28 October 2008; Zhang and Lee, 'Untangling China's Energy Policy', p. 59; and Rosen and Houser, 'China Energy', pp. 18–19.

it needs to effectively anage the energy sector were still in the hands of the NDRC. Further, the NEA may still lack the manpower to satisfy its mandate and the political power to balance the interests of various stakeholders in the energy sector.[64]

Moreover, the NEA does not necessarily enjoy equal political status with the NOCs within the CCP and the state. For instance, in 2008 when the NEA was established, Jiang Jiemin who was President and Chairman of the Board of CNPC Group was an alternate member of the CCP 17th Central Committee. Su Shulin who was President and Chairman of the Board of Sinpoec Group had been an alternate member of the 16th and the 17th CCP Central Committees. In comparison, the NEA Commissioner Zhang Guobao was not in the Central Committee. In the future, even if the energy administration function is completely separated from the NDRC and an independent ME is re-established and headed by a more important political figure in the CCP, it may still face more or less similar problems, because the NOCs' economic and political power has been growing quickly and they have substantial resources to influence the government, as will be discussed later in this chapter.

6.2.2 *Other measures for the government to control the NOCs*

Apart from those oil supervising organisations holding relevant administrative power over the NOCs, the CCP government has other important measures to administrate and control the NOC, such as the nomenclatural system and the investment approval process.[65]

6.2.2.1 *The nomenclature system*

The nomenclature system is widely regarded as a key approach for the CCP government to manage and control the SOEs. Here, it simply

---

[64] Downs, 'China's Energy Policies and Their Environmental Impacts'; and Zhang and Lee, 'Untangling China's Energy Policy', pp. 59–61.

[65] Apart from these two key measures, the government also manages the NOCs through taxation and subsidy.

means that the CCP Central Committee has the power to appoint or dismiss the heads of major central SOEs including the NOCs through its organisational organs and institutions. When the SASAC was established in 2003, there were originally 198 central SOEs under its administration. However, due to the ongoing SOE reforms which aim to promote the overall strength of the SOEs and create some strong SOEs through restructuring and mergers, the total number of central SOEs has been reduced to 112.[66] The remaining central SOEs are generally large holding companies owning multiple operating companies as subsidiaries.

The central SOEs' managers are not only business leaders but also important political figures within the CCP framework. Fifty of them are directly appointed by the CCP Central Committee's Organisational Department and have ministerial-level ranking; their appointments are finally approved by the Standing Committee of the CCP Central Committee's Political Bureau. The other one hundred or so of them are appointed by the SASAC. Although their status is not so exalted, they are also important people.[67] Actually, it is more likely that the leading SOEs' bosses are nominated and decided directly by the top CCP leaders, such as members of the Standing Committee, then appointed formally by the CCP Central Committee's Organisational Department according to the official procedure.

Nevertheless, the nomenclatural system also has its limits in managing the NOCs. In particular, appointing the heads of the NOCs is not equal to managing the NOCs' corporate affairs and business operation. Anyone appointed as heads of SOEs have to serve the CCP government's interests as well as the company's interests, or at least make a balance between the two interests. If the bosses do not serve

[66] From 2002 to 2015, the number of central SOEs has been reduced from 196 to 112, due to the ongoing SOE restructuring and reform. Website of the SASAC, the People's Republic of China, 'Yangqi minglu' (List of Central State-owned enterprises), 9 February 2015, http://www.sasac.gov.cn/n86114/n86137/c1725422/content.html, accessed 15 March 2015.

[67] Barry Naughton, 'SASAC and Rising Corporate Power in China', China Leadership Monitor, No. 24, Hoover Institution, Stanford University, http://media.hoover.org/documents/CLM24BN.pdf, accessed 9 May 2008.

the firms' interests, it would be very difficult for them to manage these huge conglomerates and gain support from their subordinates. They have to effectively manage the company and realise business achievements which are in fact their personal political achievements, paving the way for them to be appointed by the CCP in the future to more important or higher ranking positions with more power.

6.2.2.2 *The investment approval process*

Another critical approach for the CCP government to supervise and control the NOCs is the investment approval process. The power of approval is a key power in China's political and economic life. Any governmental department possessing the power of approval over a particular industrial sector is the most powerful administrative superior in this sector. In theory, proposals from companies to carry out important business operations like constructing some projects and making big investments at home or abroad are subject to the approval by such administrative agencies before implementation. Some projects of particular importance or those that exceed certain amounts of money are required to go further up to be approved by higher level governmental bodies such as the State Council.

For instance, in the past, as the most powerful department in the State Council, the SPC held the investment approval authority over many industries including the oil sector and was the essential planner and administrator in this field. At present, such authority is largely held by the NDRC and its NEA. Hence, the NOCs' domestic and outward investments must be approved by the NDRC or the State Council; and for those ODI proposals, they must be approved by the MFA as well.

According to the regulation, any ODI projects proposed by the NOCs and valued USD 30 million or above are required to be approved by the NDRC; and any proposed ODI projects valued USD 200 million or above should be firstly reviewed by the NDRC and then submitted to the State Council for approval. When more than one NOC wants to bid for the same asset and requests the approval, the NDRC usually approves the proposal submitted first. Also, the State Council usually relies on recommendations from the NDRC.

Sometimes, an NOC may be given preference due to its good relationship with the NDRC.

Some Western analysts wrote that in practice, the NOCs at least in some cases have carried out ODI activities first and informed the NDRC and the State Council after the fact. The reason is that the cumbersome investment approval system, just as the situation in many other ODI applications by Chinese enterprises, would cost the NOCs important investment opportunities.[68] Considering the NOCs' powerful networks and connection within the CCP government and their significant influence over the government, as will be explained in the next section, it is not difficult to understand that the NOCs might be able to *xianzhan houzou* (act first and request approval afterwards). If this is true, some cases of the government's investment approval could be merely the acceptance of accomplished facts.

However, some senior managers in Chinese NOCs disagree with such perspective. They maintained that all of the NOCs' proposal for ODI projects must be approved by the government especially the MFA. The NOCs cannot just invest and operate in anywhere in the world as they wish. Their overseas business plans have to be approved by the MFA before implementation. After all, they are SOEs and their money is the state's wealth.[69]

6.3 The NOCs' influence on the government

This section discusses the NOCs' resources to influence the government's policy making. The reason for the NOCs' capability to exert considerable impact upon the government is that they have significant

---

[68] For example, see Downs, 'China'; and Zhongqing Zaixian (China Youth Daily On-line), 'Henduo duiwai touzi jihui bei shenpi "zheteng" meile' ('Many opportunities to carry out overseas investment were lost due to the investment approval process'), 20 June 2006, http://chuangye.cyol.com/content/2006-06/20/content_1422733.htm, accessed 20 May 2008.

[69] Author's interview with Academician Zeng Hengyi, who is Academician of the Chinese Academy of Engineering, member of China's National Expert and Consultant Committee for Energy as well as Vice Chief Engineer of China National Offshore Oil Corporation (CNOOC), Beijing, China, 30 June 2014.

material, institutional and ideological resources. Among them, material resources refer to the NOCs' growing financial strength; institutional resources include their monopoly in China's energy market, high political status and close relationship with the CCP government; and their ideological resources are mainly derived from the government's call for 'going global' (*Zou chuqu*) and the popular discourse within China that the NOCs' overseas expansion safeguards China's energy security. To some extent, the three resources are interrelated and support one another. This section consists of three sub-sections, dealing with the three resources respectively.

6.3.1 *Material resources*

In the recent years, the NOCs have quickly increased their financial capacity. They are among the largest and most profitable SOEs in China. Their monopoly in China's domestic market — a fast expanding market with huge potential — brings them windfall income and profit, giving them prominent positions among all central SOEs.

In particular, various rankings reflect the three major NOCs' status as the leading companies not only in China but also in the world. For example, in 2014, Sinopec and CNPC were ranked the largest and second largest companies in China on the basis of their operating income.[70] Also in 2014, Sinopec was ranked the third among the Fortune Global 500, with an operating income of USD 457.2 billion, only after Wal-Mart Stores (USD 476.29 billion) and Royal Dutch Shell (USD 459.6 billion). It is the first time that a Chinese company ranked among the top three companies in the world.[71] CNPC was ranked the fourth after Sinopec, with an operating income of USD

[70] Xinhua, '2014 zhongguo 500qiang qiye paihangbang: qianshiming junwei guoqi' (The ranking of China's top 500 companies: all of the top ten companies are state-owned enterprises), 14 July 2014, http://www.bj.xinhuanet.com/bjyw/2014-07/14/c_1111600716.htm, accessed 16 March 2015.

[71] Xinhua, '2014 shijie 500qiang paihangbang: zhongguo gongsi shoujin sanqiang' (The ranking of the world's top 500 companies in 2014: the first time a Chinese company is among the top three), 8 July 2014, http://news.xinhuanet.com/overseas/2014-07/08/c_126722764.htm, accessed 16 March 2015.

Table 6.4: Ranking of the three major Chinese NOCs among the Fortune Global 500, 2014 (Ranked by operating income, USD billions)

Rank	Company	Operating income (USD billions)
3	Sinopec	457.2
4	CNPC	432.01
79	CNOOC	95.97

Source: Cankao xiaoxi, '2014 nian shijie 500qiang, zhongshihua shangbang gongsi da baijia (mingdan)' (The world's top 500 companies in 2014: 100 Chinese companies are among them), 8 July 2014, http://finance.cankaoxiaoxi.com/2014/0708/416735_2.shtml, accessed 16 March 2015.

432.01 billion, which is CNPC's highest ranking among the Fortune Global 500 in history.[72] Meanwhile, CNOOC was ranked 79th, with an operating income of USD 95.79 billion, which is also its highest ranking in the list (see Table 6.4).[73]

Also in 2014, Sinopec was ranked the first among all the enterprises in China with an operating income of RMB 2945.07 billion; CNPC was ranked the second with an operating income of RMB 2759.3 billion and CNOOC was ranked the 10th with an operating income of RMB 590.07 billion (see Table 6.5).[74]

Moreover, like many other central SOEs, the NOCs still enjoy soft budget constraints to a large extent. The SOEs' keep of all after-tax

---

[72] Cankao xiaoxi, '2014 nian "caifu" shijie 500qiang chulu, zhongshihua paiming disan' (The 2014 Fortune Global 500 released: Sinopec Group was ranked the third), 8 July 2014, http://finance.cankaoxiaoxi.com/2014/0708/417053.shtml, accessed 16 March 2015.

[73] Cankao xiaoxi, '2014 nian shijie 500qiang, zhongshihua shangbang gongsi da baijia (mingdan)' (The world's top 500 companies in 2014: 100 Chinese companies are among them), 8 July 2014, http://finance.cankaoxiaoxi.com/2014/0708/416735_4.shtml, accessed 16 March 2015.

[74] Xinhua, '2014 zhongguo qiye 500 qiang mingdan fabu' ('The list of the top 500 enterprises in China in 2014 released'), 2 September 2014, http://news.xinhuanet.com/2014-09/02/c_126945173.htm, accessed 16 March 2015.

Table 6.5: Ranking of the three major Chinese NOCs among the top 500 enterprises in China, 2014 (Ranked by operating income, RMB billions)

Rank	Company	Operating income (RMB billions)
1	Sinopec	2945.07
2	CNPC	2759.3
10	CNOOC	590.07

Source: Xinhua, '2014 zhongguo qiye 500 qiang mingdan fabu' ('The list of the top 500 enterprises in China in 2014 released'), 2 September 2014, http://news.xinhuanet.com/2014-09/02/c_126945173.htm, accessed 16 March 2015.

profits without remitting a penny to the government during 1994 and 2008 have contributed to their financial strength. In 1994 when the government was implementing tax reforms, it made a decision that SOEs would no longer remit their after-tax profits to the government. Since then, they only remit taxes to the government while retaining all of their after-tax profits until 2008 when the central government implemented a new policy initiative to collect a small proportion of after-tax profits from central SOEs. But the unification and decrease of the corporate income tax rate for Chinese invested companies from 33 per cent to 25 per cent is an important offsetting factor. The drop in tax rates would be of the same order of magnitude as the increase in the after-tax profits. Hence, central SOEs will have about the same amount of money they would have had without any changes in law.[75]

When the corporate income tax for Chinese invested firms was 33 per cent, unlike some companies that previously enjoyed preferential rates due to various reasons, oil companies had few chances to reduce their actual taxes. Thus, the new corporate income tax rate directly mitigates the oil companies' tax burdens and increases their profits.[76]

[75] Naughton, 'SASAC and Rising Corporate Power in China'.

[76] Xinhua, 'Xin qiye suodeshui fa huiji guonei shiyou qiye' ('The new Corporate Income Tax Law benefits Chinese oil enterprises'), 21 May 2007, http://news.xinhuanet.com/fortune/2007-05/21/content_6130570.htm, accessed 18 May 2008.

Similarly, some observers pointed out that from 1998 to 2005, the total profits of those profitable non-financial SOEs including the NOCs was nearly five trillion RMB. But they did not turn the profits over to the state. On the contrary, the state constantly invested in and subsidised those firms, resulting in a heavy burden on the state.[77]

Therefore, many SOEs including the NOCs still enjoy soft budget constraints, even in the post-planned economy era. If they make money, they retain the after-tax profits; if they lose money, the state is ready to subsidise them. Within the NOCs, if their downstream sections (e.g. refining) suffer losses due to the gap between the soaring international oil prices and the domestic prices for oil products set by the NDRC, they can gain the government's subsidy for such money-losing sections, although they may still have a surplus of overall profits without the state's subsidy due to their other profitable sections such as the upstream section. Once their downstream section loses money, they tend to apply for a subsidy from the state, and the government has to satisfy their demand to a certain degree, although many ordinary people are unhappy with the appetite of these monopoly firms.

The Chinese government has been subsidising CNPC and Sinopec for a long time. From 2004 to 2013, CNPC and Sinopec together received government subsidies for RMB 125.88 billion, with PetroChina Ltd. receiving RMB 48.44 billion and Sinopec Ltd. receiving RMB 77.45 billion. For instance, in 2013, PetroChina Ltd. received RMB 10.35 billion of government subsidy, making the company the largest recipient of government subsidy among all companies listed on China's A stock market for three years in consecutive (2011–2013). In 2013, Sinopec Ltd. received RMB 2.39 billion of government subsidy. Also, from 2011 to 2013, Sinopec Ltd. received government subsidy for RMB 6.73 billion, RMB 9.41 billion and RMB 10.35 billion respectively, making the company the largest recipient of government subsidy during the three years. Moreover,

[77] Kong Shanguang, 'Wu wanyi de guoyou qiye lirun nail qu le?' ('Where has gone the SOEs' 5 trillion RMB profit?'), available at Guangming wang (Bright Daily On-line), 11 February 2007, http://guancha.gmw.cn/show.aspx?id=3445, accessed on 24 May 2008.

Sinopec Ltd. was the largest recipient of government subsidy from 2005 to 2008, receiving RMB 9.42 billion, RMB 5.16 billion, RMB 4.86 billion and RMB 50.34 billion respectively. Notably, when receiving large amounts of government subsidies, PetroChina Ltd. and Sinopec Ltd. produced windfall profits. For example, in 2008, when receiving substantial government subsidies, the net profits of PetroChina Ltd. and Sinopec Ltd. were RMB 113.8 billion and RMB 28.4 billion respectively. In 2013, PetroChina Ltd. and Sinopec Ltd. respectively gained RMB 129.6 billion and RMB 67.2 billion of net profits.[78]

In addition, Chinese NOCs' listing of some of their subsidiaries on overseas stock markets has enabled them to secure more financial resources for ODI and transnational operations. Since the late 1990s, China has sought aggressively to list parts of its largest and most profitable SOEs on international stock markets. In 2000 alone, all of the three leading Chinese NOCs — CNPC, Sinopec and CNOOC — listed their subsidiaries in Hong Kong and New York and raised a total of USD 15 billion. Such overseas listings created a new platform and enabled the pooling of financial strength and managerial expertise for the listed firms and their parent companies to launch aggressive cross-border mergers and acquisitions (M&A).[79]

6.3.2 *Institutional resources*

The main institutional resources supporting the NOCs' significant influence are their monopoly in China's market and their close ties with the party-state. First of all, an important factor for the NOCs' critical role in China's energy regulation and policy making is their monopoly in the domestic market and their possession of many subsidiaries across the country. They are *de facto* regulators in China's oil market, especially in the areas of wholesale and retail. In this sense, they are both

[78] Xinhua, 'Meiti cheng zhongshihua zhongshiyou shinian huo zhengfu butie chao qianyi' (Media reports maintain that Sinopec Ltd. and PetroChina Ltd. have received government subsidies for more than RMB 100 billion during the past decade), 15 April 2014, http://news.xinhuanet.com/legal/2014-04/15/c_126391736.htm, accessed 16 March 2015.

[79] Zhang Yongjin, *China Goes Global* (London: The Foreign Policy Centre, 2005), p. 12.

players and regulators in the market. Their monopoly is not only in exploration and development, but also in import and export. Although the Chinese government has allowed the non-state-owned oil companies to participate in the trade of crude oil and oil products, it also sets a quota for them to import less than ten per cent of the country's overall import of crude oil and 14 per cent of the import of oil products. As a result, the non-state-owned oil companies are squeezed brutally by the three major NOCs. Since most of them do not have their own sources of crude oil and oil products, they have to rely on the NOCs in particular CNPC and Sinopec for the supply of such commodities, therefore maintaining the NOCs' monopoly in China.[80]

Moreover, the NOCs have a great incentive to pursue political influence within the state, because the state's policies have a fundamental impact on their fate. As some scholars noted, China's state energy companies have considerable power not only in the market but also with respect to the government.[81] Many central SOEs are large holding corporations that evolved from former governmental ministries. They have hundreds of subsidiary companies, control large sums of money and exercise strategic leverage over the government's policy making.[82] The NOCs are among those SOEs that have formed the so-called corporation interest group seeking privileges or favourable policies through the government's policy making process, or through the administrative departments and supervising agencies via their special government-business relationship.[83] Central

[80] Bo Kong, 'An Anatomy of China's Energy Insecurity and Its Strategies', Pacific Northwest National Laboratory, December 2005, http://pnwcgs.pnl.gov/Newsletter/OtherDocs/AnatChinaEnergy.pdf, accessed 16 May 2008.

[81] Philip Andrews-speed, Stephen Dow and Zhiguo Gao, 'The Ongoing Reform to China's Government and State Sector: the case of the energy industry', *Journal of Contemporary China* 9: 23, 2000, pp. 5–20.

[82] Naughton, *The Chinese Economy*, p. 317.

[83] For instance, the influence and power of those companies is reflected by the delay of the birth of China's *Anti-monopoly Law* for 13 years by some cartel-type firms and their government protectors. Even in the final draft of the law, an article explicitly states the superiority of laws or administrative regulations for a sector over the *Anti-monopoly Law*. See Yang Guangbin, 'Interest groups in China's politics and governance', EAI Background Brief, No. 361, East Asian Institute, National University of Singapore, 13 December 2007.

SOE managers are deeply embedded in the system of personal patronage which is the essence of the CCP's rule in China. The SASAC indirectly reinforces central SOEs' elite status by forcing them either to become one of the top two or three firms in their sector or to be merged.[84] The special status of the NOCs especially CNPC and Sinopec, had not only been guaranteed institutionally, as reflected by their high bureaucratic ranking, but also endorsed by the personal patronage of China's top leadership. Their high political status enabled them to become increasingly autonomous and powerful.

Particularly, CNOOC, Sinopec and CNPC were successively separated from the previous MPI that had tremendous economic and political clout in China. To some degree, the NOCs especially CNPC inherited the influence of the MPI and have remained influential throughout the reform era.[85] Also, the heads of the NOCs have direct access to China's top leadership. As in other countries, there is a revolving door between the government and oil companies. Some of China's top leaders previously worked in the oil industry.[86]

Notably, such revolving door with Chinese characteristics not only revolves between the government and the NOCs, but also among the three major NOCs themselves. Meanwhile, the exchange or revolve of personnel is not restricted at the level of senior managers or leaders of the NOCs, but also among personnel and employees with lower ranking. For the revolve or exchange of personnel between the government and the NOCs, this often happened between the

---

[84] Naughton, 'SASAC and Rising Corporate Power in China'.

[85] According to some experts, among those NOCs, CNPC Group is the only corporation that is able to exert significant influence on the government's policy making. Since its predecessor is the MPI, CNPC has substantial connections and networks to influence the government. Author's interview with Dr Jiang Yong, Beijing, China, July 2008.

[86] Downs, 'China'. However, due to the reform of China's civil servant system, the low-level personnel exchange between SOEs and the government is not as easy and frequent as before.

NOCs and some oil supervising governmental institutions like the SASAC and the NDRC and some provincial governments as well.[87]

Indeed, some large SOEs including NOCs have become more and more important not only in the economic sector but also in the political arena. In accordance with their key economic status, their political interests have to be satisfied by the CCP government to a certain degree. The NOCs' corporate interests have been increasingly represented in the personnel arrangements at the top level and reflected in the government's policy making process. Some NOCs bosses have risen to important positions in the the CCP Central Committee and the PRC central government, becoming prominent figures on China's political stage.

For example, Sheng Huaren and Zhou Yongkang, both of whom were previously NOC bosses, later took key positions in the central government, National People's Congress and the CCP Central Committee. When Zhu Rongji was the Chinese Premier, many state entrepreneurs were promoted into the SETC which was considered as Zhu's personal power base. Among them, Sheng Huaren who was former President of China Petroleum and Chemical Industry General Corporation was appointed the Chairman of the SETC. He had 40 years of experiences in the oil sector

---

[87]There are both advantages and disadvantages with the revolving door phenomenon. For the former, a manager or an official would not working at a same post for a prolonged period, therefore easily forming vested interests and his or her own sphere of influence or even gang that might be difficult for the superior to supervise or control. Moreover, working at various posts and in different environments provides of the enterprises managers or governmental officials with more opportunities to learn and practice, which is beneficial for their career development. However, the biggest disadvantage of such revolving door or exchange of personnel is that some of the managers or officials may try their best to realise most economic and political achievements within their terms at certain posts, regardless of the potential negative effects in the future or long-term development of the companies. Author's interview with Academician Zeng Hengyi, who is Academician of the Chinese Academy of Engineering, member of China's National Expert and Consultant Committee for Energy as well as Vice Chief Engineer of China National Offshore Oil Corporation (CNOOC), Beijing, China, June 2014.

(1958–1998). 'Sheng had been a successful entrepreneur managing one of the largest SOEs in China. The object of the reorganisation and expansion of SETC was to implement Zhu's plan of SOE reform. Certainly, Sheng's promotion was aimed at achieving this goal.'[88] After serving as head of the SETC, Sheng later became Vice-chairman and Secretary General of the Standing Committee of the 10th National People's Congress.

A typical example was Zhou Yongkang, who graduated from Beijing Petroleum College and worked in the oil sector for 31 years (1967–1998). He was the MPI Vice-minister, and later President of China Petroleum and Natural Gas General Corporation, before appointed the MLR Minister. Then, he became Secretary of the CCP Sichuan Provincial Committee, Minister of Public Security, State Councillor, member of the CCP 16th Central Committee's Political Bureau and finally member of the Standing Committee of the CCP Seventeenth Central Committee's Political Bureau and Secretary of the CCP Central Commission for Politics and Law. In 2012 he retired from the critical post in the CCP. In 2014, he was formally arrested and was under investigation for severe violation of the CCP disciplines and was later expelled from the CCP.

Also, some officials previously worked in the oil sector including the government's oil supervising agencies and state oil enterprises had finally made themselves among China's top leaders. Zeng Qinghong and Wu Yi were two examples.

Zeng Qinghong became the most powerful figure among those officials with a background in the oil sector, although he had experiences in the sector for merely five years (1979–1984). His status as a princeling (*taizidang*) of the CCP (his father Zeng Shan was a revolutionary veteran and an important figure in the CCP history) did help him a lot. From the late 1970s to the early 1980s, he was secretary to Yu Qiuli in the SPC and later in the SEC, before working in the MPI, some state oil enterprises and then the CCP Shanghai Municipal Committee. He returned to Beijing in 1989 to take important posts in the central government as well as the CCP Central

---

[88] Zheng Yongnian, *Globalization and State Transformation in China*, pp. 105–106.

Committee. As an alternate member of the CCP 15th Central Committee's Political Bureau, his promotion into the Standing Committee of the CCP 16th Central Committee's Political Bureau was a rare case, making him the number five leader in China from 2002 to 2007 (after Hu Jintao, Wu Bangguo, Wen Jiabao and Jia Qingli), President of the CCP Central Party School during this period and the PRC Vice-president from 2003 to 2008.

Wu Yi who is widely known as the 'iron lady' of China was another example. After graduating from Beijing Petroleum College, she worked in the oil industry for 26 years (1962–1988) before becoming a Vice-mayor of Beijing Municipality. Then, she worked in the central government as the Minister of Foreign Trade and Economic Cooperation, State Councillor and then Vice-premier. In 2002, she became member of the CCP 16th Central Committee's Political Bureau, the first female Political Bureau member since the CCP 13th National Congress in 1987. Hence, she was regarded by some people as the most powerful women in China.

Apart from the above four officials, there are another two top officials with previous experience in the oil sector. One is He Guoqiang, who had 19 years of experience in the sector (1967–1986). He is member of the Standing Committee of the CCP 17th Central Committee's Political Bureau and Secretary of the CCP Central Commission for Discipline Inspection. The other is Zhang Gaoli who had 15 years of experience in the sector (1970–1985). He is member of the Standing Committee of the Political Bureau of the CCP 18th Central Committee and Executive Vice-premier of the PRC. He is largely in charge of China's energy affairs and the NDRC.

Table 6.6 briefly shows the profile of these top officials in China with experiences in the oil sector. All of them are/were active members in China's political circle in the 2000s and early 2010s, as they hold/held important positions in the party-state system. Some of them just retired in the past few years and still have more or less influence, while some of them have been arrested or under investigation for violation of the CCP disciplines or corruption as well. They are/were state leaders and/or members of the Political Bureau (China's

Table 6.6: China's top leaders with the oil sector background (The 2000s to the early 2010s)

Name	Highest positions held in the CCP and the Chinese government	Major experiences in the oil sector
Zeng Qinghong (born in 1939)	Member of the Standing Committee of the CCP 16th Central Committee's Political Bureau; PRC Vice-president (2003–2008); President of the CCP Central Party School (2002–2007)	SPC; SEC; MPI; China National Offshore Oil General Corporation; South Yellow Sea Oil Company
He Guoqiang (born in 1943)	Member of the Standing Committee of the CCP 17th Central Committee's Political Bureau; Secretary of the CCP Central Commission for Discipline Inspection (2007–2012)	Shandong provincial chemical and petroleum industry department; Ministry of Chemical Industry
Zhou Yongkang (born in 1942)	Member of the Standing Committee of the CCP 17th Central Committee's Political Bureau; Secretary of the CCP Central Commission for Politics and Law (2007–2012)	Beijing Petroleum College; Daqing Oil field; Liaohe Oil Exploration Bureau; China Petroleum and Natural Gas General Corporation
Wu Yi (born in 1938)	Member of the of the CCP 16th Central Committee's Political Bureau; PRC Vice-premier (2003–2008)	Beijing Petroleum College; Lanzhou Oil Refinery; MPI; Beijing Dongfanghong Oil Refinery; Beijing Yanshan Petroleum and Chemical Corporation
Zhang Gaoli (born in 1946)	Member of the Standing Committee of the CCP 18th Central Committee's Political Bureau (2012 to present); PRC Executive Vice-premier (2013–present)	MPI; China Petroleum and Chemical Industry General Corporation

(*Continued*)

Table 6.6: (*Continued*)

Name	Highest positions held in the CCP and the Chinese government	Major experiences in the oil sector
Sheng Huaren (born in 1935)	Member of the CCP 15th Central Committee; Vice-chairman and Secretary General of the Standing Committee of the 10th National People's Congress; Chairman of SETC (1998–2001)	Nanjing Chemical Industry Corporation; Ministry of Chemical Industry; Ministry of Fuel and Chemical Industry; Ministry of Petroleum and Chemical Industry; China Petroleum and Chemical Industry General Corporation

Source: Author's compilation.

top decision making body) in the four most recent CCP Central Committees — the 15th, 16th, 17th and 18th Central Committees.[89]

Apart from these top officials, there are some important figures with an oil sector background that recently hold/held ministry-level positions in the State Council or some provinces. Li Yizhong, Wei Liucheng and Su Shulin are three examples. All of them graduated from petroleum specialised colleges. Li who was head of Sinopec Group has been member of the CCP 16th and 17th Central Committees. He served in the oil industry for 36 years (1967–2003) before becoming the SASAC Vice-chairman. Then, he was appointed head of the State Administration of Work Safety. Then, he became the Minister of Industry and Information. Wei who was head of both CNOOC General Corporation and CNOOC Ltd. was member of the CCP 17th Central Committee. He had 33 years of experience in the oil industry (1970–2003). He was Vice-Governor and then Governor of Hainan Province before becoming Governor of Hainan Province and later Secretary of the CCP Hainan Provincial

[89]The sessions of the 15th, 16th, 17th and 18th Central Committee of the CCP are respectively 1997–2002, 2002–2007, 2007–2012 and 2012–2017.

Committee. Su was previously head of Sinopec Group. He worked in the oil sector for 28 years (1983–2011) before becoming Governor of Fujian Province. He is member of the current CCP 18th Central Committee.

In addition to the connection between the oil sector and the CCP government at the highest level (the CCP Central Committee level), the government and the NOCs has close relationship in lower levels as well. The NOCs often directly participate in the government's energy issue studies and energy policy making. The government's lack of human and financial resources — countered by the NOCs' abundance of such resources — has contributed to the government's reliance on the NOCs for policy support, providing them with opportunities to influence Beijing's energy policy making.

First of all, the government relies on the NOCs for manpower and for their expertise. For instance, a large portion of the SELSG Office was drawn from the NOCs. Employees from Sinopec were also involved in drafting China's strategic petroleum reserves law. Moreover, Chinese officials periodically meet with the NOCs to enhance their understanding of particular energy issues. For example, in 2005, the BE officials met with representatives from CNPC, Sinopec and CNOOC to gain some knowledge about China's natural gas supply and demand. Further, the NOCs often finance studies conducted by some governmental departments which do not have adequate funds for their own research.[90] Funding relevant research has become an essential approach for the rich NOCs to influence the government's policy making. In this way, the NOCs are able to intentionally direct the policy making to lean to their preferences.

---

[90] Downs, 'China'; and Rosen, Daniel H. and Trevor Houser, 'China Energy: A Guide for the Perplexed', *China Balance Sheet*, A Joint Project by the Center for Strategic and International Studies and the Peterson Institute for International Economics, May 2007, http://www.petersoninstitute.org/publications/papers/rosen0507.pdf, accessed 3 May 2007.

6.3.3 *Ideological resources*

The NOCs' significant influence is also supported by an ideological resource, which provides key legitimacy and justification for their ODI and transnational operation. Traditionally, SOEs draw public support by repeatedly claiming that they are state-owned or owned by the people, so their operations at home and abroad are in the interests of the state and the whole nation. But the effectiveness of such arguments has become weaker and weaker in the recent years, as many Chinese people have criticised these firms for continuing monopolising the market and providing costly services. In comparison, the other two relatively new and important arguments used to justify the NOCs' overseas expansion — implementing the state's 'going global' strategy as well as safeguarding China's energy security — are more likely to be accepted by the government and the people.

In the first place, the NOCs' overseas expansion is in accordance with the government's advocacy of implementing the 'going global' strategy. Since 2001, the Chinese government has been formally calling for those capable Chinese enterprises to implement the 'going global' strategy and pursue ODI and transnational operation. For example, such a policy has been reflected in the Chinese government's 10th, 11th and 12th Five-year Plans. When articulating the outline of the 10th Five-year Plan (2001–2005) in 2001, then Chinese Premier Zhu Rongji emphasised four points for China's continuous opening up in the future. One of them is implementing the 'going global' strategy. He declared that the government needed to implement a 'going global' strategy, 'encouraging enterprises with comparative advantages to make investments abroad, to establish processing operations, to exploit foreign resources with local partners, to contract for international engineering projects...'[91] Also, when reporting

[91] Zhu Rongji, 'Report on the outline of the Tenth Five-Year Plan for national economic and social development (2001)', *Delivered at the Fourth Session of the Ninth National People's Congress on March 5, 2001,* http://english.gov.cn/official/2005-07/29/content_18334.htm, accessed 30 May 2008.

the outline of the 11th Five-year Plan (2006–2010) in 2005, Premier Wen Jiabao mentioned that the government would continue to support those well-prepared enterprises to 'go global' and conduct overseas investment under the international regulations.[92] The 12th Five-year Plan (2011–2015), China's latest five-year plan, also advocated accelerating the implementation of the 'going global' strategy and enhancing the government's supervision, guiding and service for the strategy.[93]

In fact, the NOCs have their own corporate 'going global' strategy to seek profit and development through ODI and transnational operation, even without the government's 'going global' strategy. More importantly, they started to invest abroad in the early 1990s,[94] much earlier than the official announcement of the 'going global' strategy. As a researcher with CNPC noted, the Chinese government planner took little notice of CNPC's first entry into Peru, Sudan and Kazakhstan until the mid-1990s. Some Chinese top leaders did not view ODI in overseas upstream assets as a sound strategy, and instead emphasised on continued domestic investment. They finally recognised the importance of global business to China's economic and energy development and national security in the late 1990s.[95]

---

[92] Wen Jiabao, 'Guanyu zhiding guomin jingji he shehui fazhan di shiyi ge wunian guihua jianyi de shuoming' (Report on the Outline of the Eleventh Five-Year Plan for National Economic and Social Development), 8 October 2005, http://www.gov.cn/gongbao/content/2005/content_121428.htm, accessed 30 May 2008.

[93] Xinhua, 'Shouquan fabu: zhonghua renmin gongheguo guomin jingji he shehui fazhan di shier ge wunian guihua gangyao' (Authorised to release: outline of the Tenth Five-Year Plan for national economic and social development for the People's Republic of China), 16 March 2011, http://news.xinhuanet.com/politics/2011-03/16/c_121193916_28.htm, accessed 16 March 2015.

[94] For a discussion of China's foreign oil exploration and development investment in the early 1990s, see Anonymous, 'China stepping up foreign E&P investment as oil imports soar', *Oil & Gas Journal*, May 9, 1994, pp. 56–59.

[95] Xu Xiaojie, 'Chinese NOCs' Overseas Strategies: Background, Comparison and Remarks', James A. Baker III Institute for Public Policy of Rice University, 2007, http://www.rice.edu/energy/publications/docs/NOCs/Papers/NOC_ChineseNOCs_Xu.pdf, accessed 13 March 2009.

Although the NOCs actually 'go global' largely for their own corporate interests, their transnational operations are widely viewed as a response to the government's call and an active implementation of the government's strategy, adding to the legitimacy of this movement, while generating the misperception outside China that they are still agents of the government. Meanwhile, they are inclined to keep such assumptions in order to enhance the justification for their overseas expansion. For instance, China National Oil and Gas Exploration and Development Corporation (CNODC) which is the overseas investment arm of CNPC Group, declared on its website that the company has been actively pursuing internationalisation and transnational operations since the early 1990s under the government's strategic guiding principle of 'using the resources and markets at home and abroad' and 'going global'.[96]

Furthermore, the NOCs often use the discourse of safeguarding energy security to legitimise their ODI and transnational operation. They actively adopt, support and promote the popular argument in China that their overseas business is beneficial for China's energy security. Whether these firms really take into much consideration the realisation of national interests and safeguarding national security in their business activities is another question, but such energy security discourse does provide a critical justification for their global business. As Trevor Houser wrote, the NOCs' proposition that their expanding transnational business bolsters China's energy security is for the same self-interested reasons of U.S. corn-growers' call for ethanol-based 'energy independence'.[97]

6.4 How decisions for overseas investment are made

Therefore, from the above, the interaction between the government and the NOCs is unveiled. Then, there is an important question left: How

[96] CNODC's website, http://www.cnpc.com.cn/cnodc/gsgk/gsjj/, accessed 31 May 2008.

[97] Trevor Houser, 'The Roots of Chinese Oil Investment Abroad', *Asia Policy* no. 5, January 2008, the National Bureau of Asian Research, http://www.nbr.org/publications/asia_policy/AP5/AP5_Houser.pdf, accessed 28 October 2008.

are the decisions for the NOCs' ODI projects are made, or what are the government and the NOCs' respective roles in the ODI decision making process?

In practice, the ODI activities initially come from the NOCs' proposals that were later approved by the government. If needed, the NOCs would solicit the government's involvement and support for their ODI.

Firstly, the NOCs conduct study on the market situation in various countries, getting the information of the existing and potential foreign oil and natural gas assets for sale and making the judgement of which places could be suitable for them to invest and operate. Then the NOCs decide where, when and how should they go and invest and operate. This step is purely commercial and initiated and completed by the NOCs. Then, the NOCs will submit the proposals for ODI projects to relevant oil supervising agencies in the government such as the NDRC and the MFA for approval. When considering the proposals, the governmental agencies will take account of various factors including the political and diplomatic factors as well as China's relations with the host countries of the proposed investments, then finally decide whether the proposals should be approved. During this process, the balance between the NOCs' motivation and the government's consideration is reached.[98] Of course, it is understandable that during this process, the NOCs might lobby for the government's approval and support in some cases if needed.

Also, as some authors noted, in the recent years, CNPC, Sinopec and CNOOC have used their political leverage to get support from the government for their ODI and transnational operation. Such support includes supportive high-level official visits, subsidised capital and development assistance designated for infrastructure projects.[99]

[98] Author's interview with Academician Zeng Hengyi, who is Academician of the Chinese Academy of Engineering, member of China's National Expert and Consultant Committee for Energy as well as Vice Chief Engineer of China National Offshore Oil Corporation (CNOOC), Beijing, China, June 2014.

[99] Rosen and Houser, 'China Energy', p. 22.

6.5 Conclusion

The government–NOC relationship has been evolving with China's economic and enterprise reform. The three prominent NCOs were created during the reform era; and the decentralisation process has shifted substantial power from the government to these firms, enabling them to become increasingly autonomous and powerful. Both the government and the NOCs have critical approaches and resources to influence the other. The government administrates the NOCs through various oil supervising agencies in the State Council, the nomenclatural system and the investment approval procedure; while the NOCs have significant material, institutional and ideological resources to influence the government's policy making. With the rise of these firms' strength and the relative decline of the state's authority in the sector, the government faces severe challenge in managing the NOCs.

The NOCs' material resources are mainly based on their growing financial ability. Their institutional resources come from their monopoly in China's market and their close connections with the party-state. Their ideological resources are largely derived from the official slogan of 'going global' as well as the popular discourse within China that energy diplomacy and ODI in overseas energy assets help safeguard China's energy security. Of course, there are some different discourses in the country regarding energy security and energy diplomacy. But the mainstream voice echoes and supports Beijing's energy diplomacy and the NOCs' ODI and transnational operation. The different academic discourses offer an important channel to examine the Chinese thinking about energy security and energy diplomacy. They may also influence Beijing's policy making and China's energy diplomacy in the future. Therefore, these discourses need analysing in an independent chapter.

Chapter

7

Chinese Discourses of Energy Security

In the past decade or so, with China's growing energy consumption and increasing reliance on overseas energy supply, there has been increasing Chinese academic discussion on energy security and energy diplomacy. This chapter examines major Chinese discourses of energy security. It is divided into five sections. The first section explains the reasons for the examination of the Chinese discourses of energy security, before listing the main finding of this chapter. The second, third and fourth sections respectively discuss the three major Chinese discourses of energy security, namely the oil supply-focused discourse, the strategic ability-focused discourse and the energy efficiency-focused discourse. The fifth section shows the Chinese official discourse of energy security.

7.1 Why the Chinese discourses of energy security matter?

Energy insecurity has been widely recognised as an urgent challenge to China since the beginning of the 21st century, as the country experienced three low-intensity energy crises in 2003, 2005 and 2007 respectively. These crises were largely triggered by skyrocketing international oil prices and exacerbated by the problematic energy institutions in China, affecting the people's livelihoods and the country's economic development and social stability. Thus, energy security has become a hot topic in China. The wordings of energy security or energy diplomacy have often appeared in the media reports and academic analyses on China's energy issues or Chinese leaders' trips abroad. Some Chinese scholars have produced various discourses on energy security. The Chinese government and some official think tanks have also held symposia and meetings on energy security and energy diplomacy.

There are two reasons for this chapter to examine the Chinese discourses of energy security. Firstly, it helps reveal the relationship between China's energy diplomacy and the discourses of energy security within China. For instance, the previous chapter mentioned the inspiration of Cox's idea of the three elements of hegemony for the explanation of the national oil companies' (NOCs) significant leverage over the government's policy making. Among them, the material resources (their increasing financial capabilities) and the institutional resources (their monopoly status and close connections with the party-state) can be easily understood, while the ideological resources are usually invisible and tend to be neglected. Apart from the Chinese official slogan of 'going global' mentioned in the previous chapter, Chinese discourses of energy security and energy diplomacy are a critical part of such ideological resources. This chapter will show that the mainstream Chinese discourse of energy security advocates promoting energy diplomacy to safeguard China's energy security. It provides an essential theoretical justification for the NOCs' overseas expansion and the government's support for their transnational operation.

Secondly, the domestic debates and discourses not only legitimise China's ongoing energy diplomacy, but may also influence the future development of this movement. Beijing's energy diplomacy strategy

and policies to safeguard energy security are ultimately formulated by the Chinese within China, rather than outsiders. Although scholars in other parts of the world often provide various analyses and suggestions for China to improve its energy security and energy diplomacy, their discourses have much fewer opportunities to influence China's policy making, compared with their counterparts within China, in particular those affiliated with governmental institutions or with official backgrounds.[1] Therefore, in order to better understand Beijing's consideration of energy security, as well as its options to improve energy security and conduct energy diplomacy, it is necessary to examine Chinese discourses of energy security.

In this chapter, the analysis of the discourses focuses on two questions: First, what are the major challenges to China's energy security? Second, what are the key measures to safeguard China's energy security?

The findings of this chapter: There are generally three major Chinese discourses of energy security.

The first is the oil supply-focused discourse, which is also the mainstream discourse in China. In particular, some analysts affiliated with the government hold this view. They regard insufficient domestic oil supply and oil import insecurity as the key challenge to China's energy security; and advocate protecting China's energy security through Beijing's energy diplomacy and its support for the NOCs' overseas business.

The second discourse is the strategic capacity-focused one, which considers China's limited overseas strategic ability as the essential challenge to the country's energy security, advocating resorting to strategic and arms building to defend energy security and national interests.

---

[1] In general, there are five channels for scholars in China to influence the government's policy making. First, some of them directly participate in research programmes assigned by the government. Second, some scholars directly provide consultancy for governmental institutions at various levels. Third, some scholars give lectures to Chinese officials in party schools and other academic institutions. Fourth, a few of them have opportunities to give lectures to China's top leaders (e.g. members of the CCP Central Committee's Political Bureau). Fifth, they elaborate their opinions and provide suggestions in openly published articles or internal documents within the party and the government systems.

The third is the energy efficiency-focused discourse, which maintains that low energy efficiency is the vital risk to China's energy security; and advocate resolving the problem through improving energy efficiency.

Besides, there is a Chinese official discourse of energy security, as shown in a Chinese official White Paper, which incorporates some useful ideas from the first and the third discourses. Among them, the oil supply-focused discourse and the official discourse provide important justification for China's energy diplomacy.

7.2 The oil supply-focused discourse

Some analysts affiliated with the Chinese Communist Party (CCP) or the government systems focused on oil supply when examining the country's energy security. In general, they viewed insufficient domestic oil supply and oil import insecurity as key threats to China's energy security. They also noted other critical challenges to the country's energy security, such as external threats, problematic domestic energy management, low energy efficiency, etc. They mainly proposed to increase domestic energy production and secure more overseas energy supply. Other measures to improve China's energy security include establishing the Strategic Petroleum Reserves (SPR), improving energy efficiency, etc. In particular, they stressed the importance of energy diplomacy in improving energy security; and advocated that the government should support the NOCs' transnational operation. The rest of this section is divided into three sub-sections. The first and the second sub-sections discuss the major challenges to China's energy security and measures to improve energy security according to some authors holding this kind of view; the third sub-section gives comments on this discourse.

7.2.1 *Challenges to China's energy security*

The majority of Chinese discourses of energy security regard the domestic oil shortage and the insecurity of oil imports as vital

challenges to China's energy security. For example, Xia Yishan,[2] a research professor with the China Institute of International Studies (CIIS), an official think tank, pointed out that the key challenge to China's energy security is the insufficient domestic oil supply. With the development of the Chinese economy, the gap between the domestic oil supply and demand has been widening and the country's dependence on foreign oil has been rising. Also, most of its oil imports come from the Middle East where the political situation is unstable. If an emergency occurs and causes the reduction or disruption of the overseas oil supply, China's economic development and national security would face a serious threat. Moreover, the surge in international oil prices would bring negative effects to the Chinese economy and the Chinese people's daily lives. In a word, the inadequacy of China's domestic oil production is negative for the country's economic development, political stability and national security.[3]

Also, Chen Fengying[4] who is a research professor with China Institute of Contemporary International Relations (CICIR), an official think tank having direct access to China's top leadership, argued that the major challenges to China's energy security are supply

---

[2] Xia Yishan is a veteran Chinese diplomat who had been working for the MFA since 1960. He was Director of the Centre for Energy Strategic Research under CIIS, a governmental think tank under the Ministry of Foreign Affairs (MFA). He has written extensively on China's energy security. His analyses focused on China's oil security and oil supply.

[3] See Xia Yishan, 'Dangqian guoji nengyuan xingshi he zhongguo nengyuan zhanlue' ('The international energy situation and China's energy strategy'), *Heping yu fazhan (Peace and Development)* No. 2, 2002, p. 37; Xia Yishan, 'Zhongguo nengyuan anquan wenti ji jiejue qianjing' ('China's energy security issue and the prospect of its solution'), *Heping yu fazhan (Peace and Development)* No. 4, 2003, pp. 20–22; Xia Yishan, 'Lun zhongguo de nengyuan anquan zhanlue' ('A discussion about China's energy strategy'), *Zhongwai nengyuan (China Foreign Energy)* 11: 1, 2006, p. 2; and Xia Yishan, '"Zouchuqu" xiahao liubu qi' ('China's "going global" activity should include six steps'), *Zhongguo shiyou (China Petroleum)* July, 2001, p. 12.

[4] Chen Fengying is Director of the Department of World Economics under CICIR, which is an important governmental think tank with direct access to China's top leadership. She has published several journal articles and book chapters about China's economic security generally and energy security specifically.

suspension, resource shortage and price fluctuation. She also identified some factors threatening China's energy security — the country's fast increasing energy demand, its rising dependence on oil import, the risk with oil import shipping and the instability in some oil exporting regions. In particular, she maintained that China's pursuit of overseas energy supplies is facing mounting competition and pressure from the United States, Japan, Russia and some Western energy companies.[5]

Similarly, her colleague Zhao Hongtu[6] identified seven external threats to China's energy security. The first is the uneven distribution of energy supply and demand in the world. The second is the increasingly fierce competition for energy resources and markets in the world. The third is the fluctuation of the international oil prices. The fourth is the so-called 'China oil and gas threat theory' in the West and some Western countries' energy containment against China. The fifth is the potential disruption of the overseas energy supply to China. The sixth is the increasing competition from other Asian states and U.S. military presence in China's surrounding areas. The seventh is the mounting international pressure on China to use less coal and more oil and gas.[7] In addition, he pointed out that the security of energy shipping through the Straits of Malacca is threatened by piracy, terrorist attacks and transportation accidents.[8] He also held that the pattern of China's resource consumption and the low

---

[5] Chen Fengying, 'Zhongguo nengyuan anquan de zhanlue sikao' ('The strategic thinking about China's energy security'), in Zhongguo xiandai guoji guanxi yanjiuyuan (China Institute of Contemporary International Relations), *Quanqiu nengyuan daqiju (Global Energy Structure)* (Beijing: Shishi chubanshe, January 2005), pp. 327–352.

[6] Zhao Hongtu is a research professor at the Department of World Economics under CICIR. He has written extensively about energy security and energy diplomacy.

[7] For details, see Zhao Hongtu, 'Guojia nengyuan anquan' ('National energy security'), in Zhongguo xiandai guoji guanxi yanjiuyuan (China Institute of Contemporary International Relations), *Guojia Jingji Anquan (National Economic Security)* (Beijing: Shishi chubanshe, July 2005), pp. 257–281.

[8] Zhao Hongtu, '"Maliujia kunju" yu zhongguo nengyuan anquan zai sikao' ('"Malacca dilemma" and the rethinking of China's energy security'), *Xiandai guoji guanxi* (*Contemporary International Relations*) No. 6, 2007, pp. 36–42.

efficiency of China's resource utilisation are two urgent challenges in front of China.[9]

Moreover, Zhou Dadi and Zhu Xingshan[10] who are senior research fellows with the National Development and Reform Commission (NDRC) held that China has abundant deposits of coal but is relatively short of oil and natural gas. Thus, its energy security mainly refers to the security of oil and natural gas imports; and the oil supply security is particularly critical. Also, the fluctuation of the international energy prices and its impact on China's economy and society should not be ignored. But the supply security issue is more pressing than prices fluctuation.[11]

In addition, Men Honghua,[12] a professor in the CCP Central Party School, noted that from the long-term and global perspective, China's energy security is an issue of oil and natural gas consumption. Due to China's insufficient domestic oil and natural gas supply, how to guarantee the security of its oil and natural gas import is the core of its energy security.[13]

---

[9] Zhao Hongtu, 'Guoji nengyuan anquan xingshi de xin tedian' ('The new characters of international energy security situation'), *Xiandian guoji guanxi* (*Contemporary International Relations*) No. 7, 2005, p. 2; and Zhao Hongtu, 'Guanyu zhongmei nengyuan hezuo de jidian sikao' ('Some ideas about China–U.S. energy cooperation'), *Xiandai guoji guanxi* (*Contemporary International Relations*) No. 1, 2006, p. 52.

[10] Zhou Dadi and Zhu Xingshan are two experienced energy experts who have been working for the Chinese central government for a long time. They received post-graduate degrees from Qinghua University and China University of Mining and Technology respectively. Currently, they are with the Energy Research Institute under the Academy of Macroeconomic Research in the NDRC. They have co-authored several papers on China's energy security.

[11] Zhu Xingshan and Zhou Dadi, 'Ruhe kandai zhongguo de nengyuan anquan wenti' ('Some ideas on China's energy security issue'), *Guoji shiyou jingji* (*International Petroleum Economics*) 9: 10, 2001, p. 5.

[12] Men Honghua received his PhD from Beijing University and is a professor at the CCP Central Party School.

[13] Men Honghua, 'Quebao zhongguo nengyuan anquan de zhanlue yiyi' ('The strategic significance of ensuring China's energy security'), *Taipingyang xuebao (The Pacific Journal)* No. 1, 2005, pp. 35–36.

7.2.2 *Measures to safeguard China's energy security*

Analysts focusing on the oil supply factor often proposed to increase domestic and overseas energy production, building the SPR, promoting energy diplomacy, etc., in order to defend China's energy security. For instance, as Xia argued, China must implement a global energy strategy. The core of such a strategy is using both domestic and foreign energy resources, and both the domestic and the international market. Especially, China's global energy supply system should include three elements — domestic energy supply, overseas energy supply and the SPR.[14] First of all, the domestic oil supply system is the foundation of China's global energy supply system. China should devote major efforts to promoting its domestic oil exploration and production, and try its best to lessen its dependence on foreign oil, in order to reduce the risks to its energy supply and avoid being under others' control. The second aspect of the global energy supply system is the overseas energy supply which is irreplaceable in filling the widening gap between the domestic oil demand and supply. There are two methods to utilise overseas oil resources. One is the procurement of oil and oil products in the international market; the other is 'going global' to purchase equity ownership in overseas oil fields and carry out exploration and development activities in order to acquire equity oil. Currently and in the near future, the main method for China to gain foreign oil supplies is the former. But in the long run, China should mainly use the latter method to increase its investment in overseas oil fields and gradually raise the proportion of equity oil in its total oil imports. The third aspect is the SPR, which includes oil product reserves and oil field reserves. The former refers to storing oil products for future use; and the latter means sealing up some oil fields after exploration for future development.

Chen proposed that China should improve its domestic energy management, energy market and energy legislation, and reform its

---

[14] For details, see Xia Yishan, 'China's energy security issue and the prospect of its solution', pp. 22–23; Xia Yishan, 'The international energy situation and China's energy strategy', p. 37; and Xia Yishan, 'A discussion about China's energy strategy', pp. 3–4.

energy institution. The government needs to formulate an overall and long-term energy development strategy. Also, China should focus on its overseas energy development. In particular, she proposed that China should implement three strategies of 'one third' — one third of its oil import from the Middle East and Africa, one third from the Asia–Pacific and Russia and one third from Central Asia. These regions are critical for China's energy import diversification. Oil import shipping security is also important and China should consider developing alternative shipping lanes to reduce its reliance on the Straits of Malacca. Moreover, China should coordinate various domestic actors when implementing the 'going global' strategy. Further, China should establish the SPR.[15]

Zhao proposed that China needs to promote the overall strength of its energy sector, develop overseas oil resources through various ways, establish the SPR, participate in international coordination and cooperation institutions, implement energy conservation measures, improve its energy efficiency, develop substitutes for oil and promote its ability to safeguard the oil and gas shipping lanes.[16] Besides, he proposed to develop substitute sea lanes for the Straits of Malacca. Although East Asian countries have reduced transportation costs by using this strait, it is not irreplaceable, especially when it operates in full load or has security problems. He argued that the Sunda Straits and the Lombok Straits are two alternative passages. They have not been well developed and exploited. Once the passage through the Straits of Malacca is impeded, they are well conditioned to become key new transportation sea lanes.[17]

Zhou and Zhu maintained that China has to improve energy security and minimise the economic and environmental costs through making efforts in six areas: The first is reducing its reliance on oil and promote oil conservation. The country needs to encourage research and development of substitutes for oil. The second is actively carrying

---

[15] Chen Fengying, 'The strategic thinking about China's energy security', pp. 345–352.

[16] Zhao Hongtu, 'National energy security', pp. 278–281.

[17] Zhao Hongtu, '"Malacca dilemma" and the rethinking of China's energy security', pp. 39–40.

out international energy cooperation and using political, economic, diplomatic and military measures to realise a favourable environment for its utilisation of overseas oil resources.[18] The third is accelerating the marketisation reform of its oil sector, further opening up its domestic oil market for foreign participation and enhancing its cooperation with foreign oil companies. The fourth is implementing a diversification strategy, including the diversification of China's energy consumption mix and oil supply sources.[19] The fifth is reasonably using oil resources in and outside of China. The country can shelve the development of some domestic oil resources if the costs are high and import oil to fill the gap until the international oil prices are too high. The proportion of imported oil in its overall oil consumption should be kept at around 50 per cent. The sixth is establishing the SPR which includes state reserves and the NOCs' commercial reserves.[20]

---

[18] According to Zhou and Zhu, China should promote its bilateral and multilateral dialogues with both oil importing and exporting countries. Beijing should regard oil diplomacy as an essential part of its overall diplomatic strategy and enhance its ties with oil exporting states. Moreover, China should selectively join some international organisations and actively participate in the international economic, energy, financial, trade and transportation affairs.

[19] On the one hand, the diversification of China's energy consumption mix means that when exploring and developing coal, oil and gas, China should attach great importance to reproducible energies like hydropower, nuclear energy, solar energy, wind power, etc. On the other hand, the diversification of China's oil supply means that China needs to diversify its oil import sources and reduce its over-dependence on the oil from the Middle East and coming through the Straits of Malacca, diversify the category of oil products in China's domestic market and import more oil products, diversify the way to gain overseas oil to include both oil trade and equity oil from Chinese invested oil fields abroad, diversify China's oil import transportation to include both shipping and pipeline transportation, and the government should support and encourage exploration and development of the domestic oil resources as part of the overall oil supply diversification strategy.

[20] Zhu Xingshan and Zhou Dadi, 'Some ideas on China's energy security issue', pp. 6–8; and Zhou Dadi, Han Wenke, and Zhu Xingshan, 'Zai gaige kangfang zhong baozhang shiyou gongying anquan' ('Safeguarding China's oil security during the process of reform and opening up'), *Zhongguo nengyuan* (*China Energy*) No. 3, 2001, pp. 14–17.

Men held that domestically, China should focus on developing domestic energy resources such as natural gas and clean coal, reserving and sealing up some energy fields after exploration for emergency use, establishing the SPR, promoting the energy efficiency, improving the public transportation capacity, etc.; internationally, China needs to diversify its energy import, establish strategic partnership with oil exporting states, invest in overseas oil fields, enhance its cooperation with other energy importing states, set up international institutions to manage energy crises, etc.[21]

Furthermore, as for the relationship between the government and the NOCs in developing overseas energy resources, Xia argued that the Chinese government should make efforts in three areas: First, Chinese NOCs' overseas expansion should be viewed as not only a kind of corporate behaviour but also a kind of government behaviour. Such activity must be under the government's leadership. The government should coordinate and oversee the NOCs' transnational operations, in order to avoid competition among the NOCs themselves. Second, the government should improve relevant law making and issue preferential policies to encourage the NOCs' overseas investment. Third, the Chinese leaders can use their political influence to support the NOCs' transnational operation. Chinese diplomatic organisations abroad especially those in oil exporting states should provide services to the NOCs regarding energy affairs.[22]

Chen contended that the government should issue preferential policies to support the NOCs' overseas oil and gas development, strengthen its energy information gathering and research and establish a comprehensive energy service and guarantee system. Especially, she argued that the government should set up an intelligence research office to closely follow the development of the overseas oil and gas situation. China's overseas diplomatic agencies should promote their energy information gathering. The government needs to actively support the NOCs' transnational operations and recruit and train capable

[21] Men Honghua, 'The strategic significance of ensuring China's energy security', pp. 35–38.

[22] Xia Yishan, 'China's "going global" activity should include six steps', p. 13.

personnel to promote energy diplomacy and the NOCs' overseas business.[23] Zhao also mentioned that the government should encourage the NOCs to invest in overseas oil resources.[24] Zhou and Zhu held that China needs to promote the international competitiveness of Chinese NOCs, encouraging them to become independent and strong transnational corporations. In particular, the government should reduce its intervention in and restriction on the NOCs' operations. Also, China can introduce foreign oil companies' capital and technologies into the exploration and development of the domestic oil resources while giving the NOCs more freedom to 'go global'.[25]

7.2.3 *Comment*

The oil supply-focused discourse viewed the domestic oil shortage and oil import insecurity as essential challenges to China's energy security; and advocated energy diplomacy and the government's support for the NOC's global business to safeguard China's energy security. It is the mainstream voice among Chinese discourses of energy security, providing essential justification for China's energy diplomacy including the NOCs' overseas expansion. It is a significant part of the ideological resources legitimising the NOCs' pursuit of the government's support for their outward direct investment (ODI) and transnational operation. Considering the NOCs' abundant material and institutional resources as explained in the previous chapter, such pro-NOCs discourse may not only justify these companies' ongoing ODI activities but might also influence the interaction between the government and the NOCs in the future. It may push the government to provide more assistance to the NOCs' transnational business. In this sense, this discourse is largely in line with the NOCs' corporate interests.

---

[23] Chen Fengying, 'The strategic thinking about China's energy security', pp. 345–352.

[24] Zhao Hongtu, 'National energy security', pp. 278–281.

[25] Zhu Xingshan and Zhou Dadi, 'Some ideas on China's energy security issue', pp. 6–8; and Zhou Dadi, Han Wenke, and Zhu Xingshan, 'Safeguarding China's oil security during the process of reform and opening up', pp. 14–17.

Although such discourse noted the importance of energy diplomacy for China's energy security and national interests, it often overlooked the essential corporate interests behind China's energy diplomacy and the NOCs' self-motivation in ODI. It tends to view the NOCs' transnational operation as behaviour mainly for the purpose of safeguarding China's energy security and national interests, instead of pursuing their corporate interests. But as argued in Chapter 5, there are both national and corporate interests behind China's energy diplomacy. Therefore, the obvious shortcoming of this discourse is its neglect of the vital corporate interests behind the movement. As a result, such discourse might provide the people with the misperception that the NOCs are 'going global' largely for national interests, adding to the ideological resources justifying their overseas expansion.

In addition, some points in Xia's discourse seem unrealistic. For example, he argued that the NOCs' overseas expansion must be under the government's leadership and the government should coordinate their transnational operation. But as demonstrated in Chapter 6, although the NOCs can often get support from the government, the government's leverage over the NOCs is limited. Thus, it is very difficult for the government to effectively control their business activities. Somewhat different from Xia's argument, Zhou and Zhu held that the government should reduce its intervention in and restrictions on the NOCs' operation and encourage their ODI. This idea is closer to the reality of the government–NOC relationship, as the government's control over the NOCs has been weakening during the enterprise reform and the NOCs have become increasingly autonomous and powerful. Indeed, as explained in the previous chapter, the NDRC is the essential oil supervising agency in China. It has a very close connection with the NOCs. Thus, analysts in the NDRC may know these firms better.

7.3 The strategic ability-focused discourse

Some Chinese analysts tend to consider energy security from a relatively hawkish perspective and attach great importance to whether

China's strategic capability is sufficient to defend its energy security and national interests. They often advocated accelerating China's military development particularly navy building. As some observed:

> 'The recent Chinese debates on energy security have resulted in some people strongly advocating for a speedy buildup of China's own blue water navy in order to protect vital energy shipping routes. Currently, a popular Chinese online book, The Battle in Protecting Key Oil Routes, imagines a decisive sea engagement near the Straits of Malacca linking the Indian Ocean and the South China Sea, in which the Chinese navy destroys an entire US Pacific carrier group'.[26]

Typical examples are discourses produced by Zhang Wenmu[27] and Yan Xuetong,[28] two former CICIR analysts. In brief, they viewed China's increasing dependence on oil import and limited overseas strategic reach as the key challenges to China's energy security; and strongly advocated enhancing China's military building and national strategic capacity to safeguard the country's overseas interests.

7.3.1 *Challenges to China's energy security*

Zhang made three points regarding the main challenges to China's energy security. First, as for clean energy resources such as oil, the demand is much larger than the domestic supply. Second, the problem of China's energy security is not the gap between total energy demand and supply, but the structural problem of the shortage of

---

[26] Jiang Wenran, 'Beijing's "New Thinking" on Energy Security', China Brief, Volume 6, Issue 8, 12 April 2006, the Jamestown Foundation, http://www.jamestown.org/china_brief/article.php?articleid=2373181, accessed 7 April 2008.

[27] Zhang Wenmu was previously an analyst with CICIR and currently a professor with the Centre for Strategic Studies in Beihang University. He has written extensively about China's national security, energy security and sea power.

[28] Yan Xuetong received his PhD from the University of California at Berkeley and was previously an analyst with CICIR. He is now Dean and professor at the College for Contemporary International Studies at Qinghua University. He has written extensively on China's national security and international relations.

clean energy supplies. Third, oil is the most demanded but seriously short of supply among China's clean energies. Thus, oil supply shortages are the critical challenges to China's domestic energy security.[29] Also, he pointed out that the main characteristic of China's energy security situation is its dependence on oil imports has been continuously rising while its diplomatic and military approaches to protect its overseas oil interests are too limited.[30]

Moreover, Yan held that China's large amounts of raw material imports are one of the pre-conditions for China's economic development. The lack of shipping security guarantee has become a strategic threat to the country.[31] He contended that outside risks to China's security are increasing. The key reason is the underdevelopment of the Chinese military. China's demand for energy has pushed its economic interests to expand across many developing countries in Asia, Africa and Latin America. But its energy transportation security has not been guaranteed.[32]

7.3.2 *Measures to safeguard China's energy security*

Zhang argued that China's energy security policy should combine security and environmental protection. Its target is providing clean, secure and efficient energy; and its foothold is *duli zizhu* (literally, paddling its own canoe, or acting independently and on its own initiative). He provided three suggestions: First, China's energy consumption

---

[29] Zhang Wenmu, 'Zhongguo nengyuan anquan yu zhengce xuanze' ('China's energy security and policy choices'), *Shijie jingji yu zhengzhi* (*World Economics and Politics*), No. 5, 2003, pp. 1–2; and Zhang Wenmu, 'Zhongguo de nengyuan anquan yu kexing zhanlue' ('China's energy security and feasible energy strategy'), *Shuiwu yanjiu* (*Taxation Studies*), No. 10, 2005, pp. 34–38.

[30] Zhang Wenmu, 'China's energy security and policy choices', p. 6.

[31] Yan Xuetong, 'You heping buyiding you anquan' ('Peace is not necessarily security'), Huanqiu wang (Global Times On-line), 21 September 2007, http://column.huanqiu.com/yanxuetong/2007-11/5942.html, accessed 16 December 2007.

[32] Yan Xuetong, 'Anquan liyi shi zhongguo shouyao liyi' ('Security interests are China's most important interests'), Huanqiu wang (Global Times On-line), 27 September 2007, http://column.huanqiu.com/yanxuetong/2007-11/7678.html, accessed 16 December 2007.

should be based on its own energy advantages such as its abundant coal reserves to produce clean coal. Second, China needs to raise the proportion of nuclear energy in its overall energy consumption. Third, China must accelerate its navy building and promote international cooperation. Particularly, the last point is the feature of Zhang's discourse. He wrote that the modern navy was established at the same time as the appearance of the international trade. A strong navy was the precondition for countries like the United Kingdom and the United States to expand their international trade. While China has been increasingly involved in international energy activities, its military ability to defend its overseas energy interests is lagged far behind. He stressed that historical experiences have shown without a strong navy, a nation's overseas interests may easily be harmed or grabbed by other maritime powers. Military measures, especially maritime military conflicts, are the ultimate solution for international trade disputes. Thus, China needs to make early preparation to safeguard its rapidly growing overseas national interests including energy interests. Meanwhile, Zhang proposed that when making its energy security policies, China should take into consideration cooperation with other Asian countries, in particular other Northeast Asian states such as Japan, which is even more dependent on overseas energy supplies.[33] In particular, Zhang advocated that China needs to promote its national strategic capacity (*guojia zhanlue nengli*) to ensure its resource security. He contended that the development of national strategic capacity is not only an inherent issue of national politics, but also the guarantee of resource security. Moreover, the national strategic capacity is built upon the national strategic consciousness (*guojia zhanlue yishi*) and the Chinese people needs to improve their national strategic consciousness.[34]

In addition, Yan pointed out that it is imperative for China to develop an appropriate capacity for defending its expanding economic interests overseas such as energy interests. During the 1990s, China's

---

[33] Zhang Wenmu, 'China's energy security and policy choices', pp. 8–9; and Zhang Wenmu, 'China's energy security and feasible energy strategy', pp. 39–41.

[34] Zhang Wenmu, 'Guojia zhanlue nengli shi ziyuan anquan de baozhang' ('National strategic capacity is the guarantee of resource security'), *Luye* (*Green Leaf*) No. 6, 2006, pp. 18–19.

economy grew quickly but China's national defence did not develop appropriately. Meanwhile, China's economic interests have rapidly expanded overseas; however, its defensive ability is restricted within the Chinese territory. Therefore, it is an urgent task for China to safeguard its overseas energy resources.[35]

7.3.3 *Comment*

Perhaps Zhang's argument is one of the most hawkish or pro-realist among Chinese discourses of energy security. His discourse is characterised by a strong advocacy for navy building. There are not many Chinese scholars like him that have directly and unequivocally referred to China's insufficient overseas strategic reach as the essential problem of the country's energy security and called for military building. This kind of argument is often viewed by some foreign analysts and media as the sign of China's military policy and evidence of an oil-thirsty China expanding worldwide and threatening the interests of other countries. Perhaps this is the reason that the Chinese government did not incorporate those strategic ability-focused ideas into its official discourse of energy security which will be discussed later, although they sound reasonable to some Chinese people.

This discourse may be in accordance with the preference of the Chinese People's Liberation Army (PLA) which prepares to defend China's national interests by force. The PLA Navy has been participating in the multi-national patrol authorised by the United Nations in the Gulf of Aden to protect shipping transportation from piracy. Also, the strategic ability-focused discourse may cater to the nationalist sentiment among some Chinese people.

However, this discourse is not the dominant voice within the Chinese academic or policy circles. Also, as an university professor with a purely academic position, it is unlikely that Zhang has opportunities to significantly influence the government's decision making. Thus, it is doubtful that his opinions represent the official attitude.

---

[35] Yan Xuetong, 'Security interests are China's most important interests'.

7.4 The energy efficiency-focused discourse

Some Chinese scholars regard low energy efficiency as the biggest challenge to China's energy security and improving energy efficiency as the major solution. Zha Daojiong's[36] discourse is a typical example.

7.4.1 *Challenges to China's energy security*

Zha contended that China's runaway growth in energy consumption is posing a threat to the country's energy security. The key challenge is managing the domestic demand, which is more important than securing foreign supply.[37] He pointed out that the increase of energy consumption with low efficiency is negative for China's sustainable development.[38] The low-efficiency and unrestricted growth of China's oil consumption raises the degree of China's dependence on foreign oil. The ultimate purpose for the discussion of China's oil supply security is figuring out how to manage or limit China's dependence on foreign oil supply. The policy-related significance of paying close attention to China's dependence on overseas oil supply is constantly reminding the Chinese about the strategic importance of managing the increase of the country's energy consumption.[39] Additionally, he noted that China has great difficulty finding an appropriate mechanism for governing its

---

[36] Zha Daojiong received his PhD form Hawaii University and taught in Macau and Japan before returning to China. He was previously with Renmin University and is currently a professor at Beijing University. He has written extensively on China's energy affairs, energy security and international relations.

[37] Zha Daojiong, 'China's energy security: domestic and international issues', *Survival* 48: 1, 2006, pp. 179–190; and Zha Daojiong, 'Cong guoji guanxi jiaodu kan zhongguo de nengyuan anquan' ('A review on China's energy security from the perspective of international relations'), *Guoji jingji pinglun* (*International Economic Review*) No. 11–12, 2005, pp. 4–6.

[38] Zha Daojiong, 'Nengyuan yilai jinkou bu kepa' ('It is not fearful to be dependent on energy import'), *Shijie zhishi (World Knowledge)* No. 9, 2006, p. 49.

[39] Zha Daojiong, 'Xianghu yilai yu zhongguo de shiyou gongying anquan' ('Interdependence and China's oil supply security'), *Shijie jingji yu zhengzhi* (*World Economics and Politics*) No. 6, 2005, pp. 20–21.

energy industry. It still lacks a ministerial-level agency to oversee the country's energy development.[40]

7.4.2 *Measures to safeguard China's energy security*

Zha maintained that China needs to manage its oil consumption and raise the degree of energy self-sufficiency.[41] It is difficult for the country to reduce its dependence on energy import without promoting its energy efficiency. Thus, a wise choice is seeking cooperation from foreign enterprises and foreign governments and regarding the improvement of energy efficiency as the inevitable choice to address China's energy security issue. This can prevent China from resorting to the international energy market to satisfy China's energy demands, avoiding more energy diplomacy friction between China and other countries and demonstrating that China is a responsible state in the area of global energy consumption.[42] Particularly, with regard to the question why low energy efficiency is a more urgent challenge than the domestic oil shortage, he argued that the increasing energy consumption is just like the growing population. Too large a population poses a challenge for feeding the people and the solution is slowing down population growth. Similarly, promoting energy efficiency is like controlling population growth.[43]

Also, Zha maintained that China should view energy security as an economic threat which can be addressed through market approach, rather than a military threat which has to be addressed through diplomatic approach.[44] To tackle the issue of energy inefficiency, China's

---

[40] Zha Daojiong, 'China's energy security', p. 186; and Zha Daojiong, 'A review on China's energy security from the perspective of international relations', pp. 4–6.

[41] Zha Daojiong, 'Interdependence and China's oil supply security', p. 21.

[42] Zha Daojiong, 'Bie tai danyou nengyuan yilai jinkou', ('Don't be too concerned about China's dependence on energy import'), Huanqiu wang (Global Times On-line), 18 September 2007, http://column.huanqiu.com/zhadaojiong/2007-11/4606.html, accessed 16 December 2007.

[43] Author's email correspondence with Zha Daojiong, July 2008.

[44] Zha Daojiong, 'A review on China's energy security from the perspective of international relations' 5, p. 3.

policy instruments and mechanisms of the Chinese energy industry need changing. As for China's energy search overseas, he wrote that 'China must enhance its transparency in those government-business interactions associated with its pursuit of energy interests overseas, so as to increase the level of confidence the international community can have in China's geopolitical intentions.' He also argued that China should actively pursue international cooperation in order to promote its energy efficiency; and the improvement of its energy efficiency should be a priority for international cooperation.[45] Moreover, He held that China should reduce its consumption of fossil energy and increase its use of nuclear energy; develop substitutes for oil and make effort in energy conservation.[46]

7.4.3 *Comment*

Low energy efficiency not only increases China's energy demands, but also results in a heavier environmental burden for the country. There is no doubt that it is imperative for China to improve its energy efficiency and protect the environment. This is the strength of the energy efficiency-focused discourse. Like Zha, some Chinese scholars tend to argue that China is a big energy producer itself and its energy efficiency is much lower than some developed countries. Therefore, as long as China improves energy efficiency to a certain degree, such as to Japan's level, its domestic energy supply can largely satisfy its demand. In this way, China can improve energy self-sufficiency and does not need to buy large amounts of oil from the international

---

[45] Zha Daojiong, 'China's energy security: Domestic and international issues', pp. 187–188; Zha Daojiong, 'It is not fearful to be dependent on energy import', p. 49; and Zha Daojiong, 'A review on China's energy security from the perspective of international relations', p. 10.

[46] Zha Daojiong, 'Zhongguo de nengyuan fazhan zhanlue yu guoji guanxi' ('China's energy development strategy and its international relations'), *Xuexi yuekan* (*Study Monthly*) No. 10, 2006, p. 48; and Zha Daojiong, 'Kongzhi zhongguo shiyou xiaofei zengzhang' ('Controlling the growth of China's oil consumption'), Huanqiu wang (Global Times On-line), 11 December 2007, http://column.huanqiu.com/zhadaojiong/2007-12/33512.html, accessed 16 December 2007.

market, reducing its energy competition and diplomatic friction with other oil importing countries and showing a 'responsible' image in the international community. This would be beneficial for China as well as the West, as China would no longer be a global energy competitor for Western countries and would not have energy interests in some states having sour relations with the West. Nevertheless, there are two problems with this logic.

In the first place, such discourse is somewhat idealist, as it disregards China's fundamental and complex national conditions to a certain extent. There are a number of unavoidable obstacles for China to greatly promote its energy efficiency, such as the country's vast territory and long domestic energy transportation routes, huge population, uneven economic development, under-developed science and technological levels, high cost of applying advanced energy conservation technologies across the country, etc. Thus, it is unlikely that in the foreseeable future, China would be able to effectively carry out a nationwide energy technologies update. Even if the country is able to significantly improve its energy efficiency, it still has to rely on overseas energy supply because its domestic production is very limited. In other words, it is impossible for China to largely reduce its dependence on oil import just through promoting energy efficiency.

Also, empirically speaking, so far, there has not been any case of a big country greatly reducing its dependence on foreign oil supply because of the improvement of its energy efficiency. In the past, there have been two successful cases of big countries significantly improve their oil self-sufficiency. One is the United Kingdom in the 1970s; the other is the United States in the 2000s. However, both of the successes were due to the dramatic surge of domestic oil production instead of significant promotion of energy efficiency. For the United Kingdom, the discovery and exploration of North Sea oil fields changed the country from a net oil importer to a net oil exporter. For the United States, the boom of its domestic shale oil production has largely reduced its dependence on overseas oil supply.

Moreover, the call for seeking advanced technologies from foreign enterprises and governments to improve China's energy efficiency seems a little unrealistic. As mentioned in Chapter 5, China

always desires to gain advanced energy conservation technologies from some developed countries, but there are complex commercial, political and strategic reasons leading to some developed countries' reluctance to transform such technologies to China. Therefore, the advocacy of improving energy efficiency as a priority in China's international cooperation does not look pragmatic enough.

7.5 The official discourse

Although there have been some Chinese academic discourses of energy security for years, China did not have a formal official discourse of energy security until 26 December 2007, when the State Council Information Office published a White Paper entitled *China's Energy Conditions and Policies*.[47] On 24 October 2012, the State Council Information Office issued the updated Chinese official white paper on energy policy — *China's Energy Policy 2012*.[48]

7.5.1 *Challenges to China's energy security*

According to the 2012 White Paper, China faces grave challenges to its energy security:

> 'The country's dependence on foreign energy sources has been increasing in recent years. In particular, the percentage of imported petroleum in the total petroleum consumption has risen from 32 percent at the beginning of the 21st century to the present 57 percent. Marine transportation of petroleum and cross-border pipeline transmission of oil and gas face ever-greater security risks. Price fluctuations in the international energy market make it more difficult to guarantee domestic energy supply. It will not be easy for China to

---

[47] The State Council Information Office, the People's Republic of China, *China's Energy Conditions and Policies*, 26 December 2007, http://www.china.org.cn/english/environment/236955.htm, accessed 27 December 2007.

[48] The State Council Information Office, the People's Republic of China, *China's Energy Policy 2012*, October 2012, http://english.gov.cn/archive/white_paper/2014/09/09/content_281474986284499.htm, accessed 14 January 2015.

> maintain its energy security since its energy reserves are small and its emergency response capability is weak'.

This white paper also pointed out that for some time to come, China's industrialisation and urbanisation will continue expanding and its demand for energy will continue increasing, posing tougher challenges to its energy supply.[49]

7.5.2 *Measures to safeguard China's energy security*

The 2012 White Paper provided a detailed description of China's energy strategy. The core idea is 'giving priority to conservation, relying on domestic resources, encouraging diverse development, protecting the environment, promoting scientific and technological innovation, deepening reform, expanding international cooperation, and improving the people's livelihood.'

Specifically, the strategy consists of eight elements: The first is working to build an energy-saving production and consumption system, transform the patterns of economic development and household consumption and build an energy-efficient country and an energy-saving society. The second is relying on domestic resources, enhancing energy supply capability and security and improving emergency energy reserve and response systems. The third is raising the proportion of clean, low-carbon fossil energy and non-fossil energy in the country's energy portfolio and developing substitute energy resources. The fourth is coordinating the development and consumption of energy resources with the protection of the eco-environment. The fifth is promoting basic scientific research and frontier technological research in the energy field to enhance China's scientific and technological innovation capabilities. The sixth is deepening the mercerisation reform in the energy sector. The seventh is expanding international cooperation and propelling the establishment of a new international

---

[49] The State Council Information Office, the People's Republic of China, *China's Energy Policy 2012*. The description and quotations of China's energy policies and strategy in this section are from this white paper.

energy order and promotes mutually beneficial cooperation. The last is enhancing energy infrastructure and public services to eliminate energy poverty and improve civil energy-use conditions.

Moreover, the whitepaper mentioned three approaches to promote China's energy development — intensifying efforts in prospecting and exploitation of conventional oil and gas resources, promoting the development and utilisation of non-conventional oil and gas resources and enhancing the construction of energy storage and transportation facilities.

As to international energy cooperation and energy diplomacy, the white paper declared:

> 'With accelerating economic globalisation, China has forged increasingly closer ties with the rest of the world in the field of energy. China's development of energy has not only satisfied its own needs for economic and social progress, but also made great contributions to world energy security and global market stability.'

In addition, the white paper called for the international community to foster a new energy security concept[50] featuring mutually beneficial cooperation, diversified development and common energy security through coordination. Such new energy security concept includes three aspects:

The first is strengthening dialogue and communication among energy exporting, consuming and transiting countries, which is the foundation of international energy cooperation.

Secondly, various countries should work together to stabilise the prices of bulk energy commodities, meet energy needs of various countries and maintain the order of the international energy market.

---

[50]The so-called new energy security concept (*xin nengyuan anquan guan*) was firstly proposed in July 2006 by then Chinese President Hu Jintao, when he delivered a speech to the outreach session of the G8 summit in St. Petersburg in Russia, calling for the international community to make a joint effort to ensure global energy security. For details, see Xinhua, 'President Hu urges efforts to ensure global energy security', 17 July 2006, http://news.xinhuanet.com/english/2006-07/17/content_4845287.htm, accessed 16 February 2009.

In order to realise sustainable development, developed countries should 'transfer clean and highly efficient energy technology to developing and underdeveloped countries and together promote green development globally on the condition that intellectual property rights are protected'.

Thirdly, the international community should work collaboratively to maintain stability in oil producing and exporting countries especially those in the Middle East, ensure the security of international energy transport routes and reduce the negative impact of geopolitical disputes on global energy supply. 'The various countries involved should settle major international energy disputes through dialogue and consultation. Energy issues should not be politicised, and the use of force and armed confrontation should be avoided'.

7.5.3 *Comment*

In general, the official discourse is more comprehensive but a bit vague compared with the academic discourses. Notably, it incorporates some useful points from the oil supply-focused discourse and the energy efficiency-focused discourse. It stressed the importance of international cooperation to safeguard international energy security. Although its articulation of international cooperation is not very detailed, it is supposedly related to overseas energy investment and production, energy diplomacy, energy technological cooperation with other countries, etc. Thus, it is largely in line with the oil supply-focused discourse's advocacy to enhance energy diplomacy. Also, it emphasised energy conservation, echoing the energy efficiency-focused discourse.

Furthermore, the White Paper reflected the Chinese government's intention to mitigate international concerns about a rising China with an increasing appetite for energy resources. Many Chinese people believe that although China's growing energy consumption is a natural outcome of its socio-economic development, it has been widely exaggerated and even demonised by some international media. Beijing believes that some foreign media and analysts are spreading the so-called 'China energy threat theory'. There have been various versions of China's energy policies and strategies produced by

overseas writers for years. Some of them are based on subjective speculations and exaggerate the negative impact of China's increasing energy demand and energy diplomacy on the international society. Therefore, it is necessary for the Chinese government to provide the international society with a Chinese official version of China's energy condition and policies, which can counteract the 'China energy threat theory' to some extent.

Notably, the official discourse's emphasis on relying on domestic resources to meet China's growing energy demand, environmental protection, enhancing international cooperation and coordination, sustainable development, etc. are not only what China really needs to do but are also largely in accordance with the interests of the international community. Beijing has spared no effort in reiterating its 'peaceful development' strategy so as to ease other countries' concern about the rise of China. Therefore, it is not strange that the official discourse has little resemblance with the strategic ability-focused discourse.

Lastly, the official discourse presented a very ambitious plan to promote energy efficiency and save energy. However, considering China's national conditions, such as imbalanced economic development, weak industrial foundation, large population, different conditions among various areas of the country, differences between the central government and local governments, etc., whether this plan can be effectively implemented and reach its goal is still a question.

7.6 Conclusion

There are altogether four major Chinese discourses of energy security. The first is the oil supply-focused discourse, which sees insufficient domestic oil supply and oil import insecurity as the critical threat to China's energy security; and calls for promoting energy diplomacy and Beijing's support for the NOCs' overseas business. It is the mainstream voice in China, justifying China's energy diplomacy. However, it overlooks the essential corporate interests and the NOCs' self-motivation in the movement. The second discourse is the strategic ability-focused discourse. It views China's limited overseas strategic

capacity as the vital challenge to China's energy security; and advocates navy building to safeguard energy security and overseas interests. Although catering to the nationalist sentiment, it is not the dominant voice in China. The third is the energy efficiency-focused discourse, which regards low energy efficiency as the key challenge to China's energy security and calls for improving energy efficiency to solve the problem. Although it pointed to the importance of energy efficiency and energy saving, its suggestions for the solution might be not realistic enough. As to the Chinese official discourse of energy security, it incorporates some useful ideas from the oil supply-focused and the energy efficiency-focused discourses, pledging to pursue energy saving, domestic and overseas energy development, international cooperation, etc. To a large extent, it calls for international energy cooperation and aims to reduce international concerns about China's energy demands. Notably, the oil supply-focused discourse and the official discourse provide important justification for Beijing's energy diplomacy and the NOCs' overseas expansion.

Chapter

8

Conclusion: Whither China's Energy Diplomacy?

The previous chapters have explored the key domestic dynamics of China's energy diplomacy — the interaction of national and corporate interests. Now the whole picture can be pieced together here:

China's economic rise has led to the surge of its appetite for energy resources particularly oil and natural gas. Its domestic energy production is not able to satisfy its demand, making the country to increasingly reliy on overseas energy supply. In accordance, China's energy diplomacy has developed rapidly recently, seeking access to oil and natural gas resources worldwide. The Chinese government has been actively engaging energy exporting countries around the world. Meanwhile, Chinese national oil companies (NOCs) have spared no efforts in

conducting outward direct investment (ODI) and transnational operation in many energy rich countries. The footprint of China's energy diplomacy is expanding across various regions of the world.

In particular, the Chinese government and Chinese NOCs are two equally important players in the movement; and neither of them should be underestimated. There are both national interests and corporate interests behind the phenomenon. On the one hand, energy diplomacy improves China's energy security and international relations, contributing to the maintenance of the Chinese Communist Party (CCP) government's political legitimacy which is identified as the state's essential national interests. On the other hand, this movement enables the NOCs to access new investment markets abroad, generate greater profits and fulfil their long-term development strategies, which are these firms' critical corporate interests. In this way, the convergence of national and corporate interests is realised, providing the most important and sustainable momentum for China's energy diplomacy.

Also, the government and the NOCs are natural partners in the game, as they need each other's activities to reach their respective interests. Beijing's engagement with energy exporting states and financial assistance to the NOCs are beneficial for their business operation in those countries; while the NOCs' ODI and transnational operation objectively contribute to China's energy security and provide Beijing with a new platform to strengthen its relations with many countries. In the recent years, the two players have been cooperating with each other to advance China's energy diplomacy worldwide. The convergence of national and corporate interests and the partnership between the government and the NOCs are the guarantee of such rapid progress.

Moreover, behind China's energy diplomacy, the government and the NOCs interplay with each other domestically. The government–NOC relationship has been evolving with China's economic and enterprise reform. The decentralisation process has led to a shift of power from the government to the NOCs, enabling them to become increasingly autonomous and powerful. Generally, the government manages the NOCs and oversees their business activities through various oil supervising agencies in the State Council, the nomenclatural system

and the investment approval process; while the NOCs have substantial material, institutional and ideological resources to exert considerable impact on the government's policy formulation and secure the state's diplomatic and financial support for their global business venture. The recent trend is the growth of the NOCs' economic and political capability *vis-à-vis* the waning of the state's leverage on them.

In short, such interaction of national and corporate interests forms the key domestic dynamics that motivate and influence China's energy diplomacy. However, within this interaction, two problems are emerging and may increasingly affect the movement in the future. To a large extent, whither China's energy diplomacy depends on the development of such domestic dynamics. The first is the occasional divergence of national and corporate interests. The second is the government's difficulties in managing the NOCs. Notably, these two issues are in fact related to each other, as the divergence of national and corporate interests poses a challenge for the government to effectively administrate the NOCs; while the government's handling of its ties with the NOCs determines whether and how the different interests can be coordinated. These two issues might be good topics for future research. The rest of this chapter just touches upon them briefly.

8.1 Divergence of national and corporate interests

Although the convergence of national and corporate interests is the key motivation of China's energy diplomacy, there has been occasional divergence between the two interests in the past few years, as illustrated in Figure 8.1, which may cause negative implications for the partnership between the government and the NOCs, as well as the development of China's energy diplomacy.

The reason is that the NOCs always regard their corporate interests, especially profits, as the priority, while the government stresses the national interests such as energy security and international relations. These firms' profit-chasing nature makes them reluctant to construct those energy projects preferred by the government, unless they can gain enough economic benefits from the work. Actually, while the NOCs are often able to get diplomatic and financial support

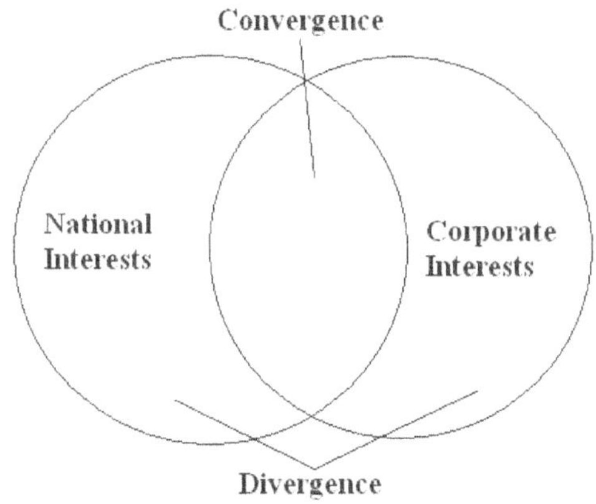

Figure 8.1: Convergence and divergence of national and corporate interests in China's energy diplomacy

Source: Author's compilation.

from the government, the government is occasionally not capable in reining these companies. This is a challenge for the Chinese leadership to coordinate different national and corporate interests behind China's energy diplomacy.

For example, in 2005 China National Offshore Oil Corporation (CNOOC) stopped talks with Chevron about purchasing liquefied natural gas (LNG) from Chevron's Gorgon project in Australia, because it did not want to pay the relatively high international price, which has reportedly angered the then National Development and Reform Commission (NDRC) Vice-Chairman Zhang Guobao who had pushed for the deal. Also, the NOCs' listing of some of their subsidiaries, which control many of their best assets, on overseas stock markets provides them with protection from state intervention. These firms always use their pursuit of profits as a justification for their resistance against projects and policies initiated by the government. Moreover, the NOCs have sometimes bid directly against each other for overseas projects, which is to the dismay of the government officials and analysts who prefer that these companies should operate

abroad as a team rather than as competitors.[1] When China National Petroleum Corporation (CNPC) and China Petroleum and Chemical Corporation (Sinopec) competed against each other for a pipeline project in Sudan, Chinese diplomats and the China International Contractors Association failed to persuade Sinopec that entered the lower bid to withdraw from the competition. Therefore, some Chinese and foreign analysts described the overseas investments of NOCs as 'each soldier is fighting his own war' (*gezi weizhan*). They criticised the poor coordination between the NOCs and the government as well as that among the NOCs. To a certain extent, the NOCs are business and political competitors with one another. They are competing not only for energy assets, but also for political status. A good investment deal may help the company secure diplomatic and financial support from the government for its subsequent investments. Also, for the NOCs bosses, business success and their firms' increasing international competitiveness are beneficial for them to obtain higher-ranking positions in the party and the government.[2]

In addition, the NOCs have some capacity to advance corporate interests at the expense of national interests. For example, they had periodically reduced their output to pressure the government to raise state-set domestic prices for oil products, which they believe did not keep pace with the rise of international oil prices. It is said that these firms have also ignored guidance from the government about where they should invest overseas.[3] On the other hand, however, these cases of the divergence between national and corporate interests can be viewed as examples to support one of the arguments by this research

---

[1] Erica S. Downs, 'China', Energy Security Series, the Brookings Foreign Policy Studies, the Brookings Institution, December 2006, http://www.brookings.edu/~/media/Files/rc/reports/2006/12china/12china.pdf, accessed 8 April 2008.

[2] Erica S. Downs, 'The Fact and Fiction of Sino–African Energy Relations', *China Security* 3: 3, 2007, pp. 42–68.

[3] Erica S. Downs, 'China's Energy Policies and Their Environmental Impacts', http://www.brookings.edu/testimony/2008/0813_china_downs.aspx, accessed 28 October 2008.

that the NOCs are not agencies of the government and their corporate interests are not subordinate to the state's national interests.

8.2 Government–NOC relationship

As showed in Chapter 6, the NOCs are increasingly powerful and exert considerable leverage on the state, while the government's administration over them is not as strong as often assumed. One important reason is that the government's administration over the NOCs is shared among different oil supervising agencies, instead of a single powerful ministry. Although some scholars and officials have called for a long time the establishment of such an agency like the Ministry of Energy (ME) that would enhance the government's supervision on these firms, such suggestion might be difficult to be realised. As long as these firms keep their monopoly status and political and economic clout in China, it would be very difficult for the state to manage them effectively. The government faces a dilemma as how to tackle its relationship with the NOCs. This section is divided into four sub-sections. The first explains how the state's authority over the NOCs is distributed among several governmental departments. The second provides two examples of the state's insufficient ability to control these firms. The third discusses whether the establishment of the ME would effectively improve the state's management over these firms. The last part shows a dilemma for the government to deal with its relationship with the NOCs.

8.2.1 *Why so many cooks in the kitchen?*

There are some major oil supervising agencies in the State Council, like the NDRC, the State-owned Assets Supervision and Administration Commission (SASAC) and the Ministry of Land and Resources (MLR), not to mention those actors playing relatively minor roles such as the Ministry of Finance (MF), the Ministry of Commerce (MOC), the Ministry of Foreign Affairs (MFA), etc.

In particular, compared with the pre-reform era, there have been more major oil supervising agencies in the State Council since the

beginning of the economic reform. From 1949 to 1978, there were two such organs; but after 1978, there were usually three or four such organs (see Table 6.3 in Chapter 6). The division of authority over energy sectors among these governmental institutions has resulted in low working efficiency and an ineffective energy management system;[4] and the state authority is fragmented among bureaucratic fiefdoms in perpetual feuds with overlapping authority and subject to reoccurring failures. As a result, the government's energy initiatives are an array of *ad hoc* policies with no central policy institution in charge of implementation. There are persistent problems with the institutionalisation of a China's energy decision making structure.[5] Therefore, an important question is why there are so many cooks in the kitchen.

First and foremost, energy or oil affairs are related to the interests of many groups in China. Since the economic reforms, the government's policy making process has been pluralised to a large extent, with more and more participants representing different interests. Moreover, the government's profit-seeking character and slowness to transform itself from an economic growth-oriented government to a service-oriented one, determines that many governmental departments want to get benefit from supervising the profitable oil industry. Thus, the state's authority over the oil sector is divided among various actors and difficult to be unified.

In practice, some governmental agencies behaved like profit-chasing corporations, as the state monopoly can easily be found in those capital-intensive and high-return sectors, like oil and natural gas, telecommunication, electric power, automobiles, etc. Meanwhile, however, the government unhesitatingly withdrew from non-profitable sectors like medical and health, education and social welfare. To a certain degree, as long as an industrial sector is profitable, some governmental departments would try to participate in the administration over the sector, so as to obtain interests. According to Jiang Yong, economic interests have increasingly combined some governmental

---

[4] Downs, 'China'.

[5] Gaye Christoffersen, 'U.S.–China energy relations and energy institution building in the Asia–Pacific', *Journal of Contemporary China* 19: 67, 2010, pp. 871–889.

bodies with the enterprises under their administration. The governmental departments grant the firms administrative monopoly, while the firms offer economic benefits to the governmental organs.[6] The government–NOC relationship is no exception. In this sense, the government may be reluctant to open up these profitable sectors to market competition and privatisation to share the interests with other players, no matter whether they are Chinese private firms or foreign companies. This is the most important reason why there are so many governmental institutions involved in the management of the oil sector while the energy market remains highly monopolised by the state-owned enterprises (SOEs) including the NOCs.

Furthermore, the CCP government has not yet been able to separate political power from direct control of economic resources. The CCP faces the task of divorcing governmental agencies from economic interests and converting them into solely service providers and regulatory bodies. Due to the governmental organisations' entrenched interests in those profitable industries, any initiative to withdraw the government from economic sectors or to shift its focus to service provision would meet strong resistance.[7]

8.2.2 *The government's difficulties in reining the NOCs*

During the economic reform, the waning of the state's authority over the oil sector and the growth of the NOCs' economic and political clout has resulted in the government's difficulties in reining these firms.

A typical example is the NOCs' capability to pressure the government to raise the prices for domestic oil products. Although

[6] Jiang Yong, 'Jingti bumen liyi pengzhang' ('Be on guard against the expansion of the interests of some governmental bodies'), quoted by Sohu, 9 October 2006, http://news.sohu.com/20061009/n245692266.shtml, accessed 13 May 2008.

[7] Zheng Yongnian and Wang Zhengxu, 'China's National People's Congress 2008: New Administration, Personnel Reshuffling and Policy Impacts', Briefing Series — Issue 38, China Policy Institute, The University of Nottingham, March 2008, http://www.nottingham.ac.uk/shared/shared_cpi/documents/policy_papers/Briefing_38_NPC_2008.pdf, accessed 3 May 2008.

CNPC and Sinopec enjoyed the monopoly status for upstream operations at home and have earned record-high profits in the recent years, they still cut production of their refineries or supply of their filling stations from time to time to minimise 'losses' whenever they perceived that the government-set domestic prices for oil products were too low. The lessons learned from the oil products shortage in China's domestic market in the 2000s, indicate that the NOCs often used the shortage of domestic oil products supply as a bargaining chip to effectively pressure the NDRC to raise the domestic prices for oil products.

For example, when international oil prices were high throughout 2006 and 2007 while China's domestic prices for oil products were kept relatively low by the NDRC, the NOCs especially Sinopec and CNPC encountered significant losses in refinery, although their overall profits were still growing. Thus, these firms actively applied pressure to the NDRC to increase the domestic prices for oil products. When refining margins drastically deteriorated in the spring of 2006, they cut oil production and sale at home to reduce losses. In consequence, there were soon many incidents of queuing at filling stations and farmers unable to buy diesel for their tractors. 'Forced to choose between the certainty of citizens angry over shortages and the possibility of citizens becoming angry over higher prices, the NDRC conceded to a 15 per cent increase in gasoline and diesel prices over the course of the next couple months'.[8]

Therefore, there is an urgent need for the state to make concrete efforts to improve its management over the oil sector and restrain the NOCs' power and authority in China's domestic market. So far, however, to improve its administration over the oil sector, the government has merely restructured its oil management structure several times throughout the reform era, as demonstrated in Chapter 6, but much remains to be done.

[8]Trevor Houser, 'The Roots of Chinese Oil Investment Abroad', Asia Policy no. 5, January 2008, the National Bureau of Asian Research, http://www.nbr.org/publications/asia_policy/AP5/AP5_Houser.pdf, accessed 28 October 2008.

8.2.3 *Why not an independent ME yet?*

Another question about the government–NOC relationship is why the government has not re-established the ME during the recent bureaucratic restructurings. Although there has been a growing voice within China to set up a powerful ME, the government merely created the Bureau of Energy (BE) under the NDRC in 2003, then the vice-ministerial level State Energy Leading Small Group (SELSG) Office in 2005, and later the vice-ministerial ranking National Energy Administration (NEA) but still under the NDRC in 2008, rather than directly establishing a fully independent ME. There are complicated reasons for it. Among them, an essential consideration is to avoid the difficult job of power redistribution within the energy bureaucracy and the State Council, so as to maintain the NDRC's leverage over the energy sector.

The NDRC was originally created in 2003 as an institution responsible for the formulation of long-term national socio-economic development strategies and the implementation of the macro-economic control. However, during the process of power recentralisation from local governments to the central government in the 2000s,[9] the authority regained by the central government was mostly power over economic and industrial sectors, which was allocated to the NDRC. As a result, the NDRC's administrative domain continued expanding; and it played a key role in not only the macro-economic control but also the micro-economic affairs and the industrial operation.

---

[9] In the 1990s, the Chinese central government decentralised much administrative power to local governments, in the hope that the latter could further decentralise the power to the society. However, local governments were able to gain much benefit during the decentralisation process, thus becoming vested interests that oppose to further decentralise the power to the society. Therefore, the central government during Hu Jintao and Wen Jiabao's era had to recentralise some power from local governments, largely for the purpose of transforming the power to the society. For detailed explanation, see Zheng Yongnian, 'Globalisation, Openness and Transformation in China', a paper and presentation for the Political Studies Association Annual Conference, Swansea University, April 2008, http://www.psa.ac.uk/2008/pps/Zheng.pdf, accessed 9 April 2008.

More importantly, the bureaucratic restructurings during the 2000s was mainly designed and watched by the NDRC which was known as the 'mini-State Council' and Premier Wen Jiabao's power base. It was highly unlikely that the NDRC would make an effort to reduce its own authority and interests in the supervision of the oil sector.

For instance, as early as in 2005 when the State Council contacted the former ME minister Huang Yicheng to solicit his opinion on what kind of an energy management agency should be established. Huang suggested setting up a ministerial-ranking administration under the NDRC headed by a NDRC Vice-Chairman. The reason was that if such an agency were under the NDRC, there would be no controversy for power redistribution; but if such an agency is separated from the NDRC, it would not have actual power and the effort would be meaningless.[10]

Just like the previous fully independent State Energy Commission (SEC) that existed for only two to three years. Although it was designed to administrate the energy sector including the oil sector, it did not have *de facto* power; the power was still in the hand of the State Planning Commission (SPC) which was the 'mini-State Council' at that time.

Another example is the re-establishment of the SEC during the 2008 bureaucratic restructuring. The new SEC was originally designed to wield significant power over the energy sector and be headed by a Vice-Premier. However, in 2010, the new SEC actually turned out to be another SELSG headed by the Premier, just under the name of the SEC. It was not a governmental institution under the State Council, does not have permanent employees under it, and of course, without any concrete administrative power in the energy sector.

Obviously, the NDRC cannot give up its grip on the energy sector. To some degree, there could be a zero-sum game between the NDRC and the proposed ME with actual power. Put it in another

---

[10] Xinlang (Sina), 'Huang Yicheng', 13 July 2005, http://finance.sina.com.cn/manage/cfrw/20050713/17011792952.shtml, accessed 26 April 2008.

way, it is impossible to create an independent and powerful ME without eroding the NDRC's authority and vested interests. If the proposed ME were established and able to enjoy major authority over the energy sector, as advocated by many scholars and officials both in and outside China, it would mean the decline of the NDRC's leverage over the energy sector or even its *de facto* dismantlement. Thus, as part of the entrenched interests and the most powerful department in the State Council, the NDRC is likely to try its best to maintain its power and interests in the profitable energy sector. In this sense, perhaps the restructuring of the NDRC itself is the pre-condition for the establishment of a powerful ME.

Further, the government's incremental efforts to restructure its energy bureaucracy were a compromise among various stakeholders and based on 'win–win' considerations. Such efforts mitigate the pressure on the Chinese leadership from both sides.

On the one hand, the leaders can demonstrate to the public that they are making efforts to address China's energy problems. Through gradually promoting the bureaucratic ranking of the key oil supervising agency from a bureau level department (the BE under the NDRC) to a vice-ministerial level department (the NEA under the NDRC) and increasing the number of employees in the agency, the government shows its willingness to adopt to some extent the strong suggestion from many experts and the society to create a high-level and more capable energy supervising agency.

On the other hand, such incremental progress or lack of real reform, to a large degree, meets the vested interests of the NDRC and energy SOEs including the NOCs. Creating a new high-level energy management agency under the NDRC can keep the NDRC's authority over the energy sector as well as its interests in the field. Also, compared with an independent ME, the NEA under the NDRC is less likely to be able to effectively check energy SOEs' activities, leaving them more room to manoeuvre. Despite the growing voices calling for the re-establishment of the ME, there are powerful ministerial and corporate interests that prefer the *status quo*. The opposition to the resumption of the ME was led by the NDRC and energy SOEs. A critical reason is that the proposed ME would deprive the NDRC

of a substantial portion of its portfolio and important tools of macro-economic control. For those energy SOEs, they are reluctant to have another political master and afraid that the ME would limit their direct access to the Chinese leadership.[11]

Therefore, with regard to the restructuring of the government's energy management organisations in the future, if the vested interests are largely protected and the NDRC still holds key authority over the energy sector, even if the ME is re-established, it is likely to be a toothless tiger or merely another cook in the kitchen.

8.3 A dilemma for the government

In the end, without a significant reform to greatly weaken the NOCs' high political status in the party-state or break their monopoly in China's energy market, the government faces a dilemma of how to manage these firms.

On the one hand, if the current trend of enterprise reform continues, the NOCs are likely to have growing autonomy and economic and political clout, becoming somewhat independent 'kingdoms', which would be an increasing headache for the party-state 'emperor'. If the divergence of national and corporate interests emerges, it would be more difficult for the state to manage or bargain with these firms. As a result, the government may encounter stronger resistance from the NOCs; and national interests may have to make way for corporate interests.

On the other hand, if the government recentralise much authority from the NOCs, these firms' autonomy would be restricted and their self-motivation and corporate interests in business operation at home and abroad might be harmed, which in turn would affect China's energy diplomacy and national interests.

Therefore, how to address the divergence of national and corporate interests and the complex government–NOC relationship is a

[11] Zhang, Libin and Jason Lee, 'Untangling China's Energy Policy', *China Security* 4: 3, 2008, pp. 58–61; and Downs, 'China's Energy Policies and Their Environmental Impacts'.

severe challenge in front of the Chinese leadership. With China's ongoing energy diplomacy and economic and enterprise reform, the interplay and bargaining between the government and the NOCs could be intensified. The outcome of this interaction will determine how national and corporate interests can be coordinated and realised. The development of such domestic dynamics will eventually have a key influence on China's energy diplomacy in the future.

Bibliography

Books, papers, comments and official publications

Alden, Chris, *China in Africa* (London and New York: Zed Books, 2007), p. 126.

Andrews-speed, Philip, Stephen Dow and Zhiguo Gao, 'The ongoing reform to China's government and state sector: The case of the energy industry', *Journal of Contemporary China* 9: 23, 2000, pp. 5–20.

Andrews-speed, Philip, Xuanli Liao and Roland Dannreuther, *The Strategic Implications of China's Energy Needs* (Oxford: Oxford University Press, 2002).

Anonymous, 'China stepping up foreign E&P investment as oil imports soar', *Oil & Gas Journal*, May 9, 1994, pp. 56–59.

Bader, Jeffrey A. and Flynt L. Leverett, 'Oil, the Middle East and the Middle Kingdom', 16 August 2005, http://www.brookings.edu/opinions/2005/0816globalenvironment_bader.aspx, accessed 24 June 2008.

Bateman, Sam and Jane Chan, 'Good order at sea in Southeast Asia', in Konrad-Adenauer Stiftung and European Union, *Maritime Security and*

Piracy: Common Challenges and Responses from Europe and Asia (Singapore: Select Books Pte. Ltd., 2014), pp. 69–85.

British Petroleum, *BP Statistical Review of World Energy June 2008*, http://www.bp.com/liveassets/bp_internet/globalbp/globalbp_uk_english/reports_and_publications/statistical_energy_review_2008/STAGING/local_assets/downloads/pdf/statistical_review_of_world_energy_full_review_2008.pdf, accessed 25 December 2008.

British Petroleum, *BP Statistical Review of World Energy June 2014*, http://www.bp.com/content/dam/bp/pdf/Energy-economics/statistical-review-2014/BP-statistical-review-of-world-energy-2014-Bob-Dudley-Group-Chief-Executive-introduction.pdf, accessed 24 August 2014.

Bustelo, Pablo, 'China and the geopolitics of oil in the Asian Pacific region', Working Paper, 38/2005, Real Instituto Elcano (Elcano Royal Institute), http://129.3.20.41/eps/othr/papers/0511/0511005.pdf, accessed 24 June 2008.

Campbell, Horace, 'China in Africa: Challenging US Global Hegemony', in Firoze Manji and Stephen Marks eds., *African Perspectives on China in Africa* (Cape Town, Nairobi and Oxford: Fahamu — Networks for Social Justice, 2007), pp. 119–137.

Cao, Rongxiang, *Jingji anquan: fazhanzhong guojia de kaifang yu fengxian* (*Economic Security: Openness and Risks of Developing Countries*) (Beijing: Shehui kexue wenxian chubanshe, June 2006), pp. 175–183.

Chen, Fengying, 'Zhongguo nengyuan anquan de zhanlue sikao' ('The strategic thinking about China's energy security'), in Zhongguo xiandai guoji guanxi yanjiuyuan (China Institute of Contemporary International Relations), *Quanqiu nengyuan daqiju* (*Global Energy Structure*) (Beijing: Shishi chubanshe, January 2005), pp. 327–352.

Chen, Gang and Ryan Clarke, 'China's diplomacy in the Middle East', in Wang Gungwu and Zheng Yongnian eds., *China: Development and Governance* (Singapore: World Scientific Publishing Co. Pte. Ltd., 2013), pp. 469–475.

Cheng, Joseph Y. S., 'A Chinese view of China's Energy Security', *Journal of Contemporary China* 17: 55, 2008, pp. 297–317.

Christensen, Thomas J., 'Windows and War: Trend Analysis and Beijing's Use of Force', in Alastair Iain Johnston and Robert S. Ross ed., *New Directions in the Study of China's Foreign Policy* (Stanford, California: Stanford University Press, 2006), pp. 50–85.

Christoffersen, Gaye, 'US–China energy relations and energy institution building in the Asia–Pacific', *Journal of Contemporary China* 19: 67, 2010, pp. 871–889.

Cox, Robert W. and Timothy J. Sinclair, *Approaches to World Order* (Cambridge: Cambridge University Press, 1996).

Cui, Minxuan, ed., *2007 Zhongguo nengyuan fazhan baogao* (*The Energy Development Report of China 2007*) (Beijing: Shehui kexue wenxian chubanshe, March 2007), p. 93.

Dannreuther, Roland, 'Asian security and China's energy needs', *International Relations of the Asia–Pacific* 3, 2003, pp. 197–219.

Department of Foreign Affairs and Trade, the Australian Government, 'PRC Country brief — May 2006', http://www.dfat.gov.au/geo/china/cb_index.html#betr, accessed 29 June 2006.

Ding, Ying, 'Seeking more than oil', *Beijing Review*, 19 June 2006, http://www.bjreview.com.cn/06-19-e/w-1.htm, accessed 25 July 2006.

Downs, Erica S., *China's Quest for Energy Security* (Santa Monica, CA: Rand Corporation, 2000).

Downs, Erica S., 'The Chinese energy security debate', *The China Quarterly* 177, 2004, pp. 21–41.

Downs, Erica S., 'China', *Energy Security Series*, the Brookings Foreign Policy Studies, the Brookings Institution, December 2006, http://www.brookings.edu/~/media/Files/rc/reports/2006/12china/12china.pdf, accessed 8 April 2008.

Downs, Erica, 'China's Quest for Overseas Oil', *Far Eastern Economic Review*, September 2007, pp. 52–56.

Downs, Erica S., 'The Fact and Fiction of Sino-African Energy Relations', *China Security* 3: 3, 2007, pp. 42–68.

Downs, Erica S., 'China's Energy Policies and Their Environmental Impacts', http://www.brookings.edu/testimony/2008/0813_china_downs.aspx, accessed 28 October 2008.

Dreyer, June Teufel, 'Sino-American Energy Cooperation', *Journal of Contemporary China* 16: 52, 2008, pp. 461–476.

Ebel, Robert E., *China's Energy Future: The Middle Kingdom Seeks Its Place in the Sun* (Washington DC: The CSIS Press, Center for Strategic and International Studies, 2005).

Engdahl, F William, 'The geopolitical stakes of 'Saffron Revolution'', Asia Times On-line, 17 October 2007, http://www.atimes.com/atimes/Southeast_Asia/IJ17Ae01.html, accessed 6 April 2008.

Eurasia Group, 'China's overseas investments in oil and gas production', 2006, http://www.uscc.gov/researchpapers/2006/oil_gas.pdf, accessed 13 March 2009.

Feldt, Lutz, Peter Roell and Ralph Thiele, 'Maritime security — perspectives for a comprehensive approach', in Konrad-Adenauer Stiftung and

European Union, Maritime security and piracy: Common challenges and responses from Europe and Asia (Singapore: Select Books Pte Ltd, 2014), pp. 23–45.

Friedberg, Aaron L., '"Going out": China's pursuit of natural resources and implications for the PRC's grand strategy', NBR Analysis, 17: 3, the National Bureau of Asian Research, September 2006, http://www.nbr.org/publications/analysis/pdf/vol17no3.pdf, accessed 20 June 2008.

Gill, Bates, Chin-hao Huang and J. Stephen Morrison, 'Assessing China's growing influence in Africa', *China Security* 3: 3, 2007, pp. 3–21.

Gilpin, Robert, *The Political Economy of International Relations* (Princeton: Princeton University Press, 1987).

Gilpin, Robert, *Global Political Economy: Understanding the International Economic Order* (Princeton: Princeton University Press, 2001).

Gilpin, Robert, 'U.S. power and the multinational corporation', in Nikolaos Zahariadis ed., *Contending Perspective in International Political Economy* (Beijing: Peking University Press and Thomson Learning, 2004).

Goldstein, Avery, *Rising to the Challenge: China's Grand Strategy and International Security* (Stanford, California: Stanford University Press, 2005).

Guizot, Armelle, *Chinese Energy Markets: Trading and Risk Management of Commodities and Renewables* (Basingstoke, UK: Palgrave Macmillan, 2007).

Guojia tongjiju (National Bureau of Statistics, the People's Republic of China), *Zhongguo tongji nianjian 2014* (*China Statistical Yearbook 2014*), online edition, http://www.stats.gov.cn/tjsj/ndsj/2014/indexch.htm, accessed 8 March 2015.

Guojia tongjiju (National Bureau of Statistics, the People's Republic of China), '2014 guomin jingji he shehui fazhan tongji gongbao' (Statistical Communiqué of the People's Republic of China on the 2014 National Economic and Social Development), 26 February 2015, http://www.stats.gov.cn/tjsj/zxfb/201502/t20150226_685799.html, accessed 8 March 2015.

Guojia tongjiju nengyuan tongjisi (Department for Energy Statistics, National Bureau of Statistics, the People's Republic of China), *Zhongguo nengyuan tongji nianjian 2013* (*China Energy Statistical Yearbook 2013*), p. 3 (Beijing: Zhongguo tongji chubanshe (Beijing: China Statistics Press), 2013), pp. 118–119.

Hatemi, Peter and Andrew Wedeman, 'Oil and conflict in Sino-American relations', *China Security* 3: 3, pp. 95–118.

He, Wenping, 'The balancing Act of China's Africa policy', *China Security* 3: 3, 2007, pp. 23–40.

Herberg, Mikkal E., 'Energy security survey 2007: The rise of Asia's national oil companies', NBR Special Report, No. 14, December 2007, the National Bureau of Asian Research, http://www.nbr.org/publications/specialreport/pdf/SR14.pdf, accessed 13 March 2009.

Houser, Trevor, 'The roots of Chinese oil investment abroad', Asia Policy no. 5, January 2008, the National Bureau of Asian Research, http://www.nbr.org/publications/asia_policy/AP5/AP5_Houser.pdf, accessed 28 October 2008, pp. 141–166.

International Energy Agency, *World Energy Outlook 2013 — Executive Summary*, OECD/IEA 2013, http://www.iea.org/publications/freepublications/publication/WEO2013_Executive_Summary_English.pdf, accessed 24 August 2014.

Jaffe, Amy Myers and Steven W. Lewis, 'Beijing's oil diplomacy', *Survival* 44: 1, 2002, pp. 115–134.

Jiang, Julie and Jonathan Sinton, *Overseas Investments by Chinese National Oil Companies: Assessing the Drivers and Impacts*, International Energy Agency, February 2011, http://www.iea.org/publications/freepublications/publication/overseas_china.pdf, accessed 24 August 2014.

Jiang, Julie and Chen Ding, Updates on overseas investments by China's national oil companies: Achievements and challenges since 2011, International Energy Agency, Partner Country Series, 2014, http://www.iea.org/publications/freepublications/publication/PartnerCountrySeriesUpdateonOverseasInvestmentsbyChinasNationalOilCompanies.pdf, accessed 24 August 2014.

Jiang, Wenran, 'Fueling the dragon: China's quest for energy security and Canada's opportunities', Asia Pacific Foundation of Canada, April 2005, http://www.asiapacific.ca/analysis/pubs/pdfs/can_in_asia/cia_fueling_dragon.pdf, accessed 28 June 2006.

Jiang, Wenran, 'Beijing's "New Thinking" on energy security', *China Brief*, 6: 8, 12 April 2006, the Jamestown Foundation, http://www.jamestown.org/china_brief/article.php?articleid=2373181, accessed 7 April 2008.

Jiang, Yong, 'Jingti bumen liyi pengzhang' ('Be on guard against the expansion of the interests of some governmental bodies'), quoted by Sohu, 9 October 2006, http://news.sohu.com/20061009/n245692266.shtml, accessed 13 May 2008.

Kambara, Tatsu and Christopher Howe, *China and the Global Energy Crisis: Development and Prospects for China's Oil and Natural Gas* (Cheltenham, UK: Edward Elgar Publishing Limited, 2008), pp. 12–67; 87–95.

Kaplan, Robert D., 'The geopolitics of energy', *Forbes*, 4 April 2014, http://www.forbes.com/sites/stratfor/2014/04/04/the-geopolitics-of-energy/, accessed 10 February 2015.

Kenny, Henry J., 'China and the competition for oil and gas in Asia', *Asia–Pacific Review*, 11: 2, 2004, pp. 36–47.

Keohane, Robert O. and Joseph S. Nye, *Power and Interdependence* (Boston: Scott, Foresman and Company, 1989), pp. 3–22.

Klare, Michael T, 'Containing China: The US's real objective,' Asia Times On-line, 20 April 2006, http://www.atimes.com/atimes/China/HD20Ad01.html, accessed 30 December 2007.

Kleine-Ahlbrandt, Stephanie and Andrew Small, 'China's new dictatorship diplomacy: Is Beijing parting with pariahs?', *Foreign Affairs* 87: 1, 2008, pp. 38–56.

Kong, Bo, 'An anatomy of China's energy insecurity and its strategies', Pacific Northwest National Laboratory, December 2005, http://pnwcgs.pnl.gov/Newsletter/OtherDocs/AnatChinaEnergy.pdf, accessed 16 May 2008.

Kong, Shanguang, 'Wu wanyi de guoyou qiye lirun nail qu le?' ('Where has gone the SOEs' 5 trillion RMB profit?'), available at Guangming wang (Bright Daily On-line), 11 February 2007, http://guancha.gmw.cn/show.aspx?id=3445, accessed on 24 May 2008.

Krasner, Stephen D., *Defending the National Interest: Raw Materials Investments and U.S. Foreign Policy* (Princeton: Princeton University Press, 1978), pp. 5–34.

Lai, Hongyi Harry, 'China's oil diplomacy: Is it a global security threat?', *Third World Quarterly* 28: 3, 2007, pp. 519–537.

Lairson, Thomas D. and David Skidmore, *International Political Economy: The Struggle for Power and Wealth* (Belmont, CA: Wadsworth/Thomson Learning, 2003).

Lanteigne, Marc, *Chinese Foreign Policy: An Introduction* (London and New York: Routledge, 2009), p. 24.

Lee, Pak K, 'China's quest for oil security: Oil (wars) in the pipeline?', *Pacific Review* 18: 2, 2005, pp. 265–301.

Leverett, Flynt and Jeffery Bader, 'Managing China–U.S. Energy competition in the Middle-East', *The Washington Quarterly* 29: 1, pp. 187–201.

Lewis, Steven W., 'Chinese NOCs and World Energy Markets: CNPC, Sinopec and CNOOC', James A. Baker III Institute for Public Policy of Rice University, 2007, http://www.rice.edu/energy/publications/docs/NOCs/Papers/NOC_CNOOC_Lewis.pdf, accessed 13 March 2009.

Li, Anshan, 'China and Africa: Policy and Challenges', *China Security* 3: 3, 2007, pp. 69–93.

Liao, Xuanli, 'Central Asia and China's Energy Security', *China and Eurasia Forum Quarterly* 4: 4, 2006, pp. 61–69.

Lieberthal, Kenneth and Michel Oksenberg, *Policy Making in China: Leaders, Structures, and Process* (Princeton: Princeton University Press, 1988).

Lieberthal, Kenneth and Mikkal Herberg, 'China's search for energy security: implications for U.S. policy', NBR Analysis, 17: 1, the National Bureau of Asian Research, April 2006, http://www.nbr.org/publications/analysis/pdf/vol17no1.pdf, accessed 20 June 2008.

Liu, Xuecheng, 'China's energy security and its grand strategy', *Policy Analysis Brief*, The Stanley Foundation, September 2006.

Lye, Liang Fook, 'China–US relations: Coping with a US pivot to the Asia–Pacific region', in Wang Gungwu and Zheng Yongnian eds., *China: Development and Governance* (Singapore: World Scientific Publishing Co. Pte. Ltd., 2013), pp. 399–404.

McGregor, Richard, 'Chinese diplomacy 'hijacked' by big companies', Financial Times on-line, 16 March 2008, http://www.ft.com/cms/s/0/28b21418-f386-11dc-b6bc-0000779fd2ac.html?nclick_check=1, accessed 18 March 18, 2009.

Men, Honghua, 'Quebao zhongguo nengyuan anquan de zhanlue yiyi' ('The strategic significance of ensuring China's energy security'), *Taipingyang xuebao* (*The Pacific Journal*) 1, 2005, pp. 33–44.

Ministry of Foreign Affairs, the People's Republic of China, 'Zhongfei hezuo luntan diwujie buzhangji huiyi — Beijing xindong jihua, 2013 nian zhi 2015 nian' (The Fifth Ministerial Meeting of the Forum on China–African Cooperation — Beijing Action Plan, 2013–2015), 23 July 2012, http://www.fmprc.gov.cn/ce/cena/chn/xwdt/t957754.htm, accessed 5 March 2015.

Naughton, Barry, *The Chinese Economy: Transition and Growth* (Cambridge, MA: The MIT Press, 2007), pp. 329–333.

Naughton, Barry, 'SASAC and rising corporate power in China', China Leadership Monitor, No. 24, Hoover Institution, Stanford University, http://media.hoover.org/documents/CLM24BN.pdf, accessed 9 May 2008.

Ni, Jianmin eds., *Guojia nengyuan anquan baogao* (*A Report of National Energy Security*) (Beijing: Renmin Chubanshe, July 2005).

Niu, Tiehang and Lye Liang Fook, 'The challenges of managing China's huge foreign reserves: From Huijin to China investment corporation', EAI Background Brief, No. 352, East Asian Institute, National University of Singapore, 1 October 2007.

Obiorah, Ndubisi, 'Who's afraid of China in Africa? Towards an African civil society perspective on China–Africa relations', in Firoze Manji and Stephen Marks eds., *African Perspectives on China in Africa* (Cape Town, Nairobi and Oxford: Fahamu — Networks for Social Justice, 2007), pp. 35–55.

O'Brien, Robert and Marc Williams, *Global Political Economy: Evolution and Dynamics* (New York: Palgrave Macmillan, 2004), pp. 21–22.

OPEC, *World Oil Outlook 2008* (Vienna: OPEC Secretariat, 2008), http://www.opec.org/library/World%20Oil%20Outlook/pdf/WOO2008.pdf, accessed 6 January 2009.

OPEC, *2014 World Oil Outlook* (Vienna: OPEC Secretariat, 2014), http://www.opec.org/opec_web/static_files_project/media/downloads/publications/WOO_2014.pdf, accessed 22 January 2015, pp. 39–41.

Paul, Joshy M., 'The role of energy security in China's foreign policy: A maritime perspective', *Maritime Affairs* 6: 2, 2010, pp. 49–71.

Pollack, Jonathan D., 'Energy Insecurity with Chinese and American Characteristics: Implications for Sino–American relations', *Journal of Contemporary China* 17: 55, 2008, pp. 229–245.

Richardson, Michael , 'A southward thrust for China's energy diplomacy in the south China sea', *China Brief* 8: 21, the Jamestown Foundation, 2008, http://www.jamestown.org/programs/chinabrief/single/?tx_ttnews%5Btt_news%5D=34140&tx_ttnews%5BbackPid%5D=168&no_cache=1, accessed 13 March 2009.

Rosen, Daniel H. and Trevor Houser, 'China energy: A guide for the perplexed', *China Balance Sheet*, A Joint Project by the Center for Strategic and International Studies and the Peterson Institute for International Economics, May 2007, http://www.petersoninstitute.org/publications/papers/rosen0507.pdf, accessed 3 May 2007, p. 5.

Shambaugh, David, 'China engages Asia: Reshaping the regional order', *International Security* 29: 3, 2005, pp. 64–99.

Stopford, John M. and Susan Strange, *Rival States, Rival Firms: Competition for World Market Shares* (Cambridge: Cambridge University Press, 1991), pp. 19–22.

Storey, Ian, 'Securing Southeast Asia's sea lanes: A work in progress', *Asia Policy* 6, 2008, the National Bureau of Asian Research, http://www.nbr.org/publications/asia_policy/AP6/AP6_E_Storey.pdf, accessed 13 March 2009, pp. 95–127.

Strange, Susan, *States and Markets* (London and New York: Continuum, 1994), p. 179; 207.

Strange, Susan, *The Retreat of the State: The Diffusion of Power in the World Economy* (Cambridge: Cambridge University Press, 1996), p. xiv.

Sutter, Robert G., *China's Rise in Asia: Promises and Perils* (Oxford: Rowman & Littlefield Publishers, Inc., 2005).

Sutter, Robert G., *Chinese Foreign Relations: Power and Policy since the Cold War* (Plymouth, UK: Rowman & Littlefield Publishers, Inc., 2008).

The State Council Information Office, the People's Republic of China, *China's Energy Conditions and Policies*, 26 December 2007, http://www.china.org.cn/english/environment/236955.htm, accessed 27 December 2007.

The State Council Information Office, the People's Republic of China, *China's Energy Policy 2012*, October 2012, http://english.gov.cn/archive/white_paper/2014/09/09/content_281474986284499.htm, accessed 14 January 2015.

The State Council Information Office, the People's Republic of China, *China–Africa Economic and Trade Cooperation*, white paper, August 2013, http://english.gov.cn/archive/white_paper/2014/08/23/content_281474982986536.htm, accessed 3 March 2015.

Tang, James, 'With the Grain or Against the Grain?: Energy Security and Chinese Foreign Policy in the Hu Jintao Era', the Brookings Institution, October 2006, http://www.brookings.edu/~/media/Files/rc/papers/2006/10china_tang/tang2006.pdf, accessed 20 June 2008.

Taylor, Ian, 'China's oil diplomacy in Africa', *International Affairs* 82: 5, 2006, pp. 937–959.

Thompson, Drew, 'China's emerging interests in Africa: Opportunities and challenges for Africa and the United States', *African Renaissance Journal*, July/August 2005, pp. 20–29.

Thomson, Elspeth and Augustin Boey, 'The role of oil and gas in China's energy strategy: An overview', *Asia Pacific Business Review* 21: 1, 2015, pp. 10–25.

Till, Geoffrey, *Seapower: A Guide for the Twenty-First Century* (London and New York: Routledge, second edition, 2009).

Tseng, Hui-Yi Katherine, 'The South China Sea disputes: Current state of play and future prospects', in Wang Gungwu and Zheng Yongnian eds., *China: Development and Governance* (Singapore: World Scientific Publishing Co. Pte. Ltd., 2013), pp. 515–526.

Tseng, Hui-Yi Katherine, 'Maritime security in Southeast Asia: Interfacing regional and extra-regional stakeholder', in Konrad-Adenauer Stiftung and European Union, *Maritime Security and Piracy: Common Challenges and Responses from Europe and Asia* (Singapore: Select Books Pte Ltd, 2014), pp. 87–105.

Tu, Jianjun, 'China's new national energy commission and energy policy', *China Brief*, 8: 7, 2008, the Jamestown Foundation, http://www.jamestown.org/programs/chinabrief/single/?tx_ttnews%5Btt_news%5D=4820&tx_ttnews%5BbackPid%5D=168&no_cache=1, accessed 13 March 2009.

United Nations Conference on Trade and Development, *World Investment Report 2006* (New York and Geneva: United Nations, 2006), http://www.unctad.org/en/docs/wir2006_en.pdf, accessed 21 March 2008.

United Nations Conference on Trade and Development (UNCTAD), *World Investment Report 2014: Investing in the SDGs: An Action Plan* (New York and Geneva: United Nations, 2014), http://unctad.org/en/PublicationsLibrary/wir2014_en.pdf, accessed 11 March 2015.

U.S. Energy Information Administration, 'Country analysis brief: China', 4 February 2014, http://www.eia.gov/countries/analysisbriefs/China/china.pdf, accessed 14 January 2015.

U.S. Energy Information Administration, 'China is now the world's largest net importer of petroleum and other liquid fuels', 24 March 2014, http://www.eia.gov/todayinenergy/detail.cfm?id=15531, accessed 14 January 2015.

U.S. Energy Information Administration, 'Short-term energy outlook', December 2014, http://www.eia.gov/forecasts/steo/pdf/steo_full.pdf, accessed 13 December 2014.

Wen, Jiabao, 'Guanyu zhiding guomin jingji he shehui fazhan di shiyi ge wunian guihua jianyi de shuoming' (Report on the outline of the eleventh five-year plan for national economic and social development), 8 October 2005, http://www.gov.cn/gongbao/content/2005/content_121428.htm, accessed 30 May 2008.

Wu, Friedrich, 'The globalization of corporate China', *NBR Analysis* 16: 3, 2005, the National Bureau of Asian Research, http://www.nbr.org/publications/analysis/pdf/vol16no3.pdf, accessed 13 March 2009.

Wu, Kang and Shair Ling Han, 'State-company goals give China's investment push unique features', *Oil & Gas Journal*, April 18, 2005, p. 20.

Wong, John, 'China's Economic growth in East Asian context', in John Wong and Wei Liu ed., *China's Surging Economy: Adjusting for More Balanced Development* (Singapore: World Scientific Publishing Co. Pte. Ltd., 2007), pp. 31–53.

World Health Organisation, 'Dr Margaret Chan nominated to be WHO Director-General', 8 November 2006, http://www.who.int/mediacentre/news/releases/2006/pr65/en/index.html, accessed 18 January 2007.

World Health Organisation, 'Dr Margaret Chan to be WHO's next Director-General', 9 November 2006, http://www.who.int/mediacentre/news/releases/2006/pr66/en/index.html, accessed 18 January 2007.

World Health Organisation, 'Dr Margaret Chan takes office as Director-General of the World Health Organization', 4 January 2007, http://www.who.int/mediacentre/news/releases/2007/pr01/en/index.html, accessed 17 January 2007.

Xia, Yishan, '"Zouchuqu" xiahao liubu qi' ('China's "going global" activity should include six steps'), *Zhongguo shiyou* (*China Petroleum*) July, 2001, pp. 12–13.

Xia, Yishan, 'Dangqian guoji nengyuan xingshi he zhongguo nengyuan zhanlue' ('The international energy situation and China's energy strategy'), *Heping yu fazhan* (*Peace and Development*) 2, 2002, pp. 36–39.

Xia, Yishan, 'Zhongguo nengyuan anquan wenti ji jiejue qianjing' ('China's energy security issue and the prospect of its solution'), *Heping yu fazhan* (*Peace and Development*) 4, 2003, pp. 20–24.

Xia, Yishan, 'Lun zhongguo de nengyuan anquan zhanlue' ('A discussion about China's energy strategy'), *Zhongwai nengyuan* (*China Foreign Energy*) 11: 1, 2006, pp. 1–5.

Xu, Xiaojie, 'Chinese NOCs' overseas strategies: Background, comparison and remarks', James A. Baker III Institute for Public Policy of Rice University, 2007, http://www.rice.edu/energy/publications/docs/NOCs/Papers/NOC_ChineseNOCs_Xu.pdf, accessed 13 March 2009.

Yahuda, Michael, *The International Politics of the Asia-Pacific* (London and New York: RoutledgeCurzon, 2004).

Yan, Xuetong, 'You heping buyiding you anquan' ('Peace is not necessarily security'), Huanqiu wang (Global Times On-line), 21 September 2007, http://column.huanqiu.com/yanxuetong/2007-11/5942.html, accessed 16 December 2007.

Yan, Xuetong, 'Anquan liyi shi zhongguo shouyao liyi' ('Security interests are China's most important interests'), *Huanqiu wang* (*Global Times On-line*), 27 September 2007, http://column.huanqiu.com/yanxuetong/2007-11/7678.html, accessed 16 December 2007.

Yang, Guangbin, 'Interest groups in China's politics and governance', *EAI Background Brief*, 361, East Asian Institute, National University of Singapore, 13 December 2007.

Yao, Yuanming Alvin, 'China's oil strategy and its implications for U.S.–China relations', *Issues & Studies* 42: 3, 2006, pp. 165–201.

You Ji, 'Dealing with the Malacca dilemma: China's effort to protect its energy supply', *Strategic Analysis* 31: 3, 2007, pp. 467–489.

You, Ji, 'Dealing with the Malacca Strait dilemma: China's efforts to enhance energy transportation security', *EAI Background Brief*, 329, East Asian Institute, National University of Singapore, 12 April 2007.

You Ji, 'China's civil-military strategies for South China Sea disputes control', EAI Background Brief, 1002, East Asian Institute, National University of Singapore, 25 February 2015.

Yu, Hong and Yang Mu, 'China as the world's largest automobile market', in Yang Mu and Yu Hong eds., *China's Industrial Development in the 21st Century*, (Singapore: World Scientific Publishing Co. Pte. Ltd., 2011), pp. 17–36.

Yu, Yongding, 'China's "twin surpluses": Causes and remedies', *EAI Background Brief*, No. 283, East Asian Institute, National University of Singapore, 27 April 2006.

Zha, Daojiong, 'Xianghu yilai yu zhongguo de shiyou gongying anquan' ('Interdependence and China's oil supply security'), *Shijie jingji yu zhengzhi* (*World Economics and Politics*) 6, 2005, pp. 15–21.

Zha, Daojiong, 'Cong guoji guanxi jiaodu kan zhongguo de nengyuan anquan' ('A review on China's energy security from the perspective of international relations'), *Guoji jingji pinglun* (*International Economic Review*) 11–12, 2005, pp. 1–10.

Zha, Daojiong, 'China's energy security: Domestic and international issues', *Survival* 48: 1, 2006, pp. 179–190.

Zha, Daojiong, 'Zhongguo de nengyuan fazhan zhanlue yu guoji guanxi' ('China's energy development strategy and its international relations'), *Xuexi yuekan* (*Study Monthly*) 10, 2006, pp. 47–49.

Zha, Daojiong, 'Nengyuan yilai jinkou bu kepa' ('It is not fearful to be dependent on energy import'), *Shijie zhishi* (*World Knowledge*) 9, 2006, pp. 47–49.

Zha, Daojiong, 'Bie tai danyou nengyuan yilai jinkou', ('Don't be too concerned about China's dependence on energy import'), *Huanqiu wang* (*Global Times On-line*), 18 September 2007, http://column.huanqiu.com/zhadaojiong/2007-11/4606.html, accessed 16 December 2007.

Zha, Daojiong, 'Kongzhi zhongguo shiyou xiaofei zengzhang' ('Controlling the growth of China's oil consumption'), Huanqiu wang (Global Times On-line), 11 December 2007, http://column.huanqiu.com/zhadaojiong/2007-12/33512.html, accessed 16 December 2007.

Zhang, Chi, 'China's oil diplomacy in Africa: The convergence of national and corporate interests', in Christopher M. Dent eds., *China and Africa Development Relations* (London and New York: Routledge, 2011).

Zhang, Chi, 'Changes in the global energy system and their implications for China', *EAI Background Brief*, 1006, East Asian Institute, National University of Singapore, 12 March 2015.

Zhang, Libin and Jason Lee, 'Untangling China's Energy Policy', *China Security* 4: 3, 2008, pp. 58–61.

Zhang, Wenmu, 'Zhongguo nengyuan anquan yu zhengce xuanze' ('China's energy security and policy choices'), *Shijie jingji yu zhengzhi* (*World Economics and Politics*), 5, 2003, pp. 1–9.

Zhang, Wenmu, 'Zhongguo de nengyuan anquan yu kexing zhanlue' ('China's energy security and feasible energy strategy'), *Shuiwu yanjiu* (*Taxation Studies*), 10, 2005, pp. 34–41.

Zhang, Wenmu, 'Guojia zhanlue nengli shi ziyuan anquan de baozhang' ('National strategic capacity is the guarantee of resource security'), *Luye* (*Green Leaf*) 6, 2006, pp. 18–19.

Zhang, Xuegang, 'Southeast Asia and energy: Gateway to stability', *China Security* 3: 2, 2007, pp. 18–35.

Zhang, Yongjin, *China's Emerging Global Business: Political Economy and Institutional Investigations* (New York: Palgrave Macmillan, 2003).

Zhang, Yongjin, *China Goes Global* (London: The Foreign Policy Centre, 2005).

Zhao Hong, 'China's efforts to enhance its energy security', in Wang Gungwu and Zheng Yongnian eds., *China: Development and Governance* (Singapore: World Scientific Publishing Co. Pte. Ltd., 2013), pp. 507–514.

Zhao, Hongtu, 'Guojia nengyuan anquan' ('National energy security'), in Zhongguo xiandai guoji guanxi yanjiuyuan (China Institute of Contemporary International Relations), *Guojia Jingji Anquan* (*National Economic Security*) (Beijing: Shishi chubanshe, July 2005), pp. 257–281.

Zhao, Hongtu, 'Guoji nengyuan anquan xingshi de xin tedian' ('The new characters of international energy security situation'), *Xiandian guoji guanxi* (*Contemporary International Relations*) 7, 2005, pp. 1–6.

Zhao, Hongtu, 'Guanyu zhongmei nengyuan hezuo de jidian sikao' ('Some ideas about China–US energy cooperation'), *Xiandai guoji guanxi* (*Contemporary International Relations*) 1, 2006, pp. 47–53.

Zhao, Hongtu, '"Maliujia kunju" yu zhongguo nengyuan anquan zai sikao' ('"Malacca dilemma" and the rethinking of China's energy security'), *Xiandai guoji guanxi* (*Contemporary International Relations*) 6, 2007, pp. 36–42.

Zhao, Suisheng, 'China's global search for energy security: Cooperation and competition in Asia–Pacific', *Journal of Contemporary China* 17: 55, 2008, pp. 207–227.

Zhao, Suisheng, 'The Making of Chinese Foreign Policy: Actors and Institutions', in Kweku Ampiah and Sanusha Naidu ed., *Crouching Tiger, Hidden Dragon?: Africa and China* (South Africa: University of KwaZulu-Natal Press, 2008), pp. 39–52.

Zheng, Yongnian, *Discovering Chinese Nationalism in China: Modernization, Identity, and International Relations* (Cambridge: Cambridge University Press, 1999).

Zheng, Yongnian, *Globalization and State Transformation in China* (Cambridge: Cambridge University Press, 2004).

Zheng, Yongnian, 'Meiguo zhuanxing waijiao dui zhongguo de hanyi' ('The impact of the US transformational diplomacy on China'), Lianhe zaobao wang (United Morning Post On-line), 18 April 2006, http://www.zaobao.com/special/china/sino_us/pages6/sino_us060418a.html, accessed 30 December 2007.

Zheng, Yongnian, 'Zhongguo dang lixing huiying yazhouban "beiyue",' ('China should appropriately react to the Asian "NATO"'), Lianhe zaobao wang (United Morning Post On-line), 27 March 2007, http://www.zaobao.com/special/forum/pages5/forum_zp070327.html, accessed 30 December 2007.

Zheng, Yongnian and Zhengxu Wang, 'China's national people's congress 2008: New administration, personnel reshuffling and policy impacts', *Briefing Series* — Issue 38, China Policy Institute, The University of Nottingham, March 2008, http://www.nottingham.ac.uk/shared/shared_cpi/documents/policy_papers/Briefing_38_NPC_2008.pdf, accessed 3 May 2008.

Zhou, Dadi, Han Wenke, and Zhu Xingshan, 'Zai gaige kangfang zhong baozhang shiyou gongying anquan' ('Safeguarding China's oil security during the process of reform and opening up'), *Zhongguo nengyuan* (*China Energy*) 3, 2001, pp. 14–17.

Zhou Fuqiu, 'Energy conservation policy development in China', in Elspeth Thomson, Youngho Chang and Jae-Seung Lee eds., *Energy Conservation in East Asia*, (Singapore: World Scientific Publishing Co. Pte. Ltd., 2011), pp. 143–159.

Zhou Zhihua, 'China launches new urbanization plan (2014–2020)', *EAI Background Brief*, 910, East Asian Institute, National University of Singapore, 16 April 2014.

Zhu, Rongji, 'Report on the outline of the Tenth Five-Year Plan for national economic and social development (2001)', Delivered at the Fourth Session of the Ninth National People's Congress on March 5, 2001, http://english.gov.cn/official/2005-07/29/content_18334.htm, accessed 30 May 2008.

Zhu, Xingshan and Zhou Dadi, 'Ruhe kandai zhongguo de nengyuan anquan wenti' ('Some ideas on China's energy security issue'), *Guoji shiyou jingji* (*International Petroleum Economics*) 9: 10, 2001, pp. 5–8.

Zweig, David, *Internationalizing China: Domestic Interests and Global Linkages* (Ithaca and London: Cornell University Press, 2002).

Zweig, David and Bi Jianhai, 'China's global hunt for energy', *Foreign Affairs* 84: 5, 2005, pp. 25–38.

Zweig, David and Shulan Ye, 'A crisis is looming: China's energy challenge in the eyes of university students', *Journal of Contemporary China* 17: 55, 2008, pp. 273–296.

Media information

BBC, 'Africa to seek UN Council seats', 3 July 2005, http://news.bbc.co.uk/2/low/africa/4645545.stm, accessed 18 January 2007.

BBC, 'Battle joined on Security Council reform', 12 July 2005, http://news.bbc.co.uk/2/hi/americas/4675823.stm, accessed 18 January 2007.

BBC, 'China oil firms buys into Nigeria', 9 January 2006, ehttp://news.bbc.co.uk/2/hi/business/4594058.stm, accessed 9 January 2006.

BBC, 'Exports fuel China's CO_2 output', 19 October 2007, http://news.bbc.co.uk/1/hi/uk/7052115.stm, accessed 20 December 2007.

Beijing Review, 'Image Improvement', http://www.bjreview.com.cn/06-24-e/w-3.htm, 24 June 2006, accessed 25 July 2006.

Cankao xiaoxi, '2014 nian "caifu" shijie 500qiang chulu, zhongshihua paiming disan' (The 2014 Fortune Global 500 released: Sinopec Group was ranked the third), 8 July 2014, http://finance.cankaoxiaoxi.com/2014/0708/417053.shtml, accessed 16 March 2015.

Cankao xiaoxi, '2014 nian shijie 500qiang, zhongshihua shangbang gongsi da baijia (mingdan)' (The world's top 500 companies in 2014: 100 Chinese companies are among them), 8 July 2014, http://finance.cankaoxiaoxi.com/2014/0708/416735_4.shtml, accessed 16 March 2015.

Dongfang wang (Eastday), 'Guoji youjia zaichuang lishi xingao, pobai bushi ouran shijian' ('The international oil price reached historical height again, the increase of the price above $100 is not an accident'), 28 February 2008, http://finance.eastday.com/m/20080228/u1a3431649.html, accessed 17 May 2008.

People's Daily On-line, 'Tebie cehua: duiyi zhanzheng zhongguo jingji sunshi you duoda?' ('Special report: to what extent has the Iraq War affected the Chinese economy?'), 25 February 2003, http://past.people.com.cn/GB/jinji/31/179/20030225/930064.html, accessed 17 May 2008.

People's Daily On-line, 'Guojia nengyuan lingdao xiaozu chengli, Wen Jiabao ren zuzhang' ('The State Energy Leading Small Group has been established and is headed by Wen Jiabao'), 30 May 2005, http://politics.people.com.cn/GB/1026/3426700.html, accessed 27 April 2008; and Downs, 'China'.

People's Daily On-line, 'Liangshui hebing: Neiwaizi qiye chongfan gongping jingzheng' ('The unification of the two tax rates: Foreign invested enterprises and domestic invested enterprises are expected to compete fairly again'), 30 October 2006, http://finance.people.com.cn/GB/4972421.html, accessed on 17 January 2007.

People's Daily On-line, 'Roundup: Africa, China usher in 50th anniversary of sincere cooperation', 31 October 2006, http://english.peopledaily.com.cn/200610/31/eng20061031_316924.html, accessed 18 January 2007.

People's Daily On-line, 'CNOOC to develop Iranian gas field', 22 December 2006, http://english.peopledaily.com.cn/200612/22/eng20061222_335088.html, accessed 13 February 2009.

People's Daily On-line, 'Xi Jinping huijian eluosi zongtnog pujing' (President Xi Jinping met with Russian President Vladimir Putin), 6 February 2014, http://politics.people.com.cn/n/2014/0206/c1024-24285729.html, accessed 2 March 2015.

People's Daily On-line, 'Xi Jinping chuxi zhonga hezuo luntan diliujie buzhangji huiyi kaimushi bing fabiao zhongyao jianghua' (President Xi Jinping attended the opening ceremony of the Sixth Ministerial-level Meeting of the China–Arab Cooperation Forum and delivered an important speech), 6 June 2014, http://politics.people.com.cn/n/2014/0606/c1024-25110595.html, accessed 3 March 2015.

People's Daily On-line, 'Xi Jinping tong agenting zongtong juxing huitan' (President Xi Jinping held talks with Argentinean President), 19 July 2014, http://politics.people.com.cn/n/2014/0719/c1024-25302513.html, accessed 2 March 2015.

People's Daily On-line, 'Xi Jinping tong weineiruila zongtong maduluo juxing huitan' (President Xi Jinping held talks with Venezuelan President Nicolas Maduro), 21 July 2014, http://politics.people.com.cn/n/2014/0721/c1024-25312095.html, accessed 2 March 2015.

People's Daily On-line, 'Xijinping he lahemeng zongtong gongtong chuxi zhongta dianli he zhongguo-zhongya tianranqi guandao hezuo xiangmu kaigong yishi' (President Xi Jinping and President Emomali Rahmon attended the commencement ceremonies of the cooperation project of China-Tajikistan power plant and Central Asia–China Gas Pipeline), 14 September 2014, http://politics.people.com.cn/n/2014/0914/c1024-25655846.html, accessed 2 March 2015.

People's Daily On-line, 'Xijinping tong tajikesitan zongtong lahemeng huitan' (President Xi Jinping held talks with Tajik President Emomali Rahmon), 14 September 2014, http://politics.people.com.cn/n/2014/0914/c1024-25655845.html, accessed 2 March 2015.

People's Daily On-line, 'Li Keqiang tong eluosi zongli meideweijiefu gongtong zhuchi zhonge zongli di shijiuci dingqi huiwu'(Premier Li Keqiang and Russian Prime Minister Dmitry Medvedev together chaired the 19th regular meeting between Chinese Premier and Russian Prime Minister), 13 October 2014, http://politics.people.com.cn/n/2014/1013/c1024-25826219.html, accessed 2 March 2015.

People's Daily On-line, 'Li Keqiang zongli tong miandian zongtong wudengsheng juxing huitanshi qiangdiao, quanmian tisheng zhongmian zhanlue hezuo shuiping, liangguo yongyuan zuo haolingju,

haopengyou, haohuoban' (When meeting with Myanmar President Thein Sein, Premier Li Keqiang proposed to comprehensively promote China–Myanmar cooperation, and stressed that the two countries should remain good neighbours, friends and partners forever), 14 November 2014, http://politics.people.com.cn/n/2014/1114/c1024-26026867.html, accessed 2 March 2015.

Quanjing Wang (Quanjing On-line), 'Jigou gaige sanshi nian jingjian weibiao, zhuanxing weili, minsheng weizhong' ('The bureaucratic restructuring in the past three decades with simplification as its surface, transformation as its substance, and the improvement of the people's livelihood as its emphasis'), 14 March 2008, http://www.p5w.net/news/gncj/200803/t1546372.htm, accessed 13 April 2008.

Renmin Ribao (People's Daily), 'Guojia nengyuanju "sanding" fang'an zhengshi huopi' ('The restructuring programme of the State Energy Bureau has formally been approved'), 30 July 2008, p. 9.

Reuters, 'Exclusive: PetroChina to join Exxon at giant Iraqi oil field', 9 August 2013, http://www.reuters.com/article/2013/08/09/us-petrochina-iraq-idUSBRE9780D520130809, accessed 20 February 2015.

Reuters, 'China buys into giant Kazakh oil field for $5 billion', 7 September 2013, http://www.reuters.com/article/2013/09/07/us-oil-kashagan-china-idUSBRE98606620130907, accessed 20 February 2015.

Sohu, 'Zhongguo jianli "shiyou zhanlue shubei" beingfei zhiwei pingyi youjia' ('The establishment of China's "strategic petroleum reserves" does not merely aim at stabilising the oil price'), 11 October 2004, http://business.sohu.com/20041011/n222432572.shtml, accessed 17 May 2008.

Souhu, '2009 nian 1yue zhongguo qiche xiaoliang shouchao meiguo quanqiu diyi' ('In January 2009, the automobile sales volume in China overtook that of the United States for the first time, making the country the largest automobile market in the world'), 12 February 2009, http://auto.sohu.com/20090212/n262192899.shtml, accessed 2 October 2014.

Souhu, 'Zhongguo 2020 nian chengshihualv jiangda 55 per cent, chenggshibing jiang jizhong baofa' ('China's urbanisation rate will reach 55 per cent in 2020 and there will be an outbreak of city diseases'), 9 February 2012, http://news.sohu.com/20120209/n334228775.shtml, accessed 17 January 2015.

Taipei Times, "Energy diplomacy' could offer diplomatic relief", 26 March 2006, http://www.taipeitimes.com/News/taiwan/archives/2006/03/26/2003299279, accessed 7 April 2008.

Telegraph On-line, 'China Prepares to Buy up Foreign Oil Companies', 22 February 2009, http://www.telegraph.co.uk/finance/newsbysector/energy/4781037/China-prepares-to-buy-up-foreign-oil-companies.html, accessed 25 February 2009.

Wangyi, 'Shisanwu nengyuan luxiantu: Zhongguo zhongyao nengyuan zhanlue ji yingxiang fenxi' (The 13th Five Year Plan energy development roadmap: An analysis of China's key energy strategy and its implications), 30 October 2014, http://money.163.com/14/1030/04/A9PEGUML00252G50.html, accessed 15 February 2015.

Xinhua, 'Jingbing jianzheng, zhuanbian zhineng — xinzhongguo chengli yilai de lici zhengfu jigou gaige' ('Streamlining the administration and transforming the function: The bureaucratic restructurings since the establishment of the People's Republic of China'), 6 March 2003, http://news.xinhuanet.com/ziliao/2003-03/06/content_761776.htm, accessed 3 May 2008.

Xinhua, 'Guojia jingmao wei pingjing miandui "danchu"' ('State Economic and Trade Commission calmly faces its "fading out"'), 11 March 2003, http://news.xinhuanet.com/zhengfu/2003-03/11/content_770480.htm, accessed 9 April 2008.

Xinhua, 'AU rejects G4 proposal on UN reform', 5 August 2005, http://news3.xinhuanet.com/english/2005-08/05/content_3311971.htm, accessed 18 January 2007.

Xinhua, 'Japan to give up UNSC bid for now', 21 August 2005, http://news3.xinhuanet.com/english/2005-08/21/content_3385064.htm, accessed 18 January 2007.

Xinhua, 'Xi qi dong shu: Gaibian woguo nengyuan xiaofei jiegou' (West to East Natural Gas Project: Changing China's energy consumption structure), 7 September 2005, http://news.xinhuanet.com/fortune/2005-09/07/content_3457146.htm, accessed 27 January 2009.

Xinhua, 'Full text: China's African Policy', 12 January 2006, http://news.xinhuanet.com/english/2006-01/12/content_4042521.htm, accessed 3 February 2009.

Xinhua, 'Guojia nengyuan lingdao xiaozu bangongshi zuo chengli, Wen Jiabao qinren zuzhang' ('The State Energy Leading Small Group Office was established yesterday and Wen Jiabao became the Head of the Group'), 3 June 2006, http://news.xinhuanet.com/fortune/2005-06/03/content_3040477.htm, accessed 27 April 2008.

Xinhua, 'Kuozhan haiwai chuliang, zhongshihua gaojia jinghuo angela san gaochan youtian' (Sinopec won bids for three highly productive oil fields in Angola with high prices, expanding its overseas oil reserves), 13 June 2006, http://news.xinhuanet.com/fortune/2006-06/13/content_4687360.htm, accessed 27 January 2009.

Xinhua, 'Shangwubu: Yiyou 64 ge guojia chengren zhognguo wanquan shichang jingji diwei' ('Ministry of Commerce: there have been 64 countries recognising China's full status as a market economy'), 7 November 2006, http://news.xinhuanet.com/fortune/2006-11/07/content_5302214.htm, accessed on 18 January 2007.

Xinhua, 'China's Margaret Chan elected WHO chief', 10 November 2006, http://news.xinhuanet.com/english/2006-11/10/content_5311568.htm, accessed 18 January 2007.

Xinhua, 'Jinrenqing: Tongyi hou de neiwaizi qiye suodeshui shuilu dingwei 25%' ('Jinrenqing: the unified corporate income tax rate in 25%'), 8 March 2007, http://www.lianghui.org.cn/2007lianghui/2007-03/08/content_7924484.htm, accessed 23 March 2008.

Xinhua, 'Xin qiye suodeshui fa huiji guonei shiyou qiye' ('The new Corporate Income Tax Law benefits Chinese oil enterprises'), 21 May 2007, http://news.xinhuanet.com/fortune/2007-05/21/content_6130570.htm, accessed 18 May 2008.

Xinhua, 'Guowuyuan jigou gaige fangan' ('The bureaucratic restructuring programme of the State Council'), 15 March 2008, http://news.xinhuanet.com/misc/2008-03/15/content_7794932.htm, accessed 3 May 2008.

Xinhua, '08nian zhongguo waijiao baipishu shuban faxing, nengyuan anquan lie shouzhang' (The 2008 edition of China's Foreign Affairs white paper has been published; the issue of energy security is addressed in the first chapter), 18 July 2008, http://politics.people.com.cn/GB/1027/7531802.html, accessed 13 March 2009.

Xinhua, 'Full text: China's Policy Paper on Latin America and the Caribbean', 5 November 2008, http://news.xinhuanet.com/english/2008-11/05/content_10308117.htm, accessed 3 February 2009.

Xinhua, '2008 nian zhongguo nengyuan xingshi baogao: nengyuan fazhan zhanlue touchu xinyi' (The 2008 report on China's energy situation: New development in the energy development strategy), 15 December 2008, http://news.xinhuanet.com/fortune/2008-12/15/content_10506475.htm, accessed 16 February 2009.

Xinhua, 'Chinese president continues visit in Saudi Arabia', 12 February 2009, http://news.xinhuanet.com/english/2009-02/12/content_10806379.htm, accessed 16 February 2009.

Xinhua, 'Wen Jiabao zhuchi zhaokai guojia nengyuan weiyuanhui diyici huiyi' (Premier Wen Jiabao chaired the first meeting of the State Energy Commission), 22 April 2010, http://news.xinhuanet.com/photo/2010-04/22/c_1250287.htm, accessed 15 March 2015.

Xinhua, 'Shouquan fabu: Zhonghua renmin gongheguo guomin jingji he shehui fazhan di shier ge wunian guihua gangyao' (Authorised to release: Outline of the Tenth Five-Year Plan for national economic and social development for the People's Republic of China), 16 March 2011, http://news.xinhuanet.com/politics/2011-03/16/c_121193916_28.htm, accessed 16 March 2015.

Xinhua, 'Zhongfei hezuo luntan diwujie buzhangji huiyi longzhong kaimu' (The fifth ministerial meeting of the forum on China–African cooperation opened), 19 July 2012, http://news.xinhuanet.com/world/2012-07/19/c_112481778.htm, accessed 5 March 2015.

Xinhua, 'Zhongfei hezuo luntan diwujie buzhangji huiyi Beijing xuanyan — quanwen' (Beijing Declaration of the Fifth Ministerial Meeting of the Forum on China–African Cooperation — full text), 20 July 2012, http://news.xinhuanet.com/world/2012-07/20/c_112493071.htm, accessed 5 March 2015.

Xinhua, 'Guowuyuan tiaozheng guojia nengyuan weiyuanhui zucheng renyuan' (The State Council has adjusted the members in the State Energy Commission), 11 July 2013, http://news.xinhuanet.com/2013-07/11/c_116499600.htm, accessed 15 March 2015.

Xinhua, 'Xi Jinping tong hasakesitan zongtong nazhaerbayefu juxing huitan' (President Xi Jinping held talks with Kazakh President Nursultan Nazarbayev), 7 September 2013, http://news.xinhuanet.com/world/2013-09/07/c_117272640.htm, accessed 3 March 2015.

Xinhua, 'Xi Jinping zai nazhaerbayefu daxue de yanjiang, quanwen' (President Xi Jinping's speech at Nazarbayev University, full text), 8 September 2013, http://news.xinhuanet.com/world/2013-09/08/c_117273079_2.htm, accessed 3 March 2015.

Xinhua, 'Xi Jinping zai yinni guohui fabiao yanjiang: Xieshou jianshe zhongguo-dongmeng mingyun gongtongti' (Xi Jinping delivered a speech at Indonesian parliament: Jointly building the China-ASEAN community of common destiny), 13 October 2013, http://news.xinhuanet.com/world/2013-10/03/c_117591652_3.htm, accessed 4 March 2015.

Xinhua, 'Zhongguo yueju shijie diyi huowu maoyi daguo' (China has become the world's largest trading nation in goods), 2 March 2014, http://news.xinhuanet.com/fortune/2014-03/02/c_119565636.htm, accessed 10 March 2015.

Xinhua, 'Xi Jinping huijian shate wangchu jian fushouxiang, guofang dachen saleman' (President Xi Jinping met with Saudi Crown Prince, Deputy Prime Minister and Defence Minister Salman), 13 March 2014, http://news.xinhuanet.com/2014-03/13/c_119762148.htm, accessed 3 March 2015.

Xinhua, 'Zhongguo baiwan yishang renkou chengshi da142ge, 6ge chengshi renkou chaoqianwan' ('The number of cities in China with a population of above one million reaches 142, and the number of cities with a population of above 10 million reaches 6'), 19 March 2014, http://news.xinhuanet.com/fortune/2014-03/19/c_119840576.htm, accessed 17 January 2015.

Xinhua, 'Meiti cheng zhongshihua zhongshiyou shinian huo zhengfu butie chao qianyi' (Media reports maintain that Sinopec Ltd. and PetroChina Ltd. have received government subsidies for more than RMB 100 billion during the past decade), 15 April 2014, http://news.xinhuanet.com/legal/2014-04/15/c_126391736.htm, accessed 16 March 2015.

Xinhua, 'Li Keqiang zhuchi zhaokai xinyijie guojia nengyuan weiyuanhui shouci huiyi' (Premier Li Keqiang chaired the first meeting of the State Energy Commission in the new administration), 20 April 2014, http://news.xinhuanet.com/photo/2014-04/20/c_126411441.htm, accessed 15 March 2015.

Xinhua, 'Li Keqiang wu niriliya zongtong: Jiang zhanlue huoban guanxi tisheng dao xinshuiping' (Premier Li Keqiang met with Nigerian President: Promoting the level of the strategic partnership to a new height), 8 May 2014, http://www.sn.xinhuanet.com/2014-05/08/c_1110591264.htm, accessed 4 March 2015.

Xinhua, 'Li Keqiang tong angela zongtong duosi sangtuosi juxing huitan' (Li Keqiang held talks with Angolan President Jose Eduardo Dos Santos), 9 May 2014, http://news.xinhuanet.com/world/2014-05/09/c_1110623607.htm, accessed 4 March 2015.

Xinhua, '2014 shijie 500qiang paihangbang: Zhongguo gongsi shoujin sanqiang' (The ranking of the world's top 500 companies in 2014: The first time a Chinese company is among the top three), 8 July 2014, http://news.xinhuanet.com/overseas/2014-07/08/c_126722764.htm, accessed 16 March 2015.

Xinhua, '2014 zhongguo 500qiang qiye paihangbang: Qianshiming junwei guoqi' (The ranking of China's top 500 companies: All of the top ten companies are state-owned enterprises), 14 July 2014, http://www.bj.xinhuanet.com/bjyw/2014-07/14/c_1111600716.htm, accessed 16 March 2015.

Xinhua, 'China-CELAC Forum established in Brazil', 18 July 2014, http://news.xinhuanet.com/english/china/2014-07/18/c_133493520.htm, accessed 4 March 2015.

Xinhua, 'Zhang Gaoli yu tukumansitan fuzongli gongtong zhuchi zhongtu hezuo weiyuanh di sanci huiyi' (Vice-premier Zhang Gaoli chaired together with Turkmenistan Deputy Prime Minister the 3rd meeting of China–Turkmenistan Cooperation Commission), 27 August 2014, http://news.xinhuanet.com/politics/2014-08/27/c_1112255872.htm, accessed 2 March 2015.

Xinhua, 'Zhang Gaoli fu eluosi juxing zhonge nengyuan hezuo weiyuanhui di shiyici huiyi huijian eluosi zongtong pujing bing gongtong chuxi zhonge dongxian tianranqi guandao ejingneiduan kaigong yishi' (Vice-premier Zhang Gaoli visited Russia to attend the 11th meeting of China–Russia Energy Cooperation Commission, met with Russia President Vladimir Putin, and attended together with President Putin the commencement ceremony of the Russia section of the eastern route of China–Russia Gas Pipeline), 2 September 2014, http://news.xinhuanet.com/politics/2014-09/02/c_1112316107.htm, accessed 2 March 2015.

Xinhua, '2014 zhongguo qiye 500 qiang mingdan fabu' ('The list of the top 500 enterprises in China in 2014 released'), 2 September 2014, http://news.xinhuanet.com/2014-09/02/c_126945173.htm, accessed 16 March 2015.

Xinhua, 'China, CELAC deepens cooperation with major documents', 9 January 2015, http://news.xinhuanet.com/english/china/2015-01/09/c_133908824.htm, accessed 4 March 2015.

Xinhua, 'Zhongguo yu lamei he jialebi guojia hezuo guihua 2015–2019' (China-CELAC Cooperation Plan 2015–2019), 9 January 2015, http://news.xinhuanet.com/world/2015-01/09/c_1113944648.htm, accessed 4 March 2015.

Xinhua, 'Zhongguo qiche chanxiaoliang lianxu liunian quanqiu diyi' ('China's automobile production and sales volumes are both the worlds' largest for the sixth consecutive year'), 13 January 2015, http://www.nmg.xinhuanet.com/xwzx/qctd/2015-01/13/c_1113975348.htm, accessed 19 January 2015.

Xinhua, 'Woguo tianranqi duiwai yicundu shengzhi 32.2 per cent' (China's natural gas dependency on foreign supply has rose to 32.2 per cent), 19 January 2015, http://news.xinhuanet.com/energy/2015-01/19/c_127398476.htm, accessed 8 February 2015.

Xinlang (Sina), 'Huang Yicheng', 13 July 2005, http://finance.sina.com.cn/manage/cfrw/20050713/17011792952.shtml, accessed 26 April 2008.

Zhongguo shiyou xinwen zhongxin (CNPC News On-line), '"Nengyuanju" shengji "guojiaju"' ('"The Energy Bureau" will be promoted to be a "state bureau"'), http://news.cnpc.com.cn/system/2008/03/19/001163500.shtml, 19 March 2008, accessed 12 April 2008.

Zhongqing Zaixian (China Youth Daily On-line), 'Henduo duiwai touzi jihui bei shenpi "zheteng" meile' ('Many opportunities to carry out overseas investment were lost due to the investment approval process'), 20 June 2006, http://chuangye.cyol.com/content/2006-06/20/content_1422733.htm, accessed 20 May 2008.

Other sources

Various issues of annual reports of China National Petroleum Corporation (CNPC), China Petrochemical Corporation (Sinopec) and China National Offshore Oil Corporation (CNOOC).

Websites of CNPC, Sinopec and CNOOC.

Websites of various Chinese governmental departments (National Development and Reform Commission, National Energy Administration, State-owned Assets Supervision and Administration Commission, Ministry of Land and Resources, Ministry of Foreign Affairs, State Administration of Foreign Exchange, etc.).

Index

About the Author

Dr ZHANG Chi received his BA in International Politics from the University of International Relations in Beijing, MA in International Political Economy from the University of Warwick, UK and PhD in Politics from the University of Nottingham, UK. He is currently Lieutenant Colonel and Assistant Professor at the Institute for Strategic Studies, National Defence University of China. His research interests include international political economy, energy security, soft power, Chinese foreign policy, Chinese politics and the international relations of the Asia-Pacific.

His English publication include two book chapters and one policy paper: "China's Oil Diplomacy in Africa: The Convergence of National and Corporate Interests", in Christopher M Dent ed., *China and Africa Development Relations*, (London and New York: Routledge, 2011), "'Soft Power' and Chinese Soft Power" (with Zheng Yongnian), in Lai Hongyi and Lu Yiyi eds., *China's Soft Power and International Relations*, (London and New York: Routledge, 2013)

and "Changes in the Global Energy System and Their Implications for China", EAI Background Brief, No. 1006, East Asian Institute, National University of Singapore, 12 March 2015. His journal articles in Chinese appeared in China's top journals in international relations such as World Economy and Politics (Shijie Jingji yu Zhengzhi) and Contemporary International Relations (Xiandai Guoji Guanxi).

CPSIA information can be obtained at www.ICGtesting.com
Printed in the USA
BVOW06*1640170915

418324BV00004B/11/P

9 789814 69673